The Organization of Lithic Technology in Late Glacial and Early Postglacial Europe

Edited by
Sarah Milliken

BAR International Series 700
1998

Published in 2019 by
BAR Publishing, Oxford

BAR International Series 700

Organization of Lithic Technology in Late Glacial and Early Postglacial Europe

ISBN 9780860549383 paperback
ISBN 9781407350158 e-book

DOI https://doi.org/10.30861/9780860549383

A catalogue record for this book is available from the British Library

This book is available at www.barpublishing.com

BAR Publishing is the trading name of British Archaeological Reports (Oxford) Ltd. British Archaeological Reports was first incorporated in 1974 to publish the BAR Series, International and British. In 1992 Hadrian Books Ltd became part of the BAR group. This volume was originally published by John and Erica Hedges in conjunction with British Archaeological Reports (Oxford) Ltd / Hadrian Books Ltd, the Series principal publisher, in 1998. This present volume is published by BAR Publishing, 2019.

BAR
PUBLISHING

BAR titles are available from:

BAR Publishing
122 Banbury Rd, Oxford, OX2 7BP, UK
EMAIL info@barpublishing.com
PHONE +44 (0)1865 310431
FAX +44 (0)1865 316916
www.barpublishing.com

CONTENTS

CONTRIBUTORS

Pierre Bodu	Laboratoire d'Ethnologie Préhistorique, U.R.A. 275, Maison René, Ginouvès, 21 Allée de l'Université, 92023 Nanterre Cedex, France
Marie-Isabelle Cattin	Service et Musée d'Archéologie, 7 Avenue Du Peyrou, 2000 Neuchâtel, Switzerland
Angela Davidzon	Institute of Archaeology, Mount Scopus, Hebrew University, 91905 Jerusalem, Israel
Nuno Ferreira Bicho	Universidade do Algarve, Unidade de Ciências Exactas e Humanas, Campus de Gambelas, 8000 Faro, Portugal
Jesus Emilio González Urquijo	Departamento de Ciencias Históricas, Universidad de Cantabria, Facultad de Filosofía y Letras, Avenida de los Castros s/n, 39005 Santander, Spain
Nigel Goring-Morris	Institute of Archaeology, Mount Scopus, Hebrew University, 91905 Jerusalem, Israel
Juan José Ibáñez Estévez	Departamento de Prehistoria y Antropologia Social, Universidad Autónoma de Barcelona, Edifici B, 03193 Bellaterra, Spain
Fawzi Ibrahim	Institute of Archaeology, Mount Scopus, Hebrew University, 91905 Jerusalem, Israel
Christina Kempcke-Richter	Niehler Strasse 1, 50670 Köln, Germany
Ofer Marder	Israel Antiquities Authority, POB 586, Jerusalem, Israel
Sarah Milliken	Dipartimento di Scienze Geologiche e Paleontologiche, Università di Ferrara, Corso Ercole I d'Este 32, 44100 Ferrara, Italy
Xavier Terradas	Consell Superior d'Investigacions Cientifiques, Laboratori d'Arqueologia, Instituto Mila i Fontanals, Egipciaques 15, 08100 Barcelona, Spain

INTRODUCTION

Lithic technology is a complex phenomenon which has technical, economic and sociocultural implications, and the organization of technology as a framework for inquiry offers great potential for focusing on the dynamic interactions between the environment, social factors and technology (Carr 1994a). Although definitions of the organization of technology vary (e.g. Binford 1979; Kelly 1988; Koldehoff 1987; Nelson 1991), they share the view that technology is a means to solve problems posed by both the physical and the social environments, and the ultimate goal of studies of technological organization is to determine "how technological changes reflect large scale behavioural changes in prehistoric societies" (Kelly 1988: 717).

The Late Glacial and Early Postglacial period was characterised by shifts in climate, seasonality, sea levels, ice cover, vegetation, fauna and soils. Europe presented a diverse arena for human adaptations across the Pleistocene-Holocene transition, and human responses to environmental change were regional and highly variable (Straus 1996). Some aspects of environmental change were abrupt while others were more gradual, and so too were the adaptive reactions of the hunter-gatherers. The aim of this publication is to present a series of case studies which examine the organization of lithic technology in the context of this changing environment.

Most studies of technological organization have focused on the reconstruction of hunter-gatherer mobility and settlement patterns, and this is particularly the case in the research carried out in North America over the last decade (e.g. Bamforth 1991; Carr 1994b; Parry and Kelly 1987; Shott 1986). This is largely due to the influence exerted by the work of Binford (e.g. 1979, 1980), who focused on hunter-gatherer mobility and principles of organization (Carr 1994a), as well as the belief that hunter-gatherer settlement patterns reflect adaptations to both the physical and the social environment.

This same concern with settlement patterns is reflected in the majority of the papers assembled here. Although these case studies vary in the scale of approach used (site-based or regional) and in the specific methodology adopted to analyse the assemblages (microwear analysis, refitting, statistical analysis, spatial analyses, etc.), the unifying thread is that they are concerned with trying to understand the relationship between tool production, use and maintenance, and how these were incorporated into other aspects of past hunter-gatherer lifestyles.

Terradas formulates a socio-economic behavioural model of the Late Glacial settlement of the eastern Pyrenees based on the reconstruction of the reduction sequences at the site of Font del Ros, and demonstrates how the stone tool production techniques were adapted to the nature and availability of the lithic raw material in the area. The reconstruction of the organizational strategies that were implemented by these groups in the management of their raw material resources enables the reconstruction of the territorial framework within which their social activity took place.

Ibáñez Estévez and González Urquijo discuss the results of their analysis of three assemblages from the Basque country: the Epipalaeolithic assemblage from Berniollo, the final Magdalenian and Azilian assemblage from Santa Catalina, and the final Magdalenian assemblage from Laminak II. Using the data derived from a reconstruction of the reduction sequences and microwear analysis of the retouched tools, they show how the technological changes which took place at the Pleistocene-Holocene transition in this region consisted only in a change in the strategy of exploitation of the tool kit, rather than in a change in the technological concept which lay behind its production, and they suggest that this could be related to a reduction in the mobility patterns of the hunter-gatherers.

Bicho presents a detailed analysis of the morphological characteristics of the formal tools, blanks, cores and core maintenance products from a sample of Portuguese Late Glacial assemblages in order to reconstruct the reduction sequences. The application of various types of statistical analysis permitted the identification of intra- and inter-assemblage variability which clearly defined a marked temporal trend from the Magdalenian to the Mesolithic and which can be related to changes in the settlement system, land use, mobility, demography and subsistence.

Milliken confronts the question of the role of raw material availability in technological organization. By means of a detailed study of two Late Palaeolithic assemblages in south-east Italy, Grotta delle Cipolliane and Grotta Paglicci, she tests the model that hunter-gatherers in non-source areas can be expected to have employed economising strategies in lithic procurement, tool production and use. She identifies the factors operating at a regional scale that contribute to the observed assemblage variability: the availability of raw material, and the alternative strategies by which it was procured; the particular activities in which the tools were made and used; and the role of the sites within a settlement system. She thus demonstrates how assemblage variability arises from the interaction of multiple causal factors as conditioners of technology, and that the prime mover models which have been proposed in the archaeological literature are too simplistic.

Kempcke-Richter uses a combination of refitting and an analysis of the cores at the late Ahrensburgian site of

Jerxen-Orbke in Germany to reconstruct the various patterns of core exploitation, and to identify different activity areas within the site. Although the assemblage contained typical Late Palaeolithic tools which permitted a preliminary and general dating, the absence of 'fossils directeurs' did not consent a more specific cultural definition of the assemblage. The reconstruction of the various patterns of core exploitation provided a basis for making comparisons with other well dated Late Palaeolithic sites in the region, and thus for attributing the assemblage from Jerxen-Orbke to the tanged points culture group.

Cattin describes a study of the lithic raw material economy at two Magdalenian sites in Switzerland: Champréveyres and Monruz. On the basis of the reconstruction of the strategies of procurement and economic management of the different types of raw material, she proposes various intriguing hypotheses regarding the mobility patterns of these Late Glacial hunter-gatherer groups.

Bodu presents a meticulous analysis of the early Azilian level at the site of Le Closeau in the Paris Basin. At this site of the Federmesser groups tradition he reconstructs the reduction sequences and examines the spatial organization of the activity loci. On the basis of the results of this study, he suggests a filiation between the last Magdalenian groups in the Paris Basin and the first groups of Azilians.

Goring-Morris, Marder, Davidzon and Ibrahim present the preliminary results of a comprehensive program of refits conducted on 24 assemblages representing various Epipalaeolithic industries (20-10,000 BP) from the western Negev Desert, Israel. The use of this approach provides an opportunity to investigate both synchronic and diachronic variability at various levels, and furnishes insights into the mobility patterns of these hunter-gatherer groups and the intrasite spatial organisation of their activities.

Despite the fact that the geographical coverage of this volume is not as complete as might have been wished, it nevertheless constitutes a significant contribution to our understanding of human adaptations to the social and physical environment during the Late Glacial and Early Postglacial in Europe, and will hopefully provide a stimulus for further research.

REFERENCES

Bamforth, D.B.
1991 Technological organization and hunter-gatherer land use. *American Antiquity* 56: 216-235.

Binford, L.R.
1979 Organization and formation processes: looking at curated technologies. *Journal of Anthropological Research* 35: 255-273.
1980 Willow smoke and dogs' tails: hunter-gatherer settlement systems and archaeological site formation. *American Antiquity* 45: 4-20.

Carr, P.J.
1994a The organization of technology: impact and potential. In P.J. Carr (ed.) *The Organization of North American Prehistoric Chipped Stone Tool Technologies.* Ann Arbor: International Monographs in Prehistory, pp. 1-8.
1994b Technological organization and prehistoric hunter-gatherer mobility: examination of the Hayes Site. In P.J. Carr (ed.) *The Organization of North American Prehistoric Chipped Stone Tool Technologies.* Ann Arbor: International Monographs in Prehistory, pp. 35-44

Kelly, R.L
1988 The three sides of a biface. *American Antiquity* 53: 717-734.

Koldehoff, B.
1987 The Cahokia flake tool industry: socioeconomic implications for late prehistory in the Central Mississippi Valley. In J.K. Johnson & C.A. Morrow (eds.) *The Organization of Core Technology*. Boulder: Westview, pp. 187-206.

Nelson, M.C.
1991 The study of technological organization. In M. Schiffer (ed.) *Archaeological Method and Theory* Vol. 3. Tucson: University of Arizona Press, pp. 57-100.

Parry, W.J. & R.L. Kelly
1987 Expedient core technology and sedentism. In J.K. Johnson & C.A. Morrow (eds.) *The Organization of Core Technology*. Boulder: Westview, pp. 285-304.

Shott, M.
1986 Technological organization and settlement mobility: an ethnographic examination. *Journal of Anthropological Research* 42: 1-15.

Straus, L.G.
1996 The world at the end of the last ice age. In L.G. Straus, B.V. Eriksen, J.M. Erlandson & D.R. Yesner (eds.) *Humans at the End of the Ice Age. The Archaeology of the Pleistocene-Holocene Transition.* New York: Plenum Press, pp. 3-9.

Sarah Milliken
March 1998

FROM RAW MATERIAL PROCUREMENT TO TOOL PRODUCTION: RECONSTRUCTION OF THE LITHIC PRODUCTION PROCESS DURING THE LATE GLACIAL PERIOD IN THE EASTERN PYRENEES

Xavier Terradas

INTRODUCTION

The aspects related to the lithic technology implemented by the human groups that occupied the Eastern Pyrenees in the Late Glacial period have been generally addressed from a positivist perspective. This perspective is rooted in a clearly empirical and culturalist tradition. Consequently, this type of archaeological research has been based on the formal aspects of the different stone tools, i.e. their morphological description, typological classification and chronological order. Furthermore, this research has not accounted for the phenomenological singularity of these aspects.

This type of archaeological research is justified by the fact that in practice a theoretical methodological approach is absolutely atheoretrical. That is to say, it is developed without an explicit initial substantive standpoint and, hence, does not offer any type of previous theoretical construction on the causality of the lithic remains under study.

It is for this reason that, for some years now, we are working on the construction of a theoretical model of the mode of production of the hunter-gatherer communities from an interdisciplinary perspective (Estévez *et al.* in press). The aims of the construction of this model are the following:

- to make empirical generalisations, to formulate hypotheses and laws with the help of which the phenomena under study could be explained;
- to arrange the different processes that are being considered, as well as their phenomenological manifestations by means of an organic structure of categories and concepts. By doing so, we aim to systematise and transmit the abstraction of this reality in a codified way.

The construction of any theoretical model on the reality we are currently studying will always be determined by the way in which its apprehension has been organised, according to our starting substantive theory (Argelés *et al.* 1995). Therefore, even though we might use a methodology similar to that used in other theoretical methodological approaches, as is the case of the *chaînes opératoires*, the results that are obtained, as well as their interpretation, will be wholly different. This is due to the fact that we proceed on different substantive theories.

The subject of study of this paper is the lithic assemblages recovered from different Late Glacial settlements located in the highest parts of the river valleys of the Eastern Pyrenees. These sites correspond to the first human occupations ever recorded so far in that area. Their accessibility would have been improbable before 12,000 BP, the date of the great retreat of the glaciers in the Pyrenees (Bordonau *et al.* 1992; Fullola *et al.* 1995).

During the Holocene, the progressive occupation of the river basins of the Eastern Pyrenees by hunter-gatherer communities favoured the accessibility not only to biotic resources, but also to abiotic resources. As regards mineral resources, the exploitation of the raw materials became a determining factor in the qualitative and quantitative composition of the lithic assemblages recovered from the archaeological sites discussed here.

As we will explain later, the stone tool production techniques were adapted to the nature and availability of the raw materials found in the immediate environs of the sites. Therefore, as an alternative to environmental limitations, these human groups applied several particular and specific technical solutions to the exploitation of the morphologies and lithologies of the blocks of the selected raw materials.

The formal singularity of the lithic remains resulting from this process lies in the rarity and diversity of the raw materials used in the production of stone tools, as well as in the morphology of these tools. The singular character of these stone tools and their scarcity, in a first stage of research, were the cause for them being almost ignored in the Spanish-French archaeological literature. The morphotypological features of these assemblages prevented them from being ascribed to any of the facies differentiated in the chronocultural seriations that were proposed for the Western Mediterranean Late Glacial period (Fortea 1973; Fullola *et al.* 1995). As some authors sustain (Barbaza and Martzluff 1995), the technological procedures undertaken for the production of these morphotypological features were not determined by cultural factors.

We will try to explain the specific nature of these stone tools beginning with the reconstruction of the lithic production process that produced them. Even though most of the data in this paper have been gathered from the analysis of the finds from the Late Glacial site of Font del Ros (Berga, Barcelona), (Terradas 1995a, 1995b), the main features of this lithic production process can also be documented in other sites located on both sides of the Eastern Pyrenees.

THEORETICAL FRAMEWORK: RECONSTRUCTION OF THE MINERAL RESOURCE MANAGEMENT AND ESTABLISHMENT OF A SOCIO-ECONOMIC BEHAVIOURAL MODEL

Raw material procurement and tool production are part of what we have been defining as the lithic production process (Vila 1987; Terradas 1995a, in press; Martínez and Afonso in press), which has been dealt with and characterised more in depth in a recent study (Terradas 1996). The lithic production process is one of the processes through which the hunter-gatherer communities acted on the environment in order to obtain material goods of a mineral nature. This process is one of the subsistence activities which developed within the global production process of these societies, in accordance with the organisational strategies that ruled their production dynamics.

The working processes that configure the lithic production process are the procurement of lithic raw material and its transformation into consumption goods. Therefore, the aim of this production process is to obtain lithic raw material of different natures, morphologies and functions. It is necessary to point out that although in the case in question most of the production volume obtained from the exploitation of the mineral resources was intended for tool manufacture, we should not forget the existence of other production strategies with very different aims. We are referring, for example, to the procurement of mineral dyes (oxides, clays etc.), of rocks used as building materials (for housing, windshields, pavements etc.) or of those used in other subsistence activities (production of heat, food cooking, water heating etc.) among others.

The development of the lithic production process entails the articulation of successive, parallel and/or accumulative production activities implied in the transformation of raw materials into products. This transformation carries along constant and progressive raw material alterations in their original conditions and properties related, not only to their spatial context, but also to their morphological traits. Normally, these sorts of changes take place in a combined way due to the simultaneous nature of the activities to which we can attribute their causality. These alterations begin in the geological context from which the raw material was obtained, continue in the society's production cycle, and end in the archaeological context/s where the effects of its transformation are recovered. These dynamics throw light onto a lithic reduction sequence which is more or less dispersed in time and space.

The recording of these spatial and morphological transformations is made possible as a result of the particular lithological properties of the different rocks that later become raw material. At the same time, the causality of its changes can be seen from the specific nature of the alterations brought about by social and/or natural agents.

In order to obtain material goods of a mineral nature, the hunter-gatherer communities devised and applied different production strategies, the implementation of which depended on the consuming goods they needed in order to satisfy their social needs. Having socially recognised these material needs, each society devised a raw material procurement and exploitation mode, according to the tools and to the way they technically organised the production that their production forces were ready to obtain, produce and/or implement.

Therefore the development of these organisational strategies conditions the intervention of a society in the mineral resources of the land they occupy, and thus it generates a specificity in the way they manage them. According to this idea, the mode which a society adopts for the mineral resources of the environment is referred to as the management of mineral resources (Terradas 1995a; 1996). This does not only affect this type of resource, but it also affects the materials extracted from them and the various products that can be obtained from their exploitation.

The reconstruction of the strategies implemented by a society in the management of mineral resources in the land they occupy enables us to characterise a segment of their productive cycle, within which we can differentiate two closely interrelated aspects (Figure 1):

- the procurement of consumption goods of a mineral nature, which makes the characterization of the lithic production process possible;
- the integration of consumption goods of a mineral nature in other working processes. Lithic production does not constitute an aim in itself, but a previous condition for the procurement of new consumption goods for the satisfaction of other social needs.

The specificity of the intervention mode for the mineral resources chosen by a society is reflected in the strategies that are applied in their management. The thesis we propose is that the strategies that were devised and applied in the management of mineral for the production of stone tools belong to the group of organisational strategies that rule the biological and social reproduction dynamics of hunter-gatherer communities. Consequently, the explanatory model of the management type applied to the mineral resources, as well as its phenomenological singularity and its causality, cannot be left aside in the account of the other productive processes. Conversely, this model should be generated taking into account all the organisational strategies that rule the socio-economic dynamics of these societies.

The notion of organisational strategy does away or minimises any random component within the global production process of a society. As its name indicates, the term we have used to designate this concept has, implicitly, a strategic character (its aim is to attain a goal) and an organisational character (its planning and implementation

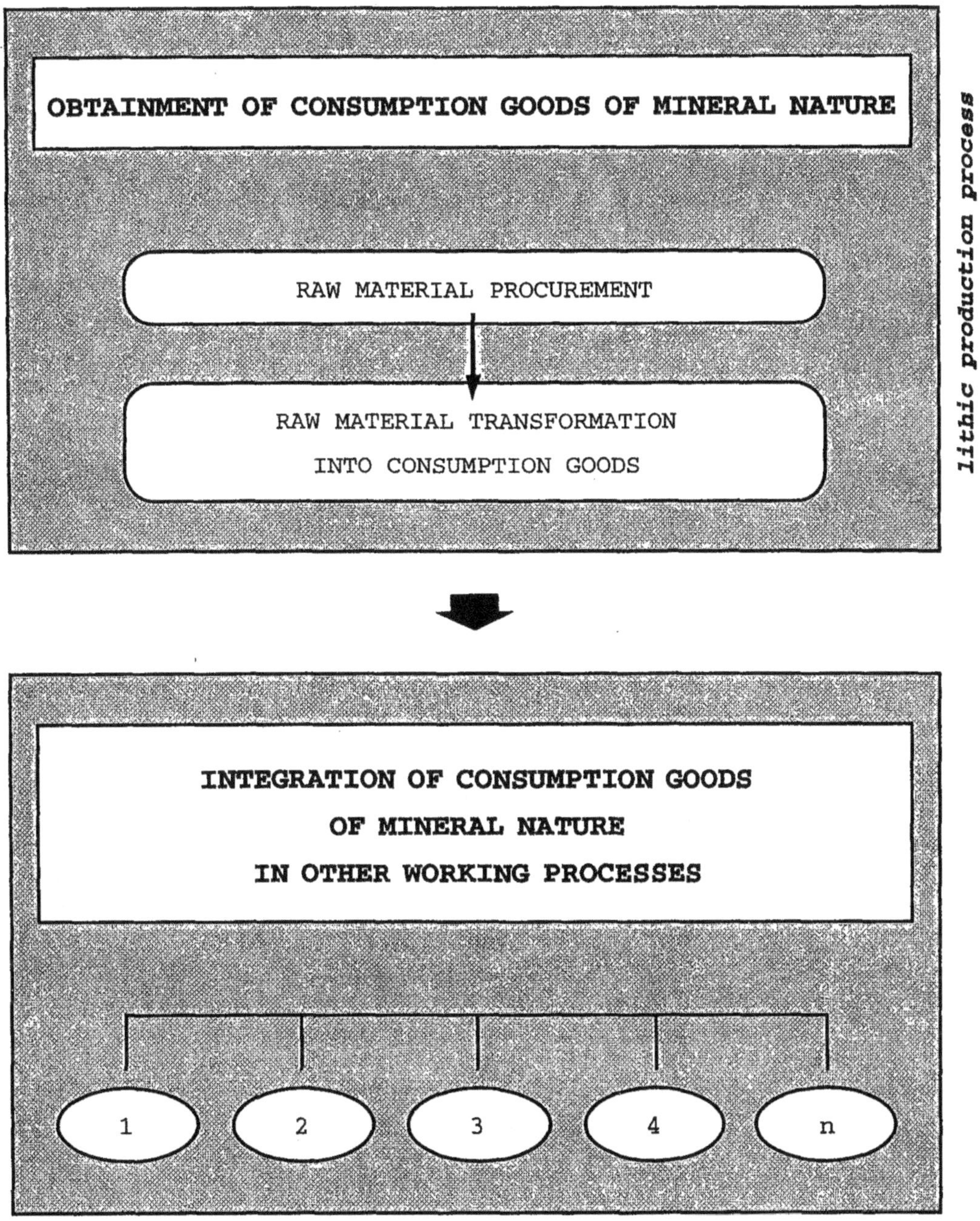

Figure 1: Segment of the productive cycle of hunter-gatherer communities involved in the development of the mineral resources management (Terradas 1996)

stems from an organic structure). Thus, the concept of organisational strategy refers to the development of an articulated and planned group of processes that were socially established, regulated and determined. These processes ruled the social production and reproduction dynamics of hunter-gatherer groups. This social organisation of the economic processes entails the following:

- a foresight, with more or less a degree of anticipation, of the social needs along its productive and reproductive cycle, in other words planning of the productive dynamics according to social requirements;
- a knowledge needed to attain major efficiency in the procurement, production and use of consumption goods.

The organisational strategies that a society applies become the coordinated and articulated development of a set of working processes which correspond to the global production process of this society. These strategies were pursued to obtain and/or elaborate the various types of goods required to satisfy its social needs, according to the level reached in the development of its production forces. Therefore, the reconstruction of the productive strategies of this society will enable us to characterise its productive dynamics and, at the same time, to identify the distinctive and particular features of their global production process and, moreover, to establish a socio-economic behavioural model.

From our theoretical-methodological research framework, we advocate that the reconstruction of the production process of any hunter-gatherer community, from the material remains of their social activity, should be carried out taking into account the management which this society developed of the various natural resources available in their environment. Therefore, the type of natural resource management applied by a human group is representative of the environment procurement process implied by the global productive process of this society.

From these theoretical premises, we try to characterise the lithic production process developed for the procurement of material goods of a mineral nature by the human groups that settled in Font del Ros during the 9th millennium BP. The reconstruction of the organisational strategies that were implemented by these communities in the management of mineral resources in the highest parts of the river valleys of the Eastern Pyrenees and its interpretation according to its temporal and spatial dynamics enable us to analyse the following aspects:

- the technical organisation of production developed by these human groups;
- the establishment of the territorial framework within which their social activity took place;
- their management of the land.

This type of social organisation of economic processes is not exclusive to the communities that settled at Font del Ros. It is a recurrent phenomenon that can also be found in other sites with similar chronological and geographic contexts. Therefore, we wish to initiate the characterisation of a socio-economic behavioural model for the hunter-gatherer communities that settled in the Eastern Pyrenees during the Late Glacial period.

METHOD AND METHODOLOGY

The reconstruction of the productive strategies involved in the procurement of mineral consumption goods by any hunter-gatherer community entails the study of two productive activities comprised in the lithic production process:

- the procurement of the raw materials used in the stone tools production;
- the procedures followed in the transformation of these raw materials into working tools.

The study of each of these thematic areas will enable us to reconstruct and characterise the technical procedures followed in stone tool production. However, our study must go beyond the identification of the techniques developed in the management of mineral resources for the procurement of mineral consumption goods. According to our criterion, the choice of the social agents to the detriment of other technical procedures that they were able to apply ought to be justified. For this reason, the design and development of specific analytical frameworks according to the initial substantive theory, that are adapted to the nature of the studied object and to the aims that we wish to attain by analysing them, is a requirement that needs to be fulfilled. The general features of the design of our proposal have been published on other occasions (Vila 1987; Pié and Vila 1992; Terradas 1995a, 1996), and therefore in this paper we only will go over its main features.

The reconstruction of the strategies of raw material procurement used in stone tool production is based on the petrological identification and characterisation of the raw material types documented in the archaeological contexts under study. Once we have this characterisation, it is possible to apply a valid contrasting method that could help to determine the geological and geographic origin of various raw materials represented in the studied archaeological contexts. Its applicability is based on the fact that the lithological features of any rock are representative of its geological origin because the specific conditions (regarding its genesis, composition, postdepositional alterations, etc.) of every geological formation conferred differential traits. This method is based on the qualitative and quantitative comparison between the results obtained from the analysed samples in the archaeological as well as in the geological contexts that could have constituted their raw materials sources (Terradas 1995a, 1996, in press; Terradas *et al.* 1991). In this case the analytical techniques used in this comparative process were polarising microscopy of thin sections and x-ray diffraction.

Stone tool production techniques may be reconstructed from the characterisation of the different phases that can be established in the transformation of raw materials into stone tools. The successive changes that take place in the original morphology of the raw material can be recorded by considering several discriminant morphotechnical characters. The hierarchical reading of those morphotechnical attributes contained in every material remain of the lithic production process enables us to determine their relative position in this process, as well as to get to know the type of technical dynamics that favoured their morphogenesis or acquisition sequence of these features (Carbonell *et al.* 1992; Terradas 1995a, 1996).

The working processes that interact in the strategies applied in the mineral resources management for the procurement of mineral consumption goods may be characterised by the material elements that take part in these processes. These are the following:

- the required product,
- the object to be transformed,
- the working tool,
- the technique they used,
- the derived products.

Henceforth, the characterisation of the working processes that constitute the strategies implemented in the mineral resources management by the interaction of production elements makes it possible to gain access to a dynamic and economic dimension of the development of these strategies and, also, their effects.

RECONSTRUCTION OF THE LITHIC REDUCTION PROCESS AT FONT DEL ROS

The SG archaeological unit at Font del Ros (Berga, Barcelona) corresponds to the material remains of a seasonal settlement of hunter-gatherers dating to the 9th millennium BP in the highest parts of the Llobregat valley, around 700 m above sea-level (Figure 3). The radiocarbon dates available for unit SG at Font del Ros, which were obtained on charcoal fragments at the Radiocarbon Dating Service of the University of Barcelona, are 8150±590 BP, 8050±150 BP and 8280±200 BP. Font del Ros is an open-air site which extends over a large surface are (1079 square meters excavated for the SG unit), (Mora *et al.* 1991; Jordá *et al.* 1992; Terradas *et al.* 1992). The most important traits of the working processes that rule stone tool production at this site are described in the chapter that follows (Terradas 1995a, 1995b).

RAW MATERIAL PROCUREMENT

The lithic raw materials selected for tool production are very varied: siliceous rocks (43.23% of the finds, n.2910), quartz (38.01%), limestone (10.79%) and quartzite are the most used. Although we refer to quartz as a rock, we should make it clear that we do not consider it as such, but rather as a mineral aggregate. The remains of this aggregate which we have studied have a magmatic origin and they are related to certain igneous rocks. As regards the quartzites, almost all the remains of these rocks which we have identified correspond to quartzites of metamorphic origin. The rest of the raw materials transported to Font del Ros are barely represented (none of them represents more than 5% of the finds). They are igneous rocks (granite, andesite, lava, rhyolite, ignimbrite) as well as metamorphic rocks (gneiss, slate) and sedimentary rocks (sandstone, loam).

This variety in the nature of the rocks brought to the settlement is translated into several differential features that are reflected not only in their suitability for knapping but also in the original morphology of the raw material blocks. Therefore, there is a differential representation of the raw materials transported to the site depending on whether we consider the number of finds or their weight (Figure 2). This can be explained by the differential response of the natures of the rocks to the type of exploitation they underwent (conchoidal fracture for siliceous rocks and granular for the rest of the rock types), obtaining morphologically differentiated products.

Most of these rocks belong to conglomeratic series from the Middle and Upper Eocene and only the siliceous rocks have a more diversified origin (Figure 3), as specified in the table below.

Mineral resources	No.	%	Grams	%
Upper Cretaceous	540	43.0	1714.33	38.6
Lower Eocene -a-	98	7.8	182.67	4.1
Lower Eocene -b-	43	3.4	127.63	2.9
Middle-Upper Eocene	446	35.5	2077.11	46.8
Indeterminate	131	10.4	342.44	7.7
Total	1258	100	4444.18	100

And so, we confirm an intensive exploitation of the mineral resources found near the settlement (within a shorter or equal distance to that covered in an hour's walk) defined as the raw material procurement zones. The resources found on the banks of the river Llobregat (Upper Cretaceous calcareous series for the procurement of silicified nodules and conglomeratic series of the Upper and Middle Eocene for the procurement of various pebbles, Figure 3), were especially exploited at the expense of other resources that are more distant and, hence, more difficult to procure. To conclude, we highlight the large amount of raw material brought to the settlement for its subsequent transformation into tools (28.5 kg approximately).

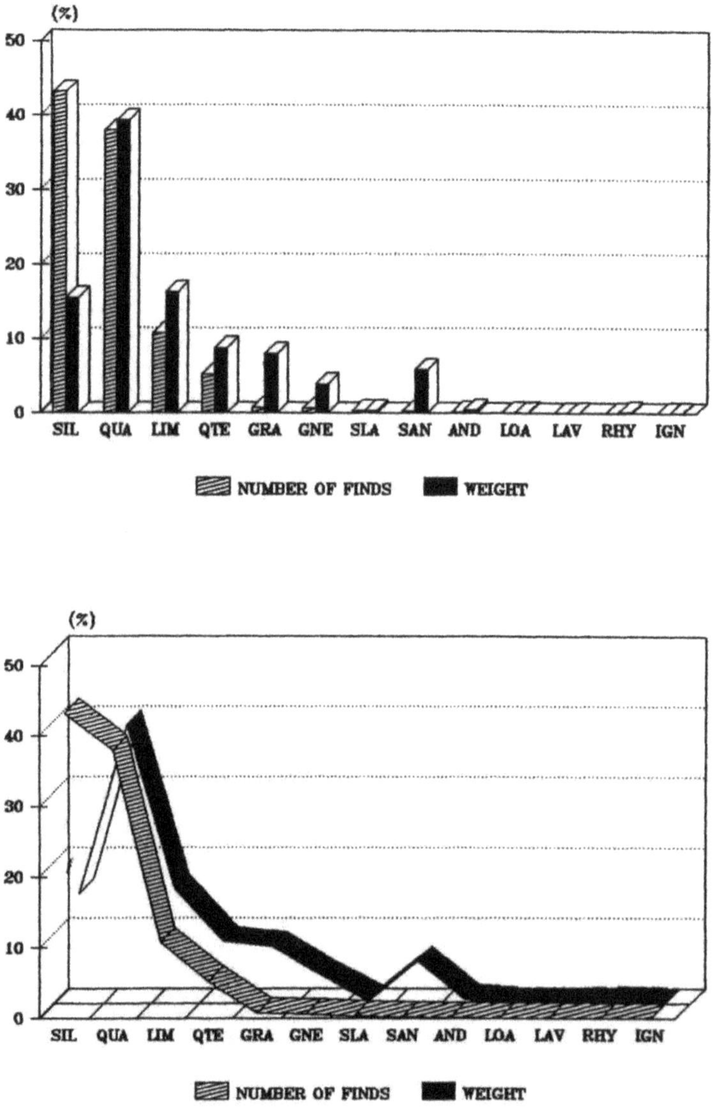

Figure 2: Histogram and tendency with the percentage of raw material types depending on whether we consider the number of finds or their weight (Terradas 1995a)

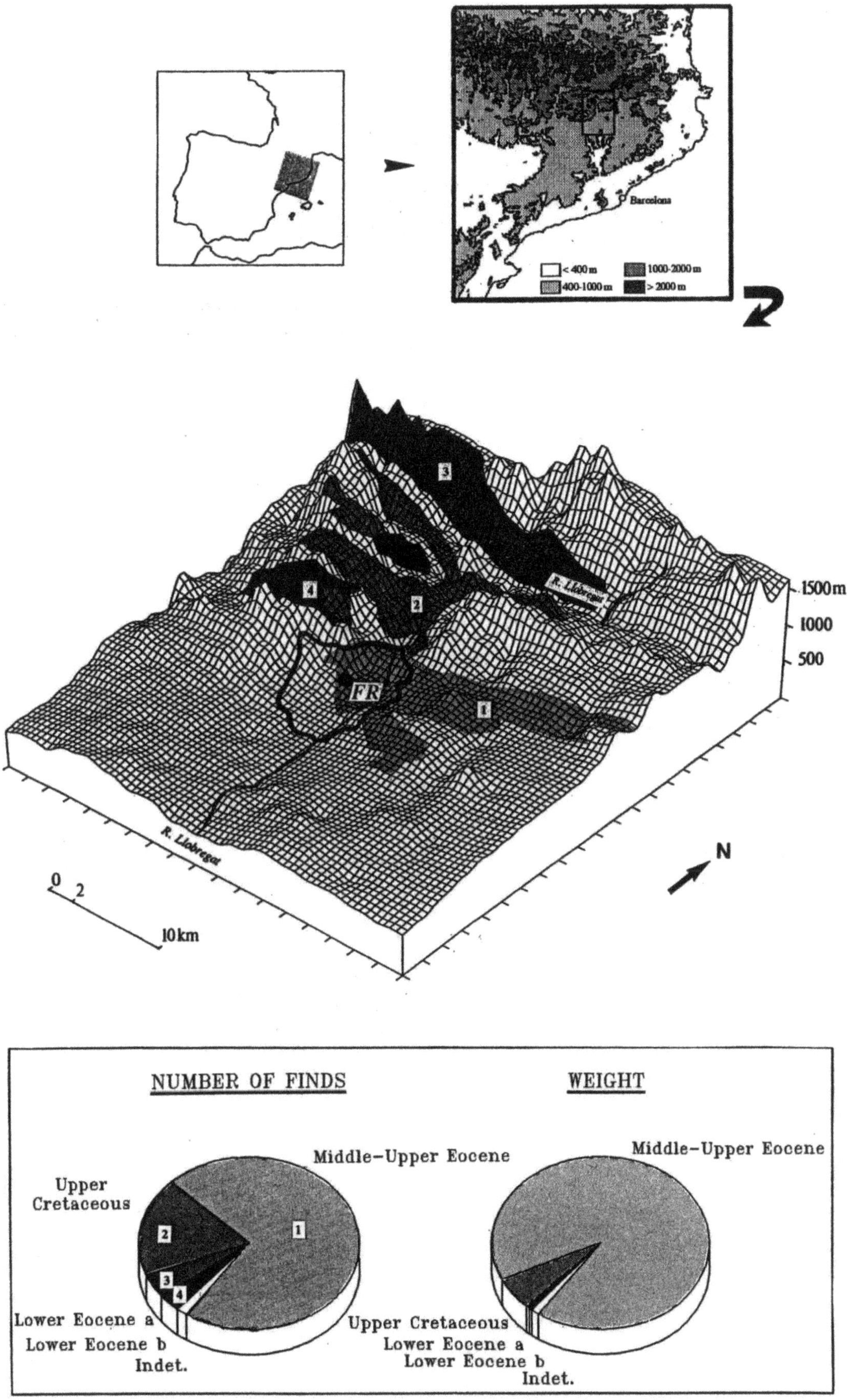

Figure 3: Exploited mineral resources and the location of the site of Font del Ros (FR) (large line delimits a shorter or equal distance to that covered in an hour's walk) (Terradas 1995a; 1995b)

STONE TOOL PRODUCTION

In this working process we distinguish three stages. These stages have their own underlying technical conceptions that can be organised in a spatial and temporal sequence that might be very variable. These are the following (Terradas 1995a, 1996):

- shaping of the raw material blocks into morphologies suitable for the procurement of supports;
- exploitation of these morphologies shaped for the procurement of supports by means of a knapping sequence;
- formatting by retouching some of the supports obtained in the previous stages.

Nevertheless, these stages do not form a liear sequence, for stone tools need not necessarily go through all the stages of their production.

Shaping the raw material blocks
The shaping carried out was minimal or non-existant because blocks with specific morphologies had already been selected in the procurement areas. For instance, in conglomeratic series from the Middle and Upper Eocene, angular pebbles were chosen which had several flattened surfaces which we later used as the striking platforms of the cores. Consequently, the decortication process of the selected raw material blocks did not correspond to a preparation of the cores consigned to the beginning of their exploitation. Therefore, the flakes obtained from the decortication of the raw material block were already part of the particular exploitation process of the core.

Consequently, the configuration of raw material blocks took place without a transformation of their original morphology. This shaping was carried out in their particular raw material procurement areas, by the selection and utilisation of the blocks with specific morphologies, which were close to the volumetric abstraction patterns that rule the types of exploitation of the cores for the procurement of supports. The selection of specific morphologies means that the preparation of striking platforms of the cores was not very elaborate, and somtimes non-existat or only minimal.

Exploitation of the cores
The exploitation of the cores seems to have bee applied to flake production. The types (Guilbaud 1985-1987; Terradas 1995a) most represented in the SG unit at Font del Ros are the following three:

- exploitation of a flaking surface by means of a conical transformation (Figure 4a). The transformation develops from a sigle striking platform. Depending on the morphology of the original block, a cortical surface or an internal fissure plan of the rock was used as a striking platform. On other occasions, the striking platform was the result of a neutral or centripetal transformation. The number of supports that were obtained was very variable, as it depends on the intensity with which the core was exploited;
- exploitation of several flaking surfaces by neutral transformations (Figure 4b). The transformation developed from several striking platforms on the cortical surfaces or on the internal fissure planes of the rock. The flaking surfaces are only slightly developed because the amount of extraction carried out on each of them was very low, and hence, the amount of flakes is very small as well;
- exploitation of flaking surfaces by centripetal transformations or discoidal exploitation (Figure 4c). This type of exploitation is based on the simultaneous exploitation of two flaking surfaces. Each surface was used as the striking platform of the other. Sometimes only one flaking surface was exploited from a cortical striking platform.

A very low number of products has shed light on the exploitation of a flaking surface by a cylindrical transformation to obtain laminar supports. The importance of this type of exploitation is very reduced in relation to the previous ones. Due to the outstanding variety observed in the nature of the selected raw materials and their lithological and morphological diversity, several technical solutions were generated. The solutions aimed at raising the profitability of the cores exploitation type by the production of a larger number of supports are as follows:

- adaptation of the exploitation performed on the nature of the selected raw materials. It was observed that each type of exploitation of the cores was performed on specific raw materials (siliceous rocks and limestones collected in the Eocene conglomerates, Figures 3 and 4a; silicified nodules from the Upper Cretaceous, Figures 3 and 4b; and quartz pebbles from Eocene conglomerates, Figures 3 and 4c). Moreover, the quartzite pebbles collected in the Eocene conglomerates represent the most adaptable lithology, for cores made on this rock are represented in all the described exploitation types;
- saving the best quality blocks (quartz and specific siliceous rocks from the Eocene conglomerates) for more complex and, at the same time, more productive types of exploitation (exploitation of two flaking surfaces by centripetal transformations, Figure 4c);
- utilisation of natural surfaces (cortex and internal fissure planes of the rock) as the striking platforms of the cores (Figures 4a and 4b). This utilisation was not casual, since a selection of these raw material blocks was made in the procurement areas;
- change or rotation of the striking platforms as new internal fissure planes emerged within the raw material block (Figure 4b). The presence of these planes prevented the performance of some exploitation types in the cores. At the same time it made their utilisation as striking platforms possible in order to exploit new

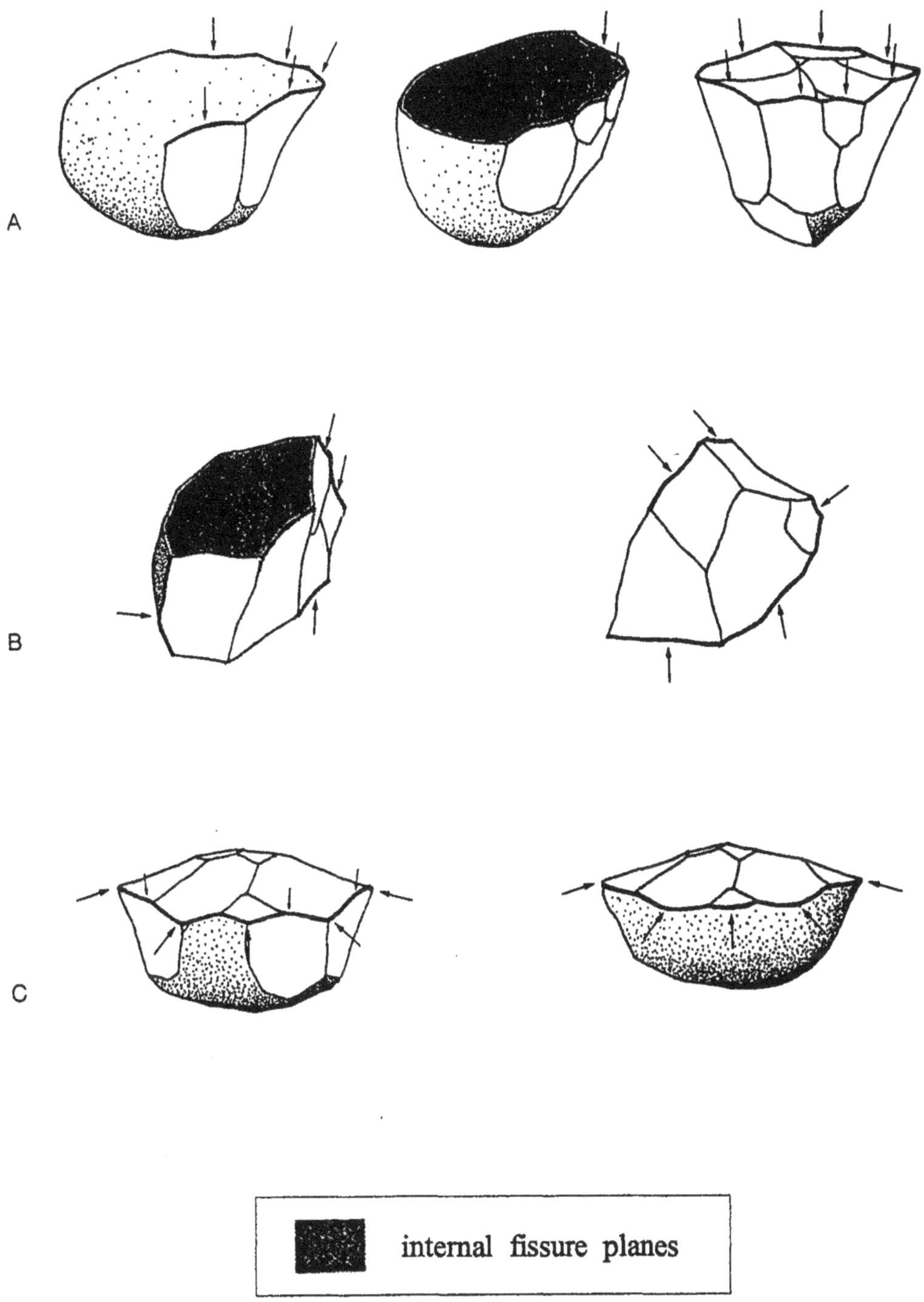

Figure 4: Scheme of the core exploitation types at Font del Ros: a: exploitation of a flaking surface by means of a conical transformation, b: exploitation of several flaking surfaces by neutral transformations, c: exploitation of flaking surfaces by centripetal transformations of discoidal exploitation (Terradas 1995a)

flaking surfaces;
- abandonment of the cores when the number of fracture platforms made it impossible to proceed with the exploitation. Therefore, their productive capacity was deemed exhausted when part of the original surface of the block was still untouched. Hence, for the production of a given number of supports, the amount of raw material needed was larger than that required by other cores exploitation types. This fact dis not pose any obstacle to the development of the stone tools production process, due to the large amount of raw material available and the proximity of the procurement areas. On the contrary, the availability of raw material could account for the large quantities in which it was transported to the site (28.5 kg)

These technical solutions were recurrent, and they costitute the core exploitation modalities (Figure 4) that already have been described. These types of exploitation were neither very standard nor homogeneous. On the contrary, they were very variable depending on the specific conditions of each raw material block. They have common primary features such as the low number of supports obtained from each core, heterogeneity, and the need for a large amount of raw material to carry out this exploitation.

Formatting the supports by means of retouch
Even though the percentage of formatted supports is very low (5.02 % of the lithic artifacts), there was a preferential selection of the supports obtained from the exploitation of siliceous rocks (82.19 % of retouched pieces) for formatting. However, there is no clear relationship between siliceous rocks and specific primary types. The most widely represented typological order is the Simple type (Laplace 1972), followed by Abrupt and Scaled pieces, while the typological groups that are most commonly documented are the denticulates, scaled pieces, abrupts and scrapers. Geometric elements are absent. Therefore, formatting techniques were simple and they seem to have been related more to the maintenance of the functionality of some given morphologies rather than to the production of specific morphologies.

DISCUSSION

If we take into account the number of supports obtained for each exploited core and the quantity of raw material needed for their production, the yield of the lithic production process could seem meagre. The fact that the production a small amount of low quality tools required a great amount of raw material would have increased their production cost. However, the proximity of the settlement to the mineral resources exploited made the transportation of the raw material blocks less costly as far as effort was concerned. Consequently, the production cost was notably reduced. Therefore, the low yield observed in stone tool production was compensated by the minimal effort required for the procurement of a large quantity of raw material. Actually, what at first might seem to be a productive process of low yield was in fact only an adaptation of the process to the raw materials of mineral nature found in the immediate environment of the Font del Ros settlement. We could affirm, therefore, that the yield of stone tool production was minimal, but that it was adapted to the transportation cost of raw materials.

This adaptation, as an alternative to the environmental limitations, resulted in some particular technical and specific solutions. These solutions were aimed at taking advantage of the exploitation of the morphologies and lithologies of the existing raw material blocks in the mineral resources of the highest parts of the Llobregat valley. Moreover, this adaptation of the technical exploitation procedures to properties related to the nature (quality and morphology) as well as to the availability (quantity and distance) of the raw material blocks of the immediate environs of the site is also reflected in the rest of productive processes bound to raw material procurement.

Figure 5 is the graphical representation of a correspondence factorial analysis (CFA) obtained by comparing the raw material types transported to the site with the categories differentiated in the morphotechnical analysis of stone tools. Three associations can be distinguished in the analysis:

- firstly, that of siliceous rocks, quartz and quartzite with the morphotechnical categories generated during tool manufacturing (cores, flakes, fragmented flakes, debris and retouched pieces);
- secondly, that of gneiss with granite and hammers;
- thirdly, that of granite with some sedimentary rocks (limestones and conglomerate) and with pebbles. There is also an association of sandstones with tabular morphologies which proves that these raw materials were itroduced onto the site in these morphologies, in other words without any particular transformation in their original morphology. In this regard, we have already outlined that these rocks were documented in the settlement in concentrations associated with the combustion areas and that most of the modifications that they underwent were due to thermal causes (Mora *et al.* 1991; Terradas *et al.* 1992). For this reason, we believe that the procurement of these rocks was closely linked to productive activities others than tool manufacturing, i.e. food preparation or heat maintenance, among others.

It has been shown, then, that some raw materials were deliberately selected for specific productive activities. We believe that this selection was aimed at obtaining a larger yield in the lithic production cost/social profit equation. We should not forget that the yield of this equation was always aimed at satisfying the social needs which motivated the application of the organisational strategies described previously in the mineral resources management in the

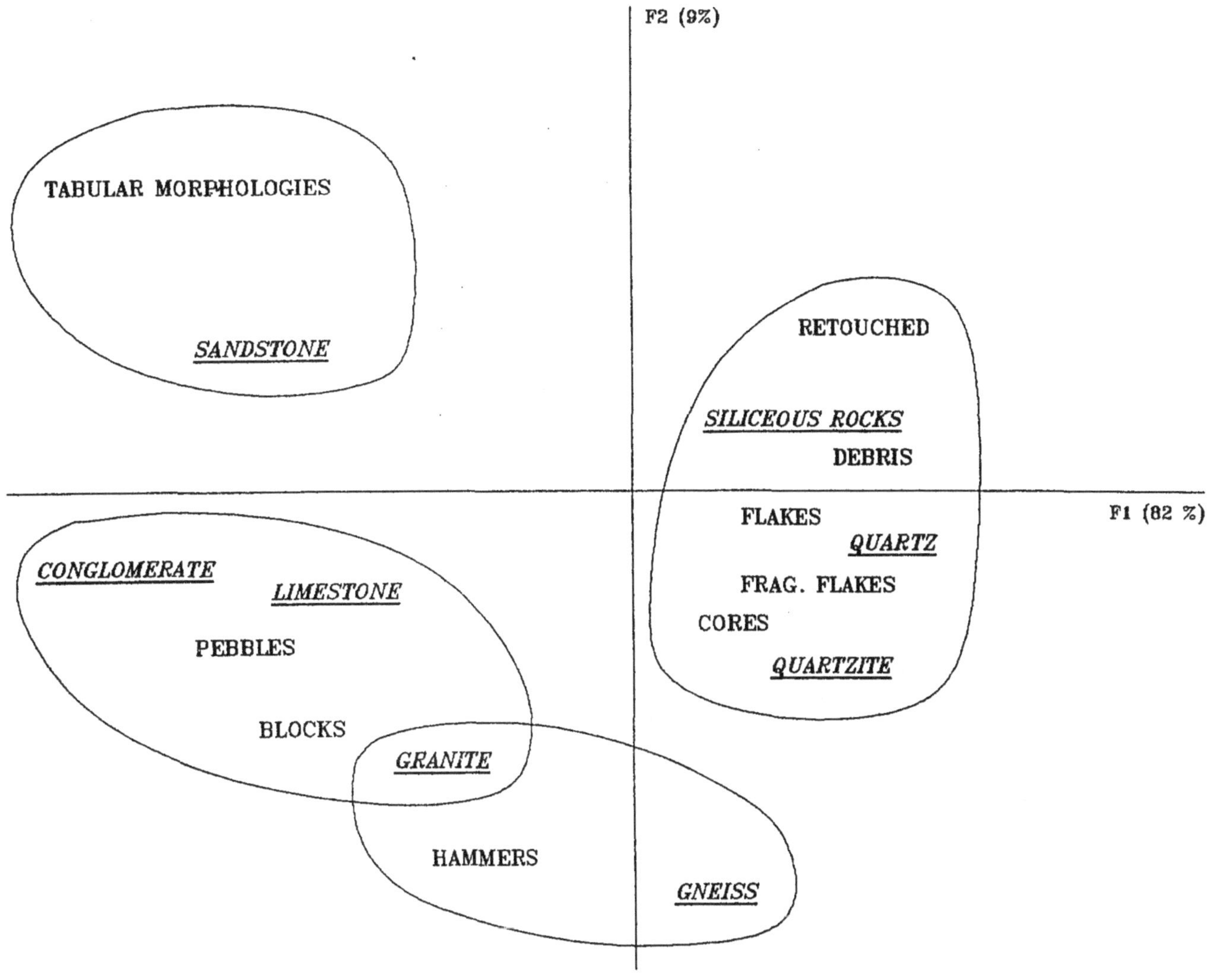

Figure 5: Graphical representation of a correspondence factorial analysis obtained comparing the raw material types with the differentiated categories in the morphological analysis (Terradas 1995a)

highest parts of Llobregat valley. The type of mineral resources management for stone tool production implemented by the human groups which settled at Font del Ros and the delimiting, and subsequent management, of the territorial framework within which their social activity took place enable us to document the existence of a short term subsistence activities organisation. This organisation had an opportunistic character and was unrelated to the specialised exploitation of a unique lithic resource. Even though the tools used in these activities had a low production cost, they still allowed for the satisfaction of immediate needs.

CONCLUSIONS

The rest of the subsistence activities documented at the settlement at Font del Ros complement the approach formulated on the mineral resources management, and were characterised by the diversification of the exploitation of the various natural resources found in the highest parts of the Llobregat valley.

As far as vegetal resources are concerned (Piqué 1995), the wood used as fuel at the site came from trees which are representative of a mild and humid climatic period (namely oak, *Quercus* sp., hazel, *Corylus avellana*, and box, *Buxus sempervirens*), and other woods particular of riparian vegetal communities (willow, *Salix* sp., and elm, *Ulmus campestris*).

The carpological analysis (López 1993) revealed that most of the carbonised fruit remains recovered from the site (232) are hazel nuts (*Corylus avellana*). However, there are also the remains of wild pear (*Pyrus piraster*), wild apple (*Malus silvestris*) and blackthorn (*Prunus spinosa*). These data do not only document the existence of wild fruit gathering, but also the season in which these fruits were gathered: late spring-early summer (blackthorn) and late summer or early autumn (rest of species). This may give a clue as to the period in which the site was occupied.

Despite their poor preservation, the bone remains indicate the presence of an eurythermal faunal association representative of the southern European Late Glacial period (Mora *et al.* 1991; Terradas *et al.* 1992). The species include animals from forest biotopes (deer, *Cervus elaphus*, and wild boar, *Sus scropha*), open and steep biotopes (goat, *Capra pyrenaica*), open and flat biotopes (an undetermined large bovid), and rabbits (*Oryctolagus cuniculus*). In spite of the proximity of the settlement to several river courses, no ichthyofaunal remains have been recovered although their absence may be due to the poor preservation of the bone remains in general.

The presence of some continental as well as marine molluscs has been documented. The continental gastropod remains are not very abundant (155), and among them there are terrestrial species (*Cepaea nemoralis*) and aquatic species (*Teodoxus fluviatilis*). However, so far there are not enough specimens to be able to relate their presence in the settlement with human consumption. Marine molluscs remains are very scarce (only 6 specimens); nevertheless, these molluscs have a greater social and economic signification. These specimens correspond to the *Dentalium sp.* species (3), *Columbella rustica* (2) and *Glycymeris violascescens* (1), and their transport to the site suggests a direct or indirect exploitation of the Mediterranean shore biotopes by the human groups that settled in Font del Ros. The current distance between the settlement and the actual Mediterranean coast (about 125 km from the headwaters, along the Llobregat valley to the delta) makes us suppose that their procurement was not exclusively related to subsistence purposes.

The palaeoecological analyses and the knowledge of the relief of the environs of the site, show that these biotopes once used to surround the place where the human settlement took place during the Boreal climatic phase. Therefore, we could say that the activities related to the management of forest resources as well as the hunting practices had developed with the aim of exploiting the animal and vegetal communities located in the immediate environs of the settlement that were an area of important ecological plurality. The strategies applied in the management of the various environmental resources, even though they are very specific, were not exclusive of this site, and they appear to be found in other human occupations of a similar chronology (9th millennium BP) located on the northern and southern side of the Eastern Pyrenees (Figure 6).

These occupations took place in rockshelters: Balma de la Margineda (Guilaine and Martzluff 1995), Balma de Guilanyà (Terradas *et al.* 1993), Bauma del Serrat del Pont (Toledo 1988), Abric del Roc del Migdia (Paz and Vila 1988; Rodríguez and Yll 1992), Abri du Roc de Dourgne (Guilaine 1994), Caune d'Arques (Sacchi 1972), Abri du Roc d'en Bertran and Grotte des Adoutx (Barbaza 1988), as well as in the open air: Sota Palou (Carbonell 1985) and La Rodona (Alcalde *et al.* 1994). All these sites present several common features according to the settlement pattern, the organisation of technical production implemented by the human groups that settled there, and the territorial frame where the social activity and the land management developed. These features are:

- their location in the highest parts of the valleys of the Pyrenees-Mediterranean river system, at a height ranging from 400 to 1100 metres, in the eastern sector of the Pyrenees physiographic unit;
- the orography of the place where the human settlement was located, in contact areas between two large physiographic units which are very differentiated: the Pyrenees, represented in this area by the Pre-Pyrenees mountain range (with very abrupt relief as a consequence of the folding of Tertiary marine series, passable through canyons open by river courses), and

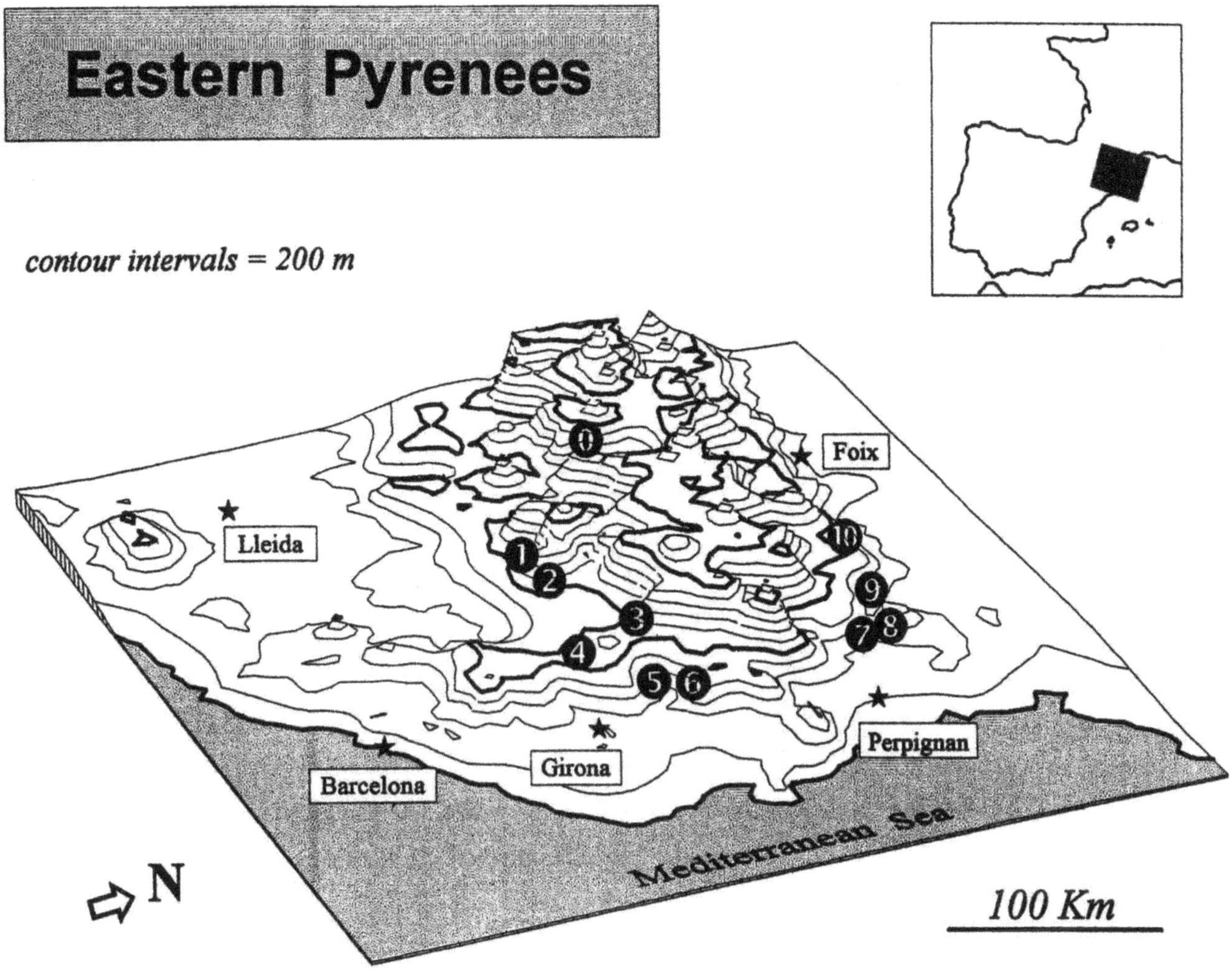

Archaeological sites:

0.- Balma Margineda (Sant Julià de Lòria, Andorra)
1.- Balma de Guilanyà (Navès, Lleida)
2.- Font del Ros (Berga, Barcelona)
3.- Sota Palou (Campdevànol, Girona)
4.- Roc del Migdia (Vilanova de Sau, Barcelona)
5.- La Rodona (Olot, Girona)
6.- Bauma del Serrat del Pont (Tortellà, Girona)
7.- Grotte des Adoutx (Caudiès de Fenouillèdes, Pyrénées Orientales)
8.- Abri du Roc d'en Bertran (Camps sur l'Agly, Aude)
9.- Caune d'Arques (Arques, Aude)
10.- Abri du Roc de Dourgne (Fontanès de Sault, Aude)

Figure 6: Location of the Eastern Pyrenean sites referred to in the text (Terradas 1995b) the plains of Rosselló, Empordà and the Great Basin of Catalunya (gentle relief corresponding to basins filled with Tertiary and Quaternary sediments);

- the direct or indirect contact with the Mediterranean shore, as proven by the discovery of different species of marine molluscs (*Columbella rustica*, *Dentalium* sp., *Glycymeris violacescens* and *Cyclonassa* sp.) at several sites (Balma de la Margineda, Balma de Guilanyà, Font del Ros, Sota Palou, Roc del Migdia, Abri du roc de Dourgne). The distance from most of these sites to the actual coastline is over 100 km along the river courses;
- the time of the year when the occupation took place (late summer or early autumn) at those sites where it has been possible to establish it (Balma de Guilanyà, Font del Ros, Sota Palou, Roc del Migdia, Abri du Roc de Dourgne);
- the diversity of the resources exploited. The orographic features of the sites resulted in the presence of ecotones or transitional zones between different biotopes. Consequentlt there was a great variety of animal and vegetal resources. As for the mineral resources, their diversity is due to the contact between several structural and physiographic units;
- the intensive use of the immediate environs of the settlements, as seen in the intensive exploitation of the mineral, animal and vegetal resources of the neighbouring environment;
- the adaptation of stone tool production techniques to the nature and availability of the raw materials existing in local mineral resources. This praxis was not exclusive to this period and it was, in fact a phenomenon that was recurrent in the Palaeolithic occupations along the Pre-Pyrenees valleys in different degrees of intensity (Soler *et al.* 1990), and which persisted until the Early Neolithic (Barbaza and Martzluff 1995).

We have outlined the mode of procurement and exploitation mode of the biotic and abiotic resources of the Eastern Pyrenees by hunter-gatherer communities that occupied these valleys during the 9th millennium BP. The organisational strategies applied by these human groups in order to meet their social needs were devised according to the tools and to the technical organisation of production that their productive forces were willing to obtain, produce and/or to implement.

The most singular element of the overall organisational strategies that ruled the socio-economic dynamics is that the general productive process they characterise was based on the exploitation of autochthonous resources of the occupied area, and no allochthonous elements have been found. As we have explained in the introduction of the theoretical framework, this fact implies, on the one hand, a planning of the productive dynamics of these human groups and, on the other, a considerable knowledge of the Pyrenees environment and of its resources, and also of the technical procedures necessary in order to attain peak efficiency in the procurement and utilisation of consumption goods. However, there is not enough evidence to assert that the human occupations of the Pyrenees valleys were continuous throughout a year. It is almost sure that in certain seasons, the climate did not favour human settlement in this area and, moreover, that a drop in the exploitable biotic resources took place, especially of vegetal resources. Nevertheless, the discovery of some marine molluscs indicates that there was a direct or indirect contact with the Mediterranean shore. Therefore, we could consider that all the production activities documented at these human settlements of the Pyrenees valleys were carried out in summer.

The success of the implementation of this overall production process lay in the diversification of the goals of the production strategies. These strategies were not aimed at the exploitation of a unique resource. The diversity of the natural resources has its origin in the ecological plurality offered by the environment where the human settlements are located, that is, in places near ecotones or transitional areas between different biotopes.

The location of human settlements in this area is not casual. It is the direct consequence of their settlement pattern. This concept refers to recurrent features in the distribution and organisation of the different segments of a society's production cycle within a territory, and its interrelationship with the environmental resources. This relationship is set in terms of complementarity, according to the specific features of the society's productive dynamics, as a reflection of the overall productive process from which the organisational strategies are generated. The existence of recurrent features that belong to the design and development of these strategies determines:

- an optimisation of the strategies in order to meet the social needs that demanded their application. This feature allows us to explain the rationality of the overall productive process of these human groups;
- the establishment of a socio-economic behavioural model for these hunter-gatherer communities that in the summer settled in the highest parts of the Eastern Pyrenees river valleys during the Late Glacial period, through the 9th millennium BP.

REFERENCES

Alcalde, G., M. Buch, J. Mateu, A. Palomo & M. Saña
1994 La intervenció arqueològica d'urgència en el jaciment a l'aire lliure de la Rodona (Olot, Garrotxa). *Tribuna d'Arqueologia* 1992-1993, 25-30.

Argelés, T., A. Bonet, I. Clemente, J. Estévez, J. Gibaja, L.G. Lumbreras, R. Piqué, M. Ríos, M.A. Taulé, X. Terradas, A. Vila & G. Wünsch

1995 Teoría para una praxis. «Splendor realitatis». *Trabalhos de Antropologia e Etnologia* 35 (1), Porto, 501-507.

Barbaza, M.
1988 La grotte des Adoutx à Caudiès-de-Fenouillèdes (Pyrénées Orientales) et le Mésolithique des Corbières. *Gallia Préhistoire* 30, 187-204.

Barbaza, M. & M. Martzluff
1995 Epipaléolithique et Mésolithique au Nord des Pyrénées. In *Cultures i medi: de la Prehistòria a l'edat mitjana.* Xè Colloqui Internacional d'Arqueologia de Puigcerdà (Girona), 177-188.

Bordonau, J., D. Serrat & J.M. Vilaplana
1992 Las fases glaciares cuaternarias en los Pirineos. In A. Cearreta and F.M. Ugarte (eds.) *The late quaternary in the western Pyrenean region.* Bilbao, 303-312.

Carbonell, E. (ed.)
1985 *Sota Palou. Un centre d'intervenció prehistòrica postglaciar a l'aire lliure.* Sèrie monogràfica del Centre d'Investigacions Arqueològiques de la Diputació de Girona 5, Girona.

Carbonell, E., M. Mosquera, A. Ollé, X.P. Rodríguez, R. Sala, M. Vaquero & J.M. Vergés
1992 *New elements of the logical analytic system.* Cahier Noir 6, Tarragona.

Estévez, J., A. Vila, X. Terradas, R. Piqué, M.A. Taulé, J.F. Gibaja & G. Ruiz
in press Cazar o no cazar, ¿es ésta la cuestión? In *1er Congreso Iberoamericano de Arqueología Social.* La Rábida (Huelva).

Fortea, J.
1973 *Los complejos microlaminares y geométricos del Epipaleolítico mediterráneo español.* Memorias del seminario de Prehistoria y Arqueología 3, Salamanca.

Fullola, J.M., P. García-Argüelles, D. Serrat & M.M. Bergadà
1995 El Paleolític i l'Epipaleolític al vessant meridional dels Pirineus catalans. Vint anys de recerca a la franja pirinenca sud; interrelacions amb les àrees circumdants. In *Cultures i medi: de la Prehistòria a l'edat mitjana.* Xè Colloqui Internacional d'Arqueologia de Puigcerdà (Girona), 159-176.

Guilaine, J. (ed.)
1994 *Dourgne: derniers chasseurs-collecteurs et premiers éleveurs de la Haute-Vallée de l'Aude.* Centre d'Anthropologie des Sociétés Rurales, Toulouse.

Guilaine, J. & M. Martzluff (eds.)
1995 *Les excavacions a la balma de la Margineda 1979-1991 (vol II).* Edicions del Govern d'Andorra.

Guilbaud, M.
1987 Dynamique du débitage. *Dialektikê. Cahiers de typologie analytique* 1985-1987, 8-15.

Jordá, J.F., R. Mora & R. Piqué
1992 La secuencia litoestratigráfica y arqueológica del yacimiento de la Font del Ros (Berga, Barcelona). *Cuaternario y Geomorfología* 6, 21-30.

Laplace, G.
1972 La typologie analytique et structurale: base rationnelle d'étude des industries lithiques et osseuses. In *Banques de données archéologiques.* Colloques nationaux du Centre National de la Recherche Scientifique 932, Marseille, 91-143.

López, V.
1993 *Estudi paleocarpològic del jaciment prehistòric de la Font del Ros (Berga, Berguedà).* Unpublished report, Universitat Autònoma de Barcelona.

Martínez, G. & J.A. Afonso
in press La producción lítica: un modelo para el análisis histórico de los conjuntos arqueológicos de piedra tallada. In J. Bernabeu *et al.* (eds.) *Recursos abióticos en la Prehistoria: caracterización, explotación e intercambio.* Universitat de València.

Mora, R., X. Terradas, A. Parpal, C. Plana, J. Martínez, A. Fígols & G. Roca
1991 Les ocupacions mesolítica i neolítica de la Font del Ros (Berga, Berguedà). *Tribuna d'Arqueologia* 1989-1990, 19-29.

Paz, M.A. & A. Vila
1988 El Roc del Migdia (Vilanova de Sau, Osona): un jaciment mesolític atípic? *Tribuna d'Arqueologia* 1987-1988, 133-143.

Pié, J. & A. Vila
1992 Relaciones entre objetivos y métodos en el estudio de la industria lítica. In R. Mora *et al.* (eds.) *Tecnología y cadenas operativas líticas.* Treballs d'Arqueologia 1, Universitat Autònoma de Barcelona, 271-278.

Piqué, R.
1995 Aproximació a l'entorn vegetal durant el Paleolític i Mesolític al vessant Sud dels Prepirineus a partir dels carbons vegetals. In *Cultures i medi: de la Prehistòria a l'edat mitjana.* Xè Colloqui Internacional d'Arqueologia de Puigcerdà (Girona), 71-78.

Rodríguez, A. & R. Yll
1992 Materias primas y cadenas operativas en el yacimiento epipaleolítico del Roc del Migdia (Vilanova de Sau, Barcelona). In R. Mora *et al.* (eds.) *Tecnología y cadenas operativas líticas.* Treballs d'Arqueologia 1, Universitat Autònoma de Barcelona, 73-82.

Sacchi, D.
1972 Datage C14 d'un gisement mésolithique des Corbières: la Caune d'Arques (Aude). *Bulletin de la Société Préhistorique Française* 69 (8), 229.

Soler, N., X. Terradas, J. Maroto & C. Plana
1990 Silex et autres matières premières au Paléolithique Moyen et Supérieur au Nord-Est de la Catalogne. In M.R. Seronie-Vivien and M. Lenoir (eds.) *Le silex:*

de sa genèse à l'outil. Cahiers du Quaternaire 17, Bordeaux, 453-460.

Terradas, X.

1995a *Las estrategias de gestión de los recursos líticos del Prepirineo catalán en el IX milenio BP: el asentamiento prehistórico de la Font del Ros (Berga, Barcelona).* Treballs d'Arqueologia 3, Universitat Autònoma de Barcelona.

1995b Les estratègies implementades en la gestió dels recursos lítics al Pirineu oriental durant el Xè millenni BP. In *Cultures i medi: de la Prehistòria a l'edat mitjana.* Xè Colloqui Internacional d'Arqueologia de Puigcerdà (Girona), 207-214.

1996 *La gestió dels recursos minerals entre les comunitats caçadores-recollectores: vers una representació de les estratègies de proveïment de matèries primeres.* Unpublished Ph.D. dissertation, Universitat Autònoma de Barcelona.

in press Estado actual de las investigaciones sobre el aprovisionamiento de materias primas líticas entre grupos cazadores-recolectores prehistóricos en el estado español. In J. Bernabeu *et al.* (eds.) *Recursos abióticos en la Prehistoria: caracterización, explotación e intercambio.* Universitat de València.

Terradas, X., F. Plana & J.S. Chinchón

1991 Aplicación de técnicas analíticas para el estudio de las materias primas líticas prehistóricas. In A. Vila (ed.) *Arqueología.* Madrid, 141-167.

Terradas, X., R. Mora, C. Plana, A. Parpal & J. Martínez

1992 Estudio preliminar de las ocupaciones del yacimiento al aire libre de la Font del Ros (Berga, Barcelona). In P. Utrilla (ed.) *Aragón/litoral mediterráneo: intercambios culturales durante la Prehistoria.* Zaragoza, 285-296.

Terradas, X., M. Pallarès, R. Mora & J. Martínez

1993 Estudi preliminar de les ocupacions humanes de la Balma de Guilanyà. *Revista d'Arqueologia de Ponent* 3, 231-247.

Toledo, A.

1988 La Bauma del Serrat del Pont, Tortellà. Un jaciment arqueològic excepcional. *Vitrina* 3, 45-52.

Vila, A.

1987 *Introducció a l'estudi de les eines lítiques prehistòriques.* Consell Superior d'Investi-gacions Científiques and Universitat Autònoma de Barcelona.

THE PRODUCTION AND USE OF LITHIC TOOLS AT THE END OF THE UPPER PALAEOLITHIC IN THE BASQUE COUNTRY

Juan José Ibáñez Estévez and Jesús González Urquijo

INTRODUCTION

The Basque country is a mountainous region situated in the north of the Iberian peninsula, and occupies the eastern spurs of the Cantabrian Cordillera and the western ones of the Pyrenees (Figure 1). The northern slope has a steep and irregular orography despite the relatively low heights of the mountains - generally between 700 and 1500 metres. The rivers which flow in this area are short, with their sources situated at less than 40 kilometres from the coast, and follow a general N-S direction. The valleys are narrow and also divided since the main hydrological network lies perpendicular to the orientation of the alpine folding which affects the region. As a result it is an area where communications are difficult and where the relief favours the appearance of particular and differentiated habitats. The southern slope forms part of the upper valley of the river Ebro. Although the relief is not flat, the valleys are more open and form plateaux at between 500 and 800 metres in altitude.

Little evidence of human occupation during the Late Glacial has been found on this southern slope, situated at a great altitude (Arribas, 1990; cf. Utrilla 1995). The site of Berniollo, which is studied here, is one of the most ancient Holocene sites in the region. In contrast abundant sites have been discovered and excavated on the northern slope, dating to both the Late Glacial and the early Holocene (Arribas 1990, Berganza 1990, González Sainz 1989). All of these are located in caves; the absence of open air sites is probably the result of the difficulty of locating them because of the dense vegetation, the scarce agriculture, the steep relief, the thickness of the sedimentary deposits on the valley bottoms and the lack of systematic surveys.

The environmental change between the Late Glacial and the Holocene is less accentuated in this region than in more northern areas of Europe. The climatic amelioration, with oscillations, started at 16,000 BP. Around 11,800 BP, the Allerød climatic amelioration installed climatic and environmental conditions in the area very similar to those of the present day and they did not deteriorate significantly during the short glacial cold period. During the Late Glacial the region was also characterised by a humid climate (Hoyos 1995), as a result of the marine influence, which permitted the development of herbaceous species and reduced the steppe in the interior zones (Peñalba 1992).

As a result the large mammal fauna did not change radically between the Late Glacial and the Holocene (cf. Altuna 1995, Castaños 1992). It was dominated by red deer (*Cervus elaphus*) and wild goat (*Capra pyrenaica*) during the whole of the Würm IV. Cold type fauna is very rare in this period; the woolly rhinoceros (*Coelodonta antiquitatis*) and the mammoth (*Mamuthus primigenius*) are only found in Isturitz, in contact with the plains of Aquitaine; the reindeer (*Rangifer tarandus*) is clearly marginal and represents less than 1% of the bones preserved. Species of open environments such as Bos, Bison or Equus are also rare. From the Allerød onwards, and above all in the Holocene, other animals better adapted to wooded environments such as wild boar (*Sus scrofa*) and roe deer (*Capreolus capreolus*) assumed greater importance although they did not replace red deer and wild goat as the dominant species.

Perhaps the environmental factor which most affected the region was the rise in sea-level, even though it is difficult to evaluate its impact. In this case too the marine transgression started long before the end of the glaciation. During the glacial maximum the coast was situated approximately at the 120 metre isobath and rose to approximately -50 metres at the beginning of the Holocene. The submerged area was much less extensive than in the rest of the Atlantic strip because of the narrowness of the Cantabrian continental platform; at the time of the maximum regression the coast was situated at 10 kilometres from the current coast and at about 10,000 BP it lay 3 or 4 kilometres away. However this transgression is comparatively very important, given the modest dimensions of the Cantabrian valleys. From an archaeological point of view, it is difficult to study the territorial organisation of the groups and a large part of their activities are unknown. From the point of view of the environmental change and the available resources, there were two main consequences: (1) it favoured the formation of estuaries and their relative marine resources and (2) it affected the availability of the lithic raw materials, given the coastal location of part of the outcrops of better quality stone.

The changes in the behaviour of the human groups did not correspond strictly with the Late Glacial/Holocene transition (cf. González Sainz 1989, Straus 1992). The main tendencies appear to change before 12,000 BP. The concentration on a single species - red deer in the sites in wooded zones and ibex in the steep and rocky ones - which took place in the middle Magdalenian gave way to a greater range of hunted species. There was a diversification of the resources exploited, with a greater importance given to fishing in the estuary zones and the collection of molluscs in the same areas. The mobility and the territories exploited were both reduced, as is indicated by the restricted distribution of resources of marine origin in the sites in the interior and the reduced presence of non local lithic raw materials. From the Allerød onwards the areas situated at higher altitudes started to be settled, with sites situated between 700 and 1000 metres. At the same time, cave art becomes more scarce; art started to be confined to mobile art.

However, the retouched stone tools in the region changed very little between 13,000 BP and 10,000 BP and it is the bone tools, and especially the harpoons, which have been used to establish the different chronological phases. The Azilian points and the nail-form scrapers are present from the lower Magdalenian onwards and they increase proportionately towards the Azilian. In later periods there is generally a higher proportion of scrapers and backed bladelets - which in the recent excavations account for 60 or 70% of the retouched tools; at the same time there is a reduction in the proportion of burins, which is attributed to the reduced importance of the bone industry (González Sainz 1989).

Studies of the stone tools have concentrated on their typology. The procurement of raw material has been studied in a very general way. Relationships have been established between the characteristics of the lithic tools during the Late Glacial and the different availability of flint along the Cantabrian coast. This region shows a greater abundance of Cretaceous formations containing flint in the east compared with the west. As a result, lithic assemblages in the west show a higher proportion of non-flint raw materials, and blade flaking is less important than in the east (González Sainz 1992). However, the precise location of the outcrops is only known in certain areas (Tarriño *et al.* 1989, Urtiaga *et al.* 1990, López Quintana 1996).

Studies of lithic technology are also rare. In the historiography of the region technology is seen as an extension of typology, with the collection of data on the types of blanks - flakes, blades and bladelets - their condition - whole or broken - the types of platforms and the measurement of the whole pieces in order to elaborate scatter plots like those proposed by Bagolini (1968); finally, the cores are classified according to their morphology. Studies which attempt to reconstruct the forms of flint working are very rare (cf. Baldeón 1985a, Arias 1987), although doctoral theses are now being written which will make a substantial contribution to this theme.

The function of the tools is hardly known. There are only the references for the work of C. Mazo at Abauntz (Utrilla & Mazo 1992), which are largely unpublished, and the analyses carried out by L.H. Keeley (1988) at nearby sites in central Cantabria.

This panorama of the research on the lithic industries obliges us to base the greater part of this article on the results of our own work on the technology and function of the tools. This analysis has been applied in recent years to the sites of Santa Catalina and Laminak II, situated in Vizcaya, on the northern slope of the Basque country, and the site of Berniollo (Álava), which is located on the southern slope (Figure 1). The occupations of these sites span between approximately 12,500 and 10,000 BP (Berganza 1989, Berganza & Arribas 1994, Baldeón 1985b).

THE PROCUREMENT AND ELABORATION OF RAW MATERIAL

An interest in the ways in which the prehistoric groups acquired raw material and manufactured and used the stone tools was born alongside the discipline itself. However, systematic studies of these aspects have only been carried out in recent decades. This is related to the demise of cultural-historical explanations and the typological analyses on which they were based.

The procurement of raw material, the manufacturing techniques and the systems of use are often studied as separate entities, especially raw material procurement and manufacture with respect to use. It is likely that part of this separation is due to the fact that those studies which are carried out require an increased specialisation of the analysts. However this separation is artificial, arbitrary and sterile. Lithic raw material was collected in order to create tools which served a certain function. If we analyse each aspect as an isolated entity, we lose the global perspective of the related behaviours. For example, the modes of manufacturing tools are not only related to the technical knowledge of the group, since this is conditioned by the characteristics of the available raw material and by the function for which the tool was intended. The procurement of raw material and the elaboration and the use of these tools together form the system designed to answer the need for tools with which to obtain and transform other resources. Many of the technical choices in each subsystem (procurement-elaboration-use) cannot be understood if the whole process is not considered.

It is precise to maintain as global a perspective as possible. For that reason we study procurement, manufacture and use as parts of a strategy of procurement of the lithic tool kit. First of all we will attempt to reconstruct the different phases through which the tools pass from the procurement of the raw material until the abandon of the tool. Once the behaviours are known, we will try to explain the process as a function of the characteristics and availability of the raw material, the technical knowledge of the group studied and the functional requirements of these groups.

The procurement of raw material was followed by the manufacture of blanks. Some of these blanks were waste products, others were used directly, while still others passed on to the phase of secondary transformation or

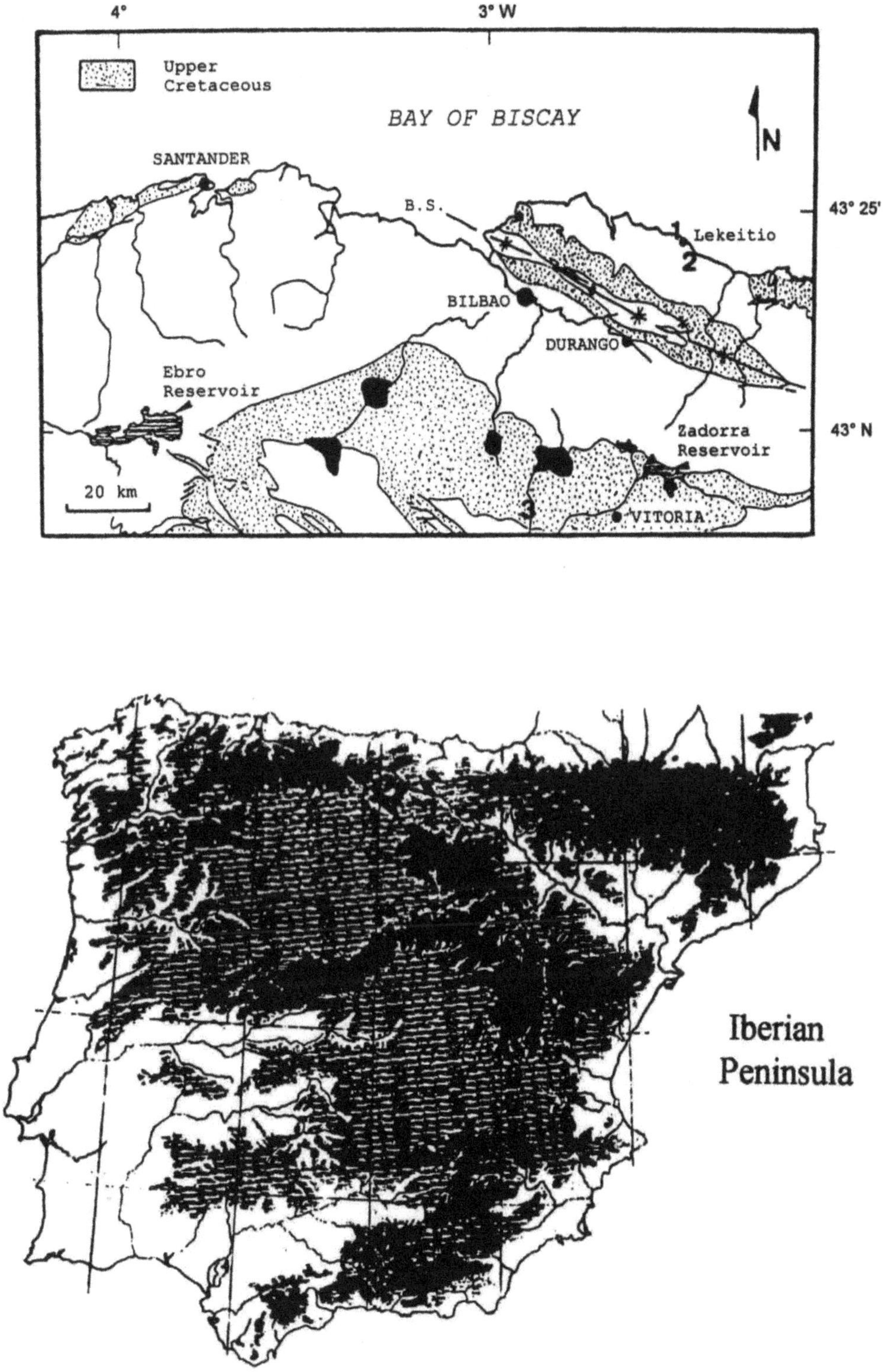

Figure 1. Lower part: the location of the Basque country in the Iberian Peninsula. Upper part (adapted from J. Elorza & M.A. Bustillo 1989): western part of the Basque Country; location of the Upper Cretaceous outcrops and the sites studied (1. Santa Catalina cave; 2. Laminak II cave; 3. Berniollo)

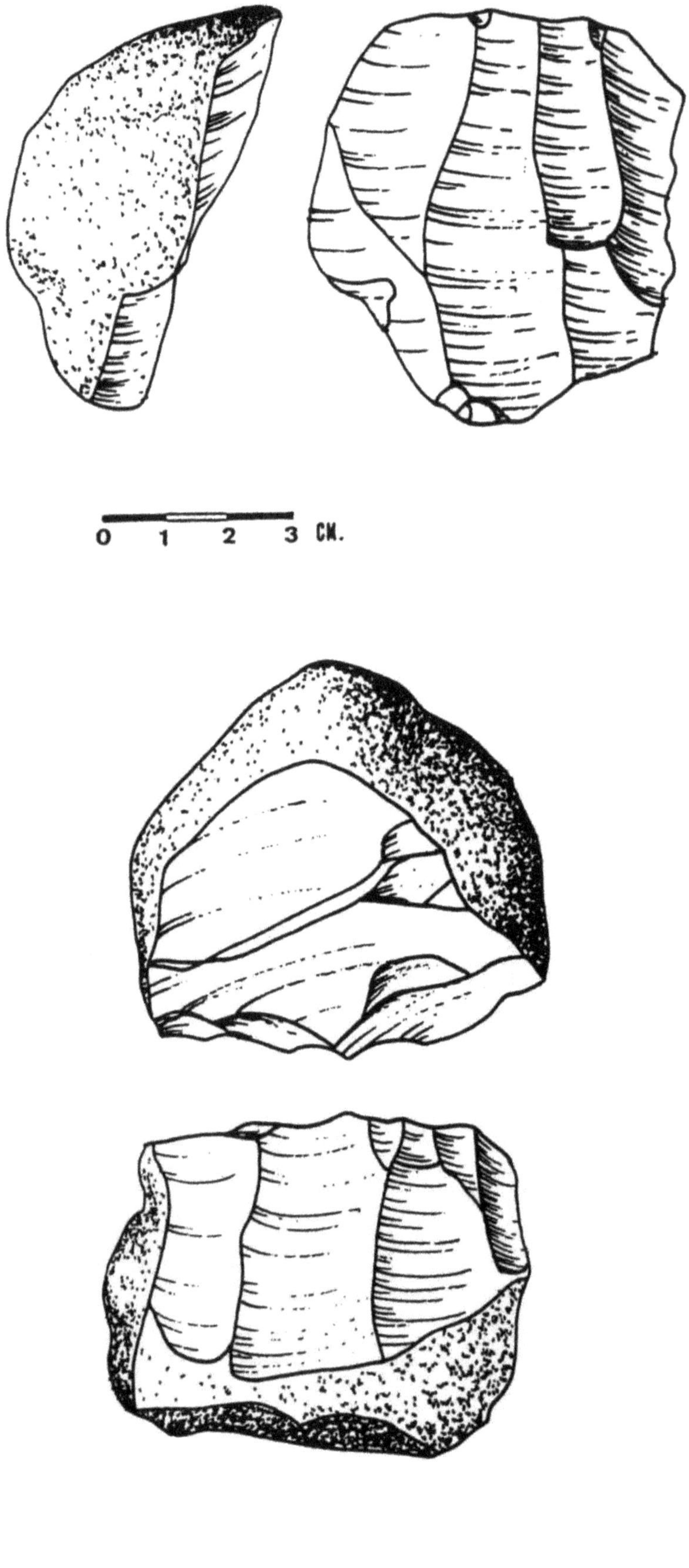

Figure 2: Cores from Berniollo, laminar production on local flint without previous preparation

transport/storage. In the phase of secondary transformation the blanks may have been retouched, broken or hafted. Before or after the phase of secondary transformation the tools may have been stored or transported to their place of use. In the use phase the matter became transformed. Resharpening implies the transformation of the active part of the tool during a single activity, reuse the utilisation of the same tool for another activity, and recycling the transformation of the morphology of the tool and its use for another activity. A complete cycle was established of the procurement of the raw material, transformation, storage/transport and use, in which the technical choices made by the human groups studied were reflected. Within the same assemblage of tools it is possible to establish different strategies of use (Ebert 1979, Nelson 1991: 65). The production-use strategy may vary as a function of the raw material from which the tool is made, the type of blank (flakes, blades or bladelets) and the fact of whether the tool is retouched or not (cf. Perles 1991).

THE SITES ANALYSED

Berniollo

The Epipalaeolithic site of Berniollo lies in the municipality of Subijana-Morillas (Álava) and it has been dated to 9940±490 B.P (Mariezkurrena 1990). It is an open air site at which we studied the manufacturing techniques and the spatial distribution of all the lithic artifacts and the use wear traces on 217 pieces (González & Ibáñez 1991, González Urquijo 1993, González & Ibáñez 1993, Ibáñez & González 1996).

The stone tools at Berniollo were made on three different kinds of flint: type 1, of a black colour, is the most abundant; type 2, of a brown colour; and type 3, also of a brown colour, with Lesengang rings. Flint similar to that of type 1 is found in the canyon which lies a few metres away from the site, while flint similar to that of types 2 and 3 has been found in the Río Rojo area, 20 kilometres to the south (Ortiz *et al.* 1990). However, in the absence of adequate analyses and a systematic survey of the flint outcrops in the area, we cannot be sure that these were the procurement sources used at Berniollo. Type 1 is of a very bad quality for flaking, since the nodules often contain natural fracture planes.

The flaking was oriented towards the production of blades and bladelets. Metrical criteria allow us to place the limit between both types of products at 40 millimetres in length and 12-13 millimetres in width. Different flaking systems were used for the production of blades:

A. The most simple system. A flake removal creates a striking platform from which production starts, without preparation of the bottom, sides or back of the core (Figure 2).

B. Flakes are removed from a large core (Figure 3) The flakes are exploited to make a burin-type core.

C. There is a main striking platform and a secondary opposed one from which the bottom of the flaking surface is rejuvenated (Figure 4). In many cases the back of the core is regularised in order to facilitate the removal or rejuvenation tablets from the platform during the flaking process.

There is a clear relationship between the type of flint and the flaking system. Types 1 and 2 were used for the production of blades and bladelets, while type 3 was used exclusively for the production of blades. Flaking systems A and B were mainly carried out using type 1 flint, while the most complex system was used with flint types 2 and 3. The blades made of type 1 flint are short and wide, while those made of type 3 flint are the longest and most well made. The nodules of type 3 flint are flat with two parallel faces, and have a good natural morphology for the production of long blades. This fact, and the use of very careful flaking techniques, leads to the production of a series of good quality laminar blanks within the general production at the site.

Blanks made of type 3 flint were transformed into retouched tools in greater proportions than the other two types of flint. The retouched tools made of type 3 flint are more regular, and the scrapers made from this type of flint were hafted to a greater degree than the rest. As a result, the tool kit entailed a higher cost of production. However, these tools, as is demonstrated by the use wear analyses, were less intensively used than the rest, which is a similar situation to that of the better quality tools at Meer (Van Noten *et al.* 1978). This implies that the aim was not to manufacture tools which could be used more intensively. The difference in effort must have had another scope.

The proportion of blades/cores is especially low for this kind of flint. The number of flake scars and the degree of exploitation of these cores leads to the conclusion that the low quantity of blades is due to a large number of them having been exported from the site. Furthermore, a high proportion of unretouched blades made from type 3 flint which were found at the site have no traces of use wear and were complete, which is in contrast with the high index of breakage of used blades made from the other flint types. The analysis of the spatial distribution of the objects revealed concentrations of unretouched blades and retouched tools made of type 3 flint outside the knapping areas.

As a result, it seems that good quality raw material was chosen, and that careful knapping was carried out with the aim of producing long and regular blades. Some of these were kept without being used, stored in specific places, while others, which were either retouched or not, were transported and used away from the site in activities which were not particularly intensive. The retouched tools were

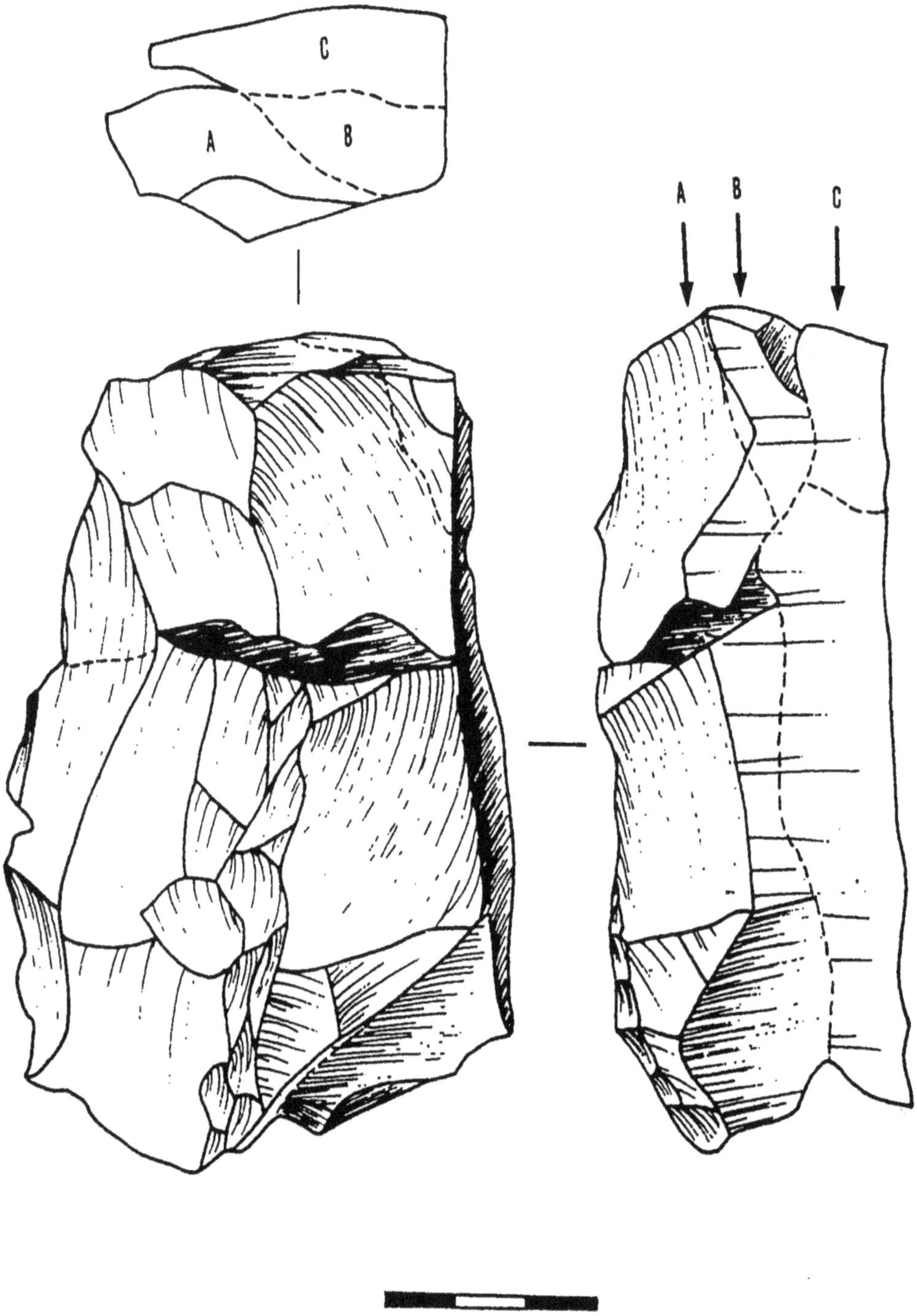

Figure 3: Berniollo. Production of large flakes which were used as blade cores

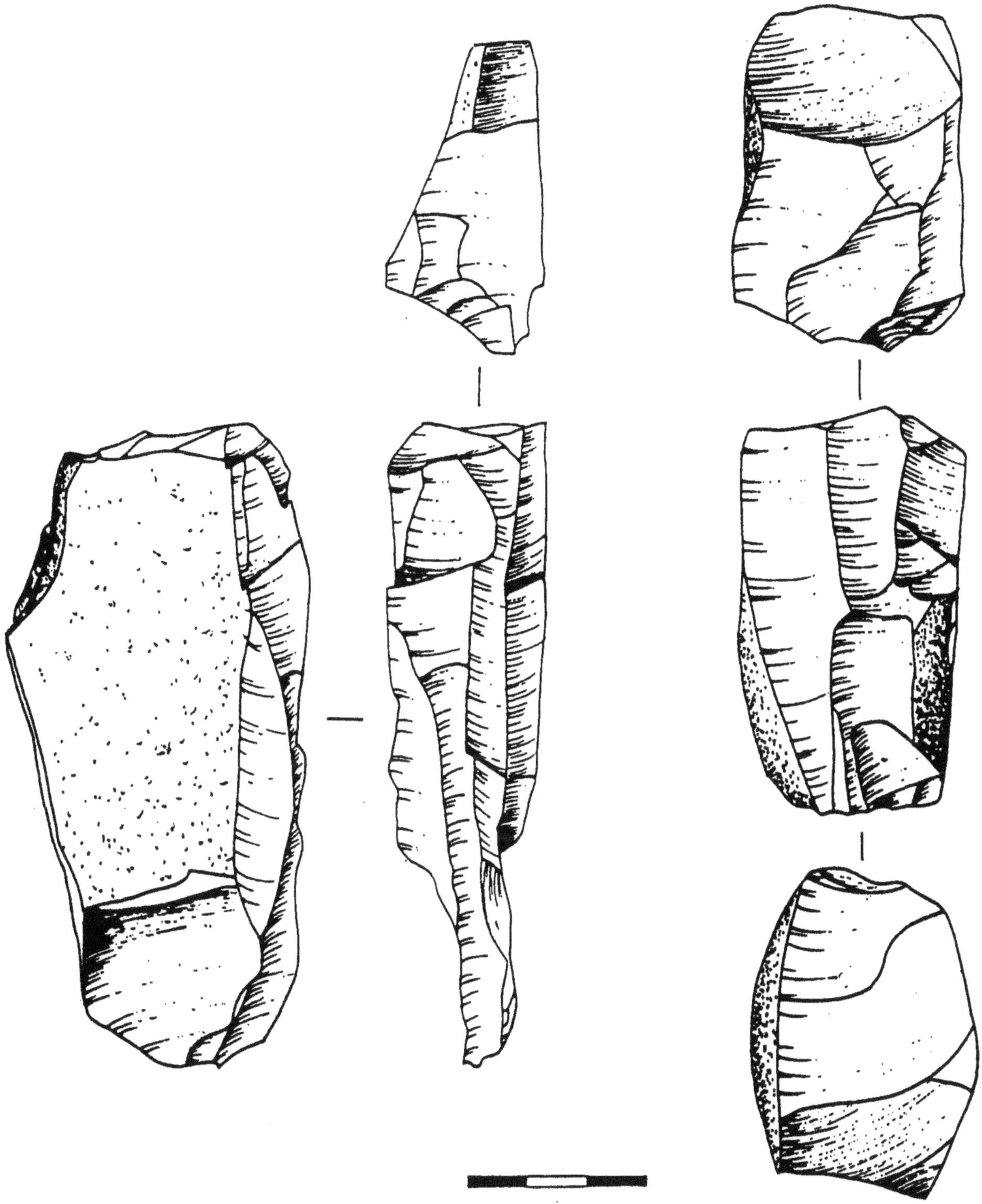

Figure 4: Berniollo. Cores with two platforms made on type 3 flint, for producing high quality blades

rejuvenated frequently, and some of them were stored at the site for future use.

At Berniollo we also detected differences in the type of exploitation of the tools made on flakes and those made on blades. Both types of blanks were also exploited without being retouched, although the blades (43%) were used to a greater extent than the flakes (23%). However, the retouched tools made on flakes were used more intensively than those made on blades.

The majority of the flakes appear to be especially associated with the knapping areas, while the blades tend to be found at some distance from these places. Broken blades tended to be used, and probably many of them were broken intentionally, while the whole unretouched blades rarely have traces of use wear. Some of these whole blades were found concentrated in small areas which we have interpreted as stores. There are also concentrations of retouched tools, the majority of which are made on blades.

The bladelets were preferentially used for making points and backed bladelets. The most important information which Berniollo offers with respect to their use is derived from the spatial distribution of the artifacts. The majority of the points and backed bladelets, which were discarded during their repair, are found in the same area as the bladelet cores and the waste products resulting from their manufacture.

Laminak II

The cave site of Laminak II lies on the northern slope of the Basque country at 3 kilometres from the current coastline and at 5 kilometres from Santa Catalina (Figure 1). There is a final Magdalenian level dated to 11,700±140 B.P. (Berganza & Arribas 1994). Here we studied the manufacturing techniques used for the lithic industry and the use wear traces on 136 tools (Ibáñez & González 1994; González & Ibáñez 1994).

The majority of the flakes found at the site derive from the production of blades and bladelets. However, some of them were made from flake cores which were exploited in order to obtain products little over 1 centimetre in length (Ibáñez & González 1988). Some of these flakes were used to make scrapers, truncated pieces or sidescrapers. Some of the larger ones of these were used without retouch for various activities, such as skinning and hide working. Other unretouched flakes, which are sometimes very small, were used for cutting fresh hide and meat or for butchery.

At Laminak II blade production was centred on the manufacture of short blanks, between 4 and 6 centimetres long, which were not very regular, and the majority of which can be classified as laminar flakes. These sizes are the maximum that can be obtained from the nodules of flint used. The procurement of these blocks probably took place in the immediate surroundings, in the lower Cretaceous marls and marly limestones which outcrop 4 to 5 kilometres away to the SW (López Quintana 1995). These blades were obtained using very simple flaking schemes. A striking platform was created by removing a decortication flake and flaking started by exploiting the natural convexity of the nodule. In some cases an opposed platform was used during flaking in order to correct the bottom of the flaking surface. The sides of the core were not prepared, and these remained covered in cortex until the core was abandoned.

There is another group of blades of longer length, more regular and with parallel edges, made from very variable types of flint. These pieces were not manufactured at the site, as is demonstrated by the absence of flaking waste which would have accompanied their production. The blanks were imported and retouched, giving the occupants at Laminak II pieces which were difficult to produce with the raw material blocks available in the immediate surroundings. These blades were used and retouched very intensively, and were employed in a greater variety of activities (working bone, hide, wood or butchery). In some cases the same blade was used to work more than one matter. The short blades of local production reflect a less intense and less varied use, and were used for working hide and bone, which were the two most common activities carried out at the cave.

Bladelets represent 80% of the laminar products found, and they were used for making backed points and backed bladelets. The size of the abandoned cores, of the unretouched blanks and of the primary bladelets, indicates that there are two assemblages of bladelets of different sizes, one between 15 and 20 millimetres in length and the other between 30 and 35 millimetres. The long ones and part of the small ones were made from block cores (Figure 5). In addition there was a specific production of small bladelets from flake cores (Figure 6). This diversity of blank types demonstrates a functional significance. In general the backed tools made from the small bladelets were used as projectile barbs while the longer ones were used to cut soft animal matter, hide and meat, or for dismembering small animals. The backed points, which show very variable sizes, were made on two types of blanks (Figure 7). In both cases the bladelets which were chosen to be retouched came from the full flaking phase, and are the most regular; those which were produced during the initial phases or which were produced after the rejuvenation of the flaking surface were systematically abandoned.

Santa Catalina

The cave site of Santa Catalina (Figure 1) is located in the coastal municipality of Lekeitio (Vizcaya); there are two occupation levels, one Azilian which dates between 10,500 and 9600 BP, and the other final Magdalenian with dates between 12,200 and 11,400 BP (Berganza 1989). We studied the function of 167 tools from the Azilian level and 300 from the Magdalenian one (Ibáñez *et al.* 1993, Ibáñez & González 1996).

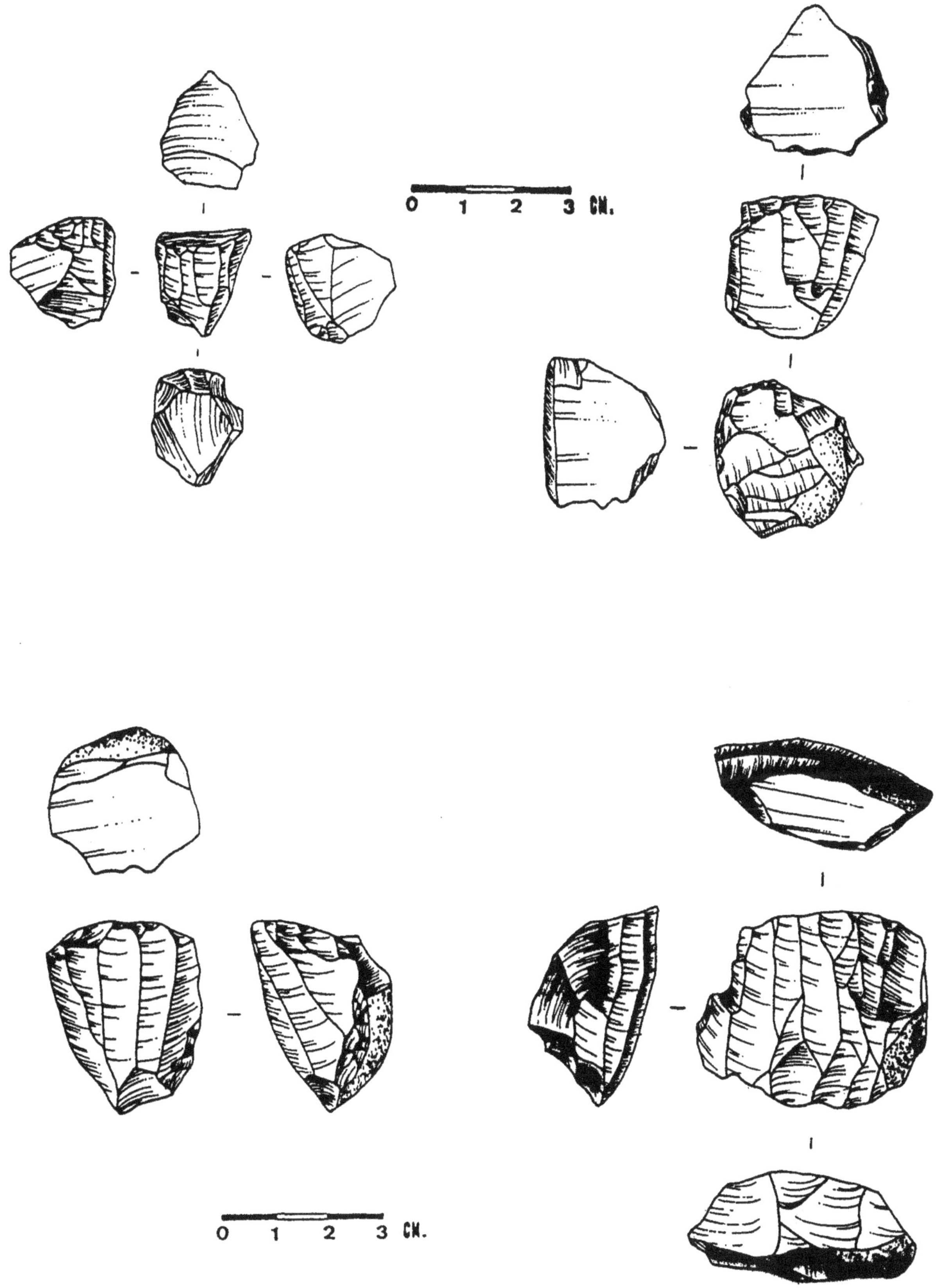

Figure 5: Laminak II. Block cores for bladelet production

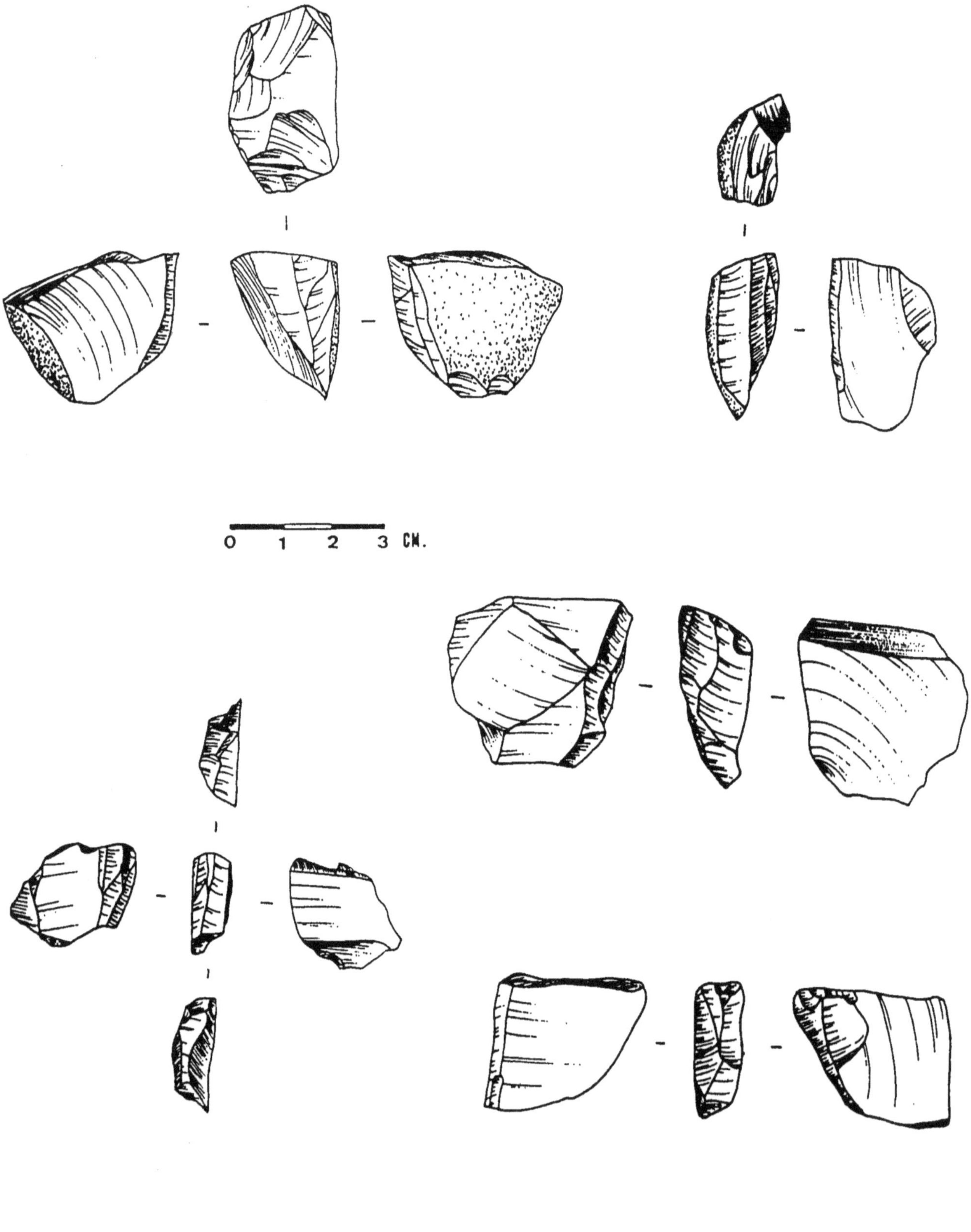

Figure 6: Laminak II. Flake cores for microbladelet production

The lithic production in the Azilian level at Santa Catalina was mainly aimed at the production of bladelets, which represent 75% of the laminar production. In the Magdalenian level, on the other hand, the proportion of blades and bladelets is very similar. Bladelets were rarely used unretouched, as they served to make backed bladelets and points. The backed tools served a double function. While the larger sized ones were used to cut soft animal matter, the smaller sized ones rarely show traces of use wear. Probably these small backed tools were used as projectile armatures.

The most common activities carried out in the Azilian level were butchery and fresh hide working, mainly by means of scraping. For this both retouched and unretouched tools were used. Among the unretouched blanks, blades were used more frequently than flakes. Among the retouched tools, those made on blades show more intensive use than those made on flakes.

In the Magdalenian level the most common activities were working antler and dry hide. In this level there is no evidence for a differential exploitation of the tools made on flakes and those made on blades, since the intensity of use of the tools made on both types of blank is very similar.

GENERAL CHARACTERISTICS OF THE MANUFACTURE AND USE OF THE TOOL KIT

As is shown in the sites analysed, the lithic production in the final Palaeolithic in the Basque country is fundamentally laminar. The distribution of the lengths and widths of the laminar products shows that the metrical differences between blades and bladelets lie at around 40 millimetres in length and 12 millimetres in width, since at these measurements there are breaks in the distributions.

A large part of the laminar production was destined to obtaining short blades, between 4 and 6 centimetres in length. For this, medium sized nodules were exploited which were procured from the surroundings of the sites. The flaking scheme was very simple. The striking platform was created by means of the removal of a decortication flake and flaking started by exploiting the natural convexity of the nodule. The sides, the back and the bottom of the core remained cortical. However, when the morphology of the nodule permitted it, the distal zone of the flaking surface was corrected from the bottom of the core. This type of production is present in all the occupation levels at Berniollo, Laminak II and Santa Catalina.

There was also another production of blades of better quality, between 7 and 12 centimetres in length, which are more regular than the previous ones. At Laminak II these blades were not produced at the site but rather they were imported from elsewhere. At Berniollo this type of blade was produced at the site and preferably with high quality flint. The natural convexity of the nodule was used for the beginning of the flaking process, as in the previous flaking scheme, while during flaking opposed platforms were prepared from which to correct the distal end of the flaking surface, and the back of the core was regularised in order to facilitate the removal of platform rejuvenation tablets.

A large part of the supports was used without being retouched first. At Laminak II and in the Azilian level at Santa Catalina, more unretouched tools were used than at Berniollo and in the Magdalenian level at Santa Catalina. The greatest use of the unretouched tools coincides with the sites in which flaking activities were less important and where they were more specialised in bladelet production.

In general unretouched blades were used more intensively than unretouched flakes. Furthermore, the blades were retouched more than the flakes. Among the blades, we observed that the longer ones were more likely to be retouched than the shorter ones.

The management of the blades is more complex than that of the flakes, especially that of the high quality blades. At Berniollo the spatial distribution of the objects consented the recognition of this differential management. The flakes appear to be more closely related to the flaking areas, while the blades are found further away in places which we interpret as being zones of tool use. Some of the whole long blades, made of high quality flint, are found concentrated in deposits, stored for future use. We also found stores of retouched tools made on the same kinds of blanks. It is likely that a large part of the laminar production was destined to be used away from the site. At Laminak II we observed a similar behaviour, since the blade tools were not produced at the site. With this we can conclude that the good quality blanks and tools had been stored and transported more frequently than the rest of the tool kit.

As regards their use, the unretouched tools were used in a wide range of activities including working bone, antler, wood and hide, and for butchery. The unretouched tools were used in the activities which the tool assemblage presents as the most important at each site. They were used for working fresh hides and for butchery in the Azilian level at Santa Catalina, for working wood and antler at Berniollo and for working dry hides and antler in the Magdalenian level at Santa Catalina and at Laminak II. The unretouched blades were used more frequently for butchery than the flakes, which can be explained by the suitability of the long and straight edges of these pieces for this kind of activity.

We found clear common elements at the different sites in the use of the retouched tools. The scrapers were mainly used for working hides, both dry and fresh. However, some of them were used for working antler and others for working wood. The use of scrapers for working antler is more evident in the Magdalenian level at Santa Catalina, where this activity was very important. The vast majority of the burins

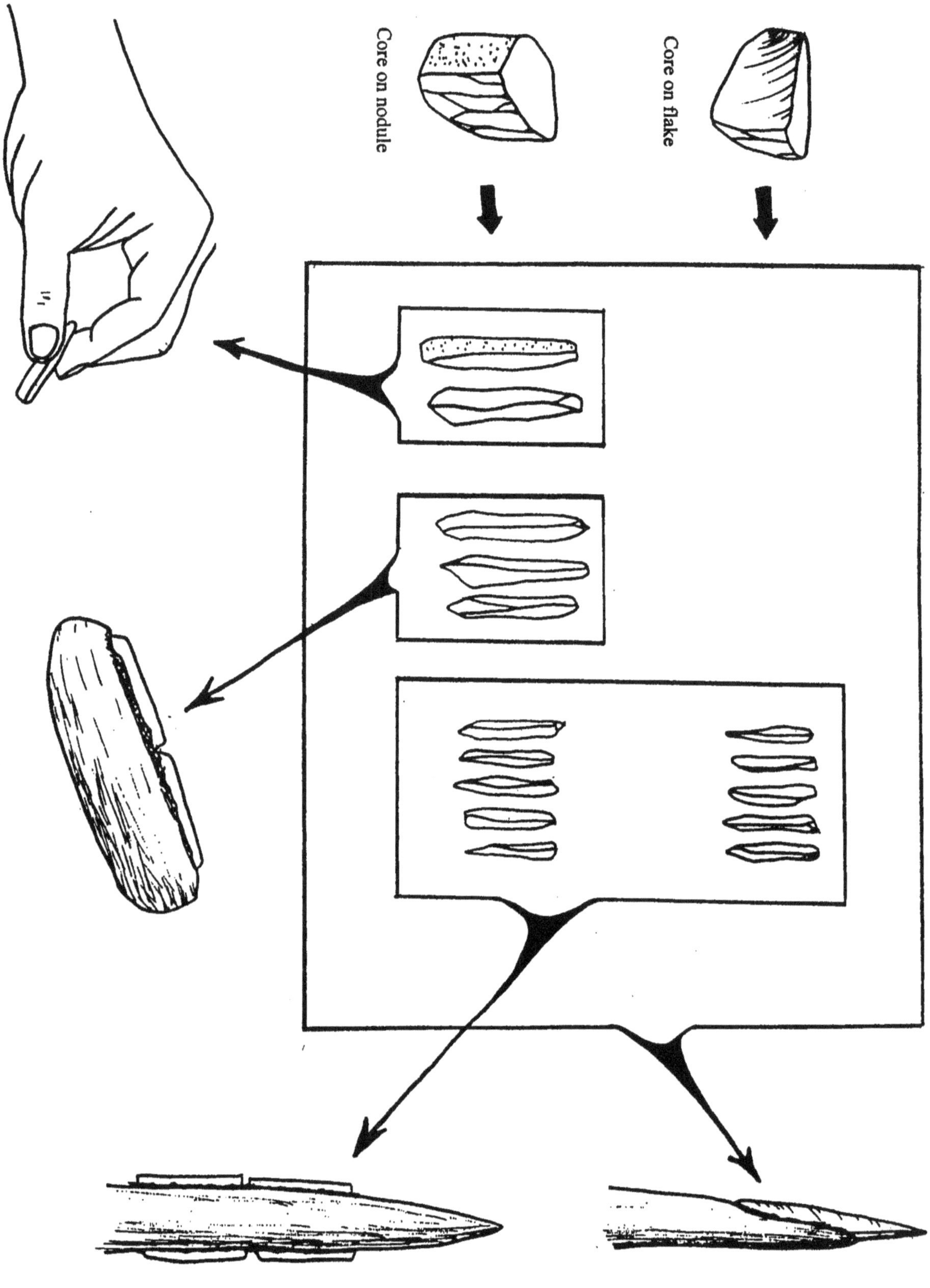

Figure 7: Laminak II. Scheme of the production, selection and use of the bladelets

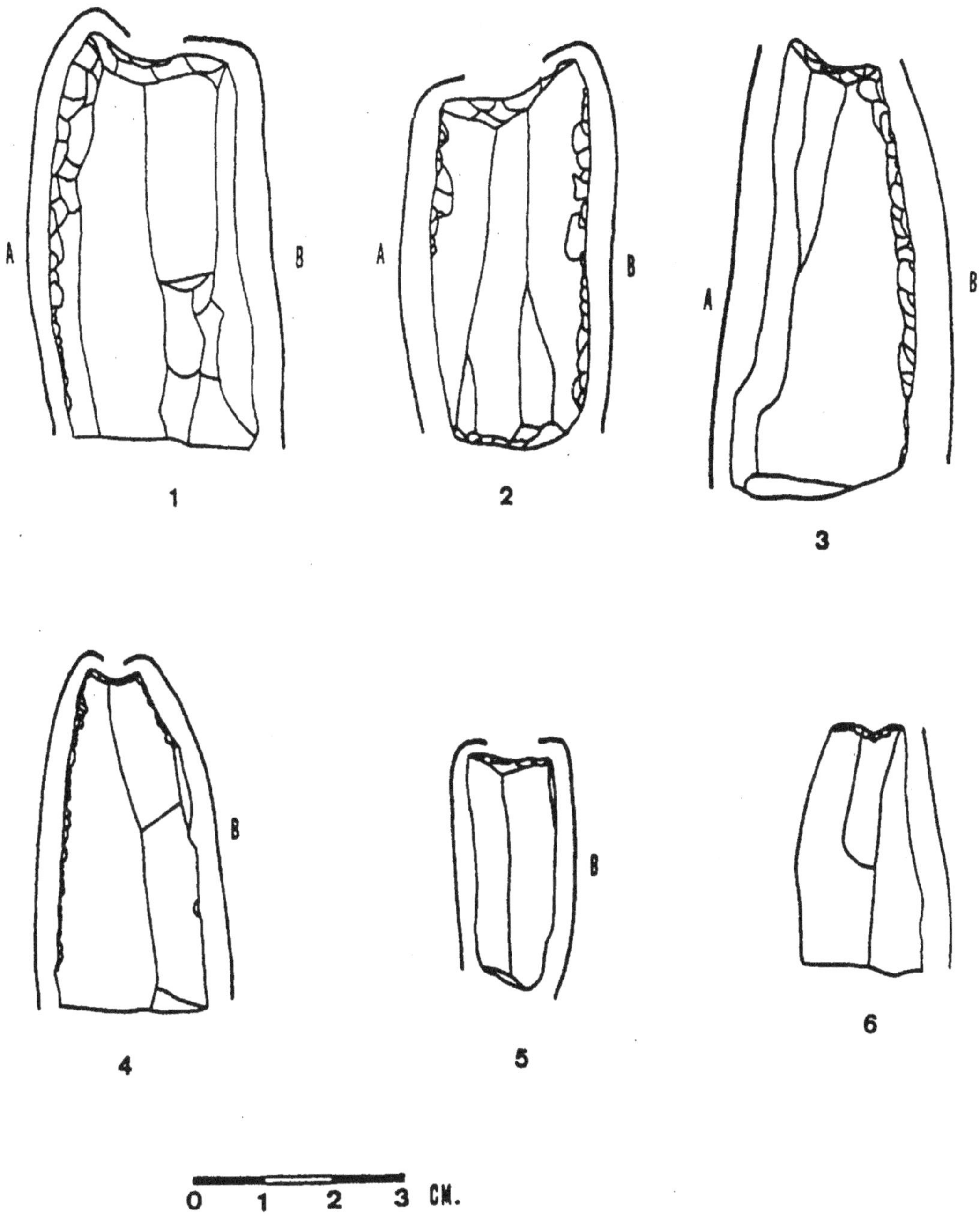

Figure 8: Blades with concave truncations which were probably hafted, from Berniollo and Santa Catalina

were used for working antlers. The tops of the burins were used to incise and the edges of the facets for scraping, while the unretouched edges were used for scraping or cutting. Two functional groups were recognised among the truncated artifacts. The short artifacts with a straight truncation were used for scraping bone and wood, while the long truncated blades, generally with concave truncations, were used for cutting soft animal matter, hides or meat. In these cases the truncation was not the active part of the tool. The use wear traces stop abruptly at the edge of the area of the truncation. For this reason and because of the position and morphology of the truncation we think that it served to facilitate the hafting of the blade (Figure 8). Other groups of tools, such as the flakes and blades with continuous retouch and the denticulate tools, were used for very diverse activities.

As regards the tools made on flakes and blades, we observed different strategies of exploitation at the sites studied. There is a tool kit made with careful knapping techniques on better quality raw material which permitted the production of long and regular blanks. These tools were destined for preferential uses, the blanks were stored, and many of these were retouched before use; these tools were then used in activities away from the site. In contrast, the unretouched tools, both on flakes and on blades and made of lower quality raw materials, tended to be used for activities which were carried out at the site.

It has been demonstrated that one of the main advantages which blades offer in comparison with less standardised tools is that of transportability, in other words their high use potential in relation to their weight. The length and regularity of the active margins and the versatility of the blanks make it possible to manufacture a large range of tools whenever necessary. With the transported laminar blanks it is possible to carry out urgent tasks, without the limitations of having to look for raw material to make the necessary tools. Therefore, in situations in which it is necessary to have a transportable tool assemblage with which to carry out planned activities in different places, a series of high quality blanks was made of a raw material type which allowed accurate flaking (Goodyear 1979: 29, Edmonds 1987: 163-165, Jeske 1989: 36, Myers 1987). "Curated" tools tend to be made of high quality raw material, hafted and frequently rejuvenated (Keeley 1982, Shott 1986, Hayden 1989). This is the case of the tools made on blades of type 3 flint at Berniollo.

Tools made on bladelets are abundant at all the sites, and are especially important in the Azilian level at Santa Catalina and at Laminak II. For manufacturing the bladelet cores, small nodules were used or sometimes flake cores. In the latter case flakes of around 4 centimetres in length and 1 centimetre in width were used; a platform was made by means of a truncation and the edge of the first bladelet was regularised by means of a semicrest. Two groups of bladelets tended to be made, one of about 30-35 millimetres in length and the other between 15 and 20 millimetres. This duality in the bladelet assemblage was found both at Berniollo and at Laminak II. The analysis of the distribution of the lengths of the bladelets at the Magdalenian site of Ekain, in Guipúzcoa, also revealed the existence of two groups of bladelets (Merino 1984: 140). As we have seen, at Laminak II the smallest bladelets tend to be made from flake cores. In general the bladelets were rarely used without being retouched, since they served to make backed points and backed bladelets. A relative exception is the case of Laminak II where part of the larger sized unretouched bladelets were used in the same activities as the unretouched blades.

The backed points, many of which have impact traces, were used as projectile points. There is a great variability in the size of these points, which was perhaps related to the use of different propulsion techniques or different prey.

A group of tools which are typologically classified as "backed points" show use wear traces resulting from cutting meat or hide, and consequently these pieces should be considered pointed backed bladelets. We distinguished two groups among the assemblage of points according to the classification of de Sonneville-Bordes: "Azilian points", which were made on relatively thick blanks and which have a curved back, and "microgravettes" which are thinner and have a straight back. 50% of the microgravettes (n=20) have impact traces, compared with 18% of the Azilian points (n=16); 10% of the microgravettes have use wear traces resulting from cutting meat or hide, compared with 30% of the Azilian points. This indicates that part of the morphological distinction between the two types of point had a functional significance. More regular supports were preferred for the projectile points and the straight backed bladelets.

The larger backed bladelets were used to make composite knives which were used for cutting soft animal matter, while the smaller ones tended to be used as projectile barbs. In the sites studied butchery activities were carried out using unretouched blades, the lateral edges of truncated blades, and backed bladelets which formed composite knives. The edges of the bladelets show less intensive use wear traces compared with the other types of tools used for butchery. The backed bladelet knives are less robust than the other types of tool. Furthermore, the edges of the backed bladelets with use wear traces of butchery are sharper (24) than those of the blades used for the same activity (38). All this may indicate that the blades, truncated or not, tended to be used for the primary butchery activities, for dismembering the larger animals, while the knives with backed bladelets were used for secondary butchery, for removing the meat and cutting it for consumption, and for dismembering the smaller animals.

The management of the tool kit made on bladelets seems quite simple: they were manufactured when and where they were needed, and no planned production existed. At Berniollo the backed bladelets, and in particular the smaller ones, were found in the same areas as the bladelet cores, the flaking waste and the unretouched bladelets; these are waste or bladelet production zones. This seems to indicate that when it was necessary to repair the hunting tool kit, they abandoned the used backed bladelets, manufactured new ones, retouched them and replaced them in the tool, all in a continuous manner. Their presence at all of the sites where there is abundant evidence of bladelet manufacture suggests the same thing.

THE EXPLOITATION OF THE TOOL KIT AND THE FUNCTION OF THE SITE

The exploitation strategy among the groups of hunter-gatherers is defined by the distribution of the human group and the activities of the acquisition, transformation and use/consumption of the resources at habitation and activity sites in a certain time and space. The study of the manufacture and use of the tool kit enables us to identify the activities carried out at each site, thus providing valuable information about the function of the site within the economy of the group.

So far we have outlined the general aspects of the exploitation of the stone tools which are common to the sites studied. However, the exploitation of the stone tools in each of the sites shows certain peculiarities which, to a large extent, are the result of the functional role which each site played within the exploitation strategy. We believe that the characteristics of tool production and use were substantially conditioned by the exploitation strategy of the tools for which they served. This strategy was itself related to the organisation and planning of the general activities of acquisition and transformation of the other resources.

The Azilian level at Santa Catalina

In this assemblage the tools which reflect hunting activities, such as the backed points and the small backed bladelets, are more abundant than in the other assemblages. The butchery and fresh hide working activities took place more intensively than at Berniollo, Laminak II or in the Magdalenian level at Santa Catalina. This indicates that the occupations which form the Azilian level were in general more oriented towards the procurement and first phases of the transformation of the animal resources.

The characteristics of the technical processes carried out confirm the specialised nature of the occupation of the Azilian level, given that they avoided carrying out activities which bore no relation to the main function of the site. Stone tool production was oriented towards the manufacture of bladelets which were then transformed into backed points and backed bladelets.

The wood and bone working activities were simple and not very intensive, as a result of which they can be related to repair works of the tool kit made of this matter. In contrast, all the phases of antler working appear to be represented. If we take into account the fact that antler was mainly used for making hunting tools, once again we see the importance of the technical processes aimed at the production of this type of tool.

As regards the hide working activities, the first phases of cleaning the hides are well represented, while the later phases of working dry hides are scarce. It is necessary to take into account the fact that cleaning the hides (scraping fresh hides) has to be done shortly after they have been removed from the animal so that they do not putrefy, while the activities of making the hide flexible and thin (scraping dry hides), which are those which require most time and considerable effort, can take place in a more convenient moment.

In his study of the function of the tools at the sites of El Juyo, Verberie and Rascaño, L.H. Keeley (1988) also discovered a high proportion of fresh hide working at Verberie and Rascaño. These sites were probably the result of seasonal occupations for a very specialised function.

The Italian site of Paglicci shows specialisation in hunting (Donahue 1985), as butchery was the predominant activity. Here stone tool production was also oriented towards the manufacture of bladelets, while the production of other types of objects and tools was much more simplified. At the French site of Cassegros, which was specialised in hide working, evidence for working other matter such as wood and bone is scarce and simple (Vaughan 1985).

The characteristics of the exploitation of the tool kit can be understood in this context. Unretouched tools were widely used, since the selection criteria were less strict compared with the other assemblages. The intensity of use of the tools is comparatively low, if we exclude the tools used for working fresh hide. The carrying out of activities which were neither very intensive nor specialised led to the use of a more expedient technology which involved fairly unspecialised tools which were abandoned at the end of the activity. However, in the most important tasks more specialised tools were used, such as scrapers for cleaning the hides and burins for working antler.

Blade production is scarcely represented at the site. It is likely that part of the blades and blade tools were imported onto the site. As a result, the cost of the blades with respect to the flakes was sufficiently high and the blade tools were used with greater intensity compared with those made on flakes.

Berniollo

This site offers a contrasting picture to that of the Azilian level at Santa Catalina. The activities of hunting, butchery and working fresh hide are relatively scarcely represented, while bone, wood and dry hide working were the dominant activities.

Activities involving a high cost of production were carried out. Tool manufacture was oriented towards the production of blades, and to a lesser extent of bladelets. Wood working was intensive and shows a certain complexity. Hide working is represented by softening and thinning the hides, which are those activities which require the most time and effort, while the first phases of cleaning and the last phases of manufacturing objects in hide were secondary activities.

The need to carry out intensive activities at the site led the production of tools of a higher production cost compared with those of the Azilian level at Santa Catalina, mainly in the form of retouched tools, and the unretouched tools were chosen with greater care.

Part of the stone tool production was destined to the manufacture of tools used in activities away from the site. This required an assemblage of transportable and efficient tools which could be easily transformed and rejuvenated. These tools were made on blades, and mainly of type 3 flint. We therefore observe a dichotomy in the tool kit, with one series of very expedient tools and another with more curated uses. When it was necessary to carry out an activity at the site unretouched tools or retouched flake tools were used, while the retouched high quality blades were reserved for use away from the site. This explains why the tools with a higher cost of production were less intensively used.

The Magdalenian level at Santa Catalina

The characteristics and the economic activities in the Magdalenian level at Santa Catalina are intermediate between those of the two other sites. We see a greater equilibrium in the group of economic activities. Hunting and the primary transformation of animal matter is better represented than at Berniollo, but less than in the Azilian level. Scraping dry hide and the manufacture of objects made of hide, wood and bone matter is more important than in the Azilian level and less so than at Berniollo. In comparison, there is no predominance of part of the economic system, since all the different phases are equally represented.

We do not find any breaks in the technical processes. Stone tool production was oriented towards the manufacture of blades and bladelets. The importance of the production of bladelets is greater than at Berniollo and less so than in the Azilian level. Wood, bone and antler working activities were complex and intensive. All the phases of hide working are represented: cleaning, softening/thinning, both of which were carried out using scrapers, finishing by rubbing with pebbles (González & Ibáñez in press), and the manufacture of objects.

The intensity of use and the degree of selection of the tools are also intermediate between those at the other two sites: greater than in the Azilian level and less than at Berniollo. The difference in use between the flake tools and blade tools is not marked.

The absence of discontinuities in the technical processes, the performance at the site of all the activities of the economic system and the absence of a differential exploitation of the tool kit with respect to the blanks indicate a lesser degree of complexity of the economic behaviour compared with the previous two assemblages.

Laminak II

The function of the site presents certain ambiguities. Backed points and backed bladelets are predominant in the tool kit. Antler working was important, since it is reflected both in the use wear traces on the stone tools as well as in the presence of spearheads and manufacturing waste of these tools (Berganza & Arribas 1994). This fact seems to indicate that it was a temporary site specialised in hunting activities. The presence of tools made on high quality blades which were imported onto the site supports this hypothesis. However, neither the butchery activities nor the scraping of fresh hides, activities which are common in a hunting camp, are particularly important.

As regards hide working, the phases of cleaning and softening/thinning are absent. This is in agreement with the low number of scrapers which were found at the site, since these are the tools which were used for these tasks. However, the phases of the maintenance of the hide and the manufacture of objects are important. The hide was cut and scraped with flint tools, ochre was added and it was rubbed in with pebbles.

Bone working activities were intensive and imply actions of cutting and scraping. Furthermore, a wide variety of animal resources was exploited. Among the mammals red deer was dominant, and wild goat, chamois (*Rupicapra rupicapra*), a large bovid (*Bos* sp.), horse and roe deer (*Capreolus capreolus*) were also present (Castaños 1994). They also hunted various species of bird (Hernández 1994), fished (Roselló *et al.* 1994) and collected molluscs (Imaz 1994). The characteristics of the hide and bone working and the diversity of the resources exploited do not concord with the characteristics of a specialised hunting site.

The red deer show that they were hunted at the end of spring and the beginning of summer (Castaños 1994: 181). It is likely that the site was formed by occupations at this time of year. In numerous ethnographic examples hide preparation is carried out in the autumn when the animal hides are in the best condition, in time for the necessities of the winter (Nelson 1986, Odell 1980: 415). The absence of conditioning of the hide is due to the fact that in the period of occupation of the site this activity was not necessary. Hide working activities were limited to the maintenance and manufacture of objects.

The Magdalenian site of Ekain at Guipúzcoa shares many similar characteristics with those at Laminak II, with a great abundance of backed points and bladelets and a scarcity of scrapers. The seasonality of this site, which has been determined on the basis of the hunted red deer, was spring and the beginning of summer (Altuna & Mariezkurrena 1984). It is likely that both sites belong to the same type of scheme of exploitation of the environment.

SYNTHESIS

The study of the systems of raw material procurement in the region still has to be carried out. It is only possible to recognise very general patterns of behaviour. In the surroundings of the sites studied on the northern slope (Santa Catalina and Laminak II), it is possible to find small sized nodules in outcrops of Aptian age about 5 and 8 kilometres away. This type of nodule is similar to those which were used for the manufacture of short blades and bladelets. The better quality flint is currently found in the upper Cretaceous outcrops, of Senonian age, in the coastal cliffs (Figure 1) slightly further away (Elorza & Bustillo 1989); its procurement largely depends on the sub-aerial and marine erosive action which make the cliffs recede with the rise in sea-level (Cearreta *et al.* 1992). On the southern slope, outcrops have been found close to Berniollo with similar flint to the poor quality one used at the site; flint types similar to the better quality ones are found in outcrops situated 20-25 kilometres away.

The flaking activities were oriented towards the production of three types of blank: bladelets, short blades and long blades. The bladelets, which were made from small nodules or from flakes, are of two types: one group with lengths comprised between 20 and 25 millimetres, and another with lengths comprised between 30 and 35 millimetres. The bladelets were rarely used without retouch, as they were destined for the manufacture of points and backed bladelets. The backed points were used as projectile elements, while among the backed bladelets there are two functional groups: the smaller ones tended to be used as projectile armatures, while the larger ones were inserted in a haft, probably forming composite knives, and used for cutting meat and hide. The management of the bladelet tools was very simple. When it was necessary to repair the hunting tool a nodule or flake was selected and the bladelets were manufactured, retouched and replaced in the tool, all in a continuous manner. It was a simple technology which could be carried out using only a few fragments of flint, as a result of which it was practised in all kinds of contexts. It is present in all of the sites studied here, although it is more common in those sites which were more specialised in hunting activities and primary activities of the transformation of the animal resources, such as the Azilian level at Santa Catalina.

For the production of short blades nodules between 6 and 8 centimetres in length were used. The flaking technique was very simple: the platform was prepared by means of the removal of a decortication flake and a series of flakes was removed from this platform, exploiting the natural convexity of the nodule. This generated quite wide and short flaking surfaces on cores which preserved the cortex on the bottom, the sides and the back.

The production of long blades can be recognised in detail at the site of Berniollo. They used flat pieces of high quality flint between about 10 and 15 centimetres in length. As flaking proceeded they used an opposed platform from which to correct the distal part of the flaking surface, and regularised the back of the core in order to facilitate the removal of platform rejuvenation tablets.

Logically blade production resulted in as many flakes as it did blades. The unretouched blades were used to a greater extent than the unretouched flakes, and a greater proportion of the blades was retouched. The unretouched flakes and blades were used for a wide range of activities (working bone, wood, hide, butchery etc). In general at each site they were used more intensively in those activities which are more common to the whole tool assemblage. One of the most common activities for which unretouched blades were used was butchery. At Berniollo the unretouched tools were mainly used for woodworking.

The management of the long blades was more complex than that of the short blades. The short blades were manufactured in all of the sites studied, and were used in a wide range of activities and tended to be abandoned after use. Once they had been made, some of the long blades were stored without being used. They were used in activities which were carried out away from the site, as a transported tool kit with many possibilities for use (versatility) and rejuvenation. This type of tool kit was made at Berniollo, where there is flaking waste and stores of unretouched blanks and tools. In contrast, at Laminak II this type of tool kit is only represented by retouched tools.

At the sites studied the flint tool kit destined for hunting activities consisted of points and backed bladelets. Butchery was carried out using unretouched blades, truncated blades and backed bladelets. The blades were probably used for the primary butchery of the hunted animals, while the composite knives made with backed bladelets were used for cutting the meat for consumption. The hide was cleaned and tempered using scrapers; it was cut using unretouched blades or the lateral edges of the scrapers. The antlers were made mainly using burins, and these tools were used for activities such as scraping, incising and cutting, necessary activities for making spearheads and harpoons. Bone was worked using different kinds of tools: burins, retouched flakes and blades, and some scrapers. Wood was worked preferentially using unretouched tools.

Part of the variability in the use of the tool kits at each site can be related to the function of the site. At sites such as the Azilian level at Santa Catalina, which was specialised in hunting and the primary transformation of the animal resources, the characteristics of the use of the tool kit are as follows:

- use of a high proportion of unretouched tools;
- scarce intensity of use of each tool, except in the main activities carried out at the site (butchery, working fresh hide);
- great importance of the production of bladelets and the use of backed points and backed bladelets.

At Berniollo, the site of a relatively stable occupation where activities of the economic system were carried out which imply greater effort and work time, the tool kit is characterised by the following characteristics:

- scarce use of unretouched pieces;
- high intensity of use of the tool kit;
- manufacture of flint tools involving high production costs.

If we compare the characteristics of the economic activities of the most recent assemblages, the site of Berniollo and the Azilian level at Santa Catalina, with the oldest one, the Magdalenian level at Santa Catalina, we observe some differences.

In the most recent sites:

- not all the phases of the economic system are represented in an equilibrated way; in the Azilian level hunting, butchery and fresh hide working prevail, and at Berniollo fresh hide working and tool making predominate;
- there are breaks in the technical processes of the different materials;
- one observes a differential use of the tool kit in relation to the type of flint from which the tools are made and the type of blank (flake, blade).

In the Magdalenian level:

- all parts of the technical system are present;
- there are no observable breaks in the working process of the different materials;
- there is no differential exploitation of the tool kit.

The segmentation of the economic activities implies a more complex economic behaviour in the recent sites, in which the economic activities seem to have been planned.

CONCLUSION

Changes in relation to the lithic industry can be understood in the historical context in which they took place. The environmental change was less marked in the Basque country than in the European regions situated further north, and the environmental modifications determined by the end of the glacial conditions took place over a relatively long period of time. Similar conditions to those of the present day characterised the Late Glacial, at least from the Allerød onwards. Some very basic environmental aspects such as the ungulate fauna available for hunting show very irrelevant variations. The marine transgression, which is an environmental change which evolves at a very slow rate, provoked the appearance of estuarine ecosystems with new exploitable resources and altered the availability of the lithic raw material on the northern slope.

Changes in the economic strategies are also observed in a period before the end of the glaciation, within the upper-final Magdalenian. This change resulted in a reduction in the territorial mobility of the groups which occupied the region and an increase in the range of resources that were exploited; it was also related to an occupation of the upland sites which previously had been of marginal importance. The emphasis seems to lead to a more intensive and territorial exploitation. This process was accompanied by a tendency towards a greater planning of the economic activities, with the appearance of more specialised sites in some parts of the economic system dedicated to the procurement and transformation of specific resources.

There do not seem to be any spectacular technological innovations in order to adapt the lithic industry to the new necessities of the economic system in which it participated. Those which existed are basically a question of degree. The procurement of the raw material was limited to more local outcrops. This led to important adaptations in the western part of Cantabria, with the scarcity of good raw material, but did not affect the Basque country where good quality flint is more widespread. The manufacturing techniques show no great variations; in general there seems to be a reduction in the production of the larger sized blades (González Sainz 1989) and an increase in the production of bladelets, with specific exploitation of flakes to make smaller ones. Neither were there any changes in the types of retouched tools produced: the final Magdalenian stone tools are practically indistinguishable from those which appear in the Azilian levels. The range of materials worked is the same across the Late Glacial/Holocene transition apart from the greater importance given to wood working in the more recent periods.

The change does not consist of new technological concepts but rather of a new organisation of the techniques which themselves hardly varied. The change lies in the strategy of exploitation of the tool kit and not in the concrete technical aspects of its procurement, transformation and use. A more complex and more segmented strategy of the exploitation of the tool kit develops: sites where the tools were predominantly made for local activities, but where tools were also produced which were to serve in other places or at other times; and other sites, with more specialised activities in the general strategy of the exploitation of the site, where part of the more complex tool kit was imported and where the manufacturing activities were reduced to the production of tools which were used directly in the activities carried out at the site.

REFERENCES

Altuna, J.
1995 Faunas de mamíferos y cambios ambientales durante el Tardiglaciar cantábrico. In A. Moure & C. González (eds.) *El Final del Paleolítico Cantábrico.* Universidad de Cantabria, 77-118.

Altuna, J. & K. Mariezkurrena
1984 Bases de subsistencia de origen animal en el yacimiento de Ekain. In J. Altuna & J. Merino *El yacimiento prehistórico de la cueva de Ekain (Deba, Guipúzcoa).* Eusko Ikaskuntza B1, 211-280.

Arias, P.
1987 Acerca de la clasificación de un tipo de cantos tallados postpaleolíticos de la región cantábrica. *Veleia* 4, 99-118.

Arribas, J.L.
1990 El magdaleniense superior/final en el País Vasco. *Munibe* 42, 55-63.

Bagolini, B.
1968 Ricerche sulle dimensione dei manufatti litici preistorice non ritocati. *Annali dell'Università de Ferrara, sez. XV, I,* 10, 195-219.

Baldeón, A.
1985a Estudio de las industrias lítica y ósea de Erralla. *Munibe* 37, 123-185
1985b Berniollo (Morillas, Alava). II campaña de excavaciones. *Arkeoikuska 85,* 12-14.

Berganza, E.
1989 Cueva de Santa Catalina. VIII campaña de excavaciones. *Arkeoikuska 89,* 58-60.
1990 El epipaleolítico en el País Vasco. *Munibe* 42, 81-89.

Berganza, E. & J.L. Arribas
1994 El asentamiento paleolítico de Laminak II. Excavación, cronología y análisis de las industrias lítica y ósea. *Kobie* 21, 9-84.

Castaños, P.
1992 Evolución de los macromamíferos durante el Tardiglaciar cantábrico. In A. Cearreta & F.M. Ugarte (eds.) *The Late Quaternary in the Western Pyrenean Region.* Universidad del País Vasco, 45-56.
1994 Estudio de los macromamíferos de Laminak II. *Kobie* 21, 173-188.

Cearreta, A., J.M. Edeso & F.M. Ugarte
1992 Cambios en el nivel del mar durante el cuaternario reciente en el golfo de Bizkaia. In A. Cearreta & F.M. Ugarte (eds.) *The Late Quaternary in the Western Pyrenean Region.* Universidad del País Vasco, 57-94.

Donahue, R.E.
1985 Landscape and site function: A preliminary report on Paglicci Cave, level 4a. In S. Stoddart & C. Malone (eds.) *Papers in Italian Archaeology IV: The Cambridge Conference.* BAR International Series 244, 28-49.

Ebert, J.I.
1979 An ethnoarchaeological approach to reassessing the meaning of variability in stone tool assemblages. In C. Kramer (ed.) *Ethnoarchaeology: Implications of Ethnography for Archaeology.* Columbia University Press, New York, 59-74.

Edmonds, M.
1987 Rocks and Risk: Problems with Lithic Procurement Strategies. In A.G. Brown & M. Edmonds (eds.) *Lithic Analysis and Later British Prehistory. Some Problems and Approaches.* BAR International Series, 162.

Elorza, J. & M.A. Bustillo
1989 Early and late diagenetic chert in carbonate turbidities of the senonian flysch, northeast Bilbao, Spain. In J.R. Hein & J. Obradovic (eds.) *Siliceous Deposits of the Tethys and Pacific Regions.* Springer-Verlag, New York, 93-105.

Goodyear, A.C.
1979 *A Hypothesis for the Use of Cryptocrystalline Raw Materials Among Paleo-Indian Groups of North Columbia.* University of South Carolina.

González Urquijo, J.E.
1993 *Tecnología lítica y organización económica.* Doctoral thesis, Universidad de Deusto.

González, J.E. & J.J. Ibáñez
1991 La tecnología de talla laminar en la ocupación epipaleolítica de Berniollo (Subijana-Morillas, Alava). In R. Mora, X. Terradas, A. Parpal & C. Plana (eds.) *Tecnología y eadenas operativas líticas.* Universitat Autònoma de Barcelona, 201-222.
1993 Utilización del instrumental lítico y funcionalidad del asentamiento en el yacimiento de Berniollo (Álava, España). In P.C. Anderson, S. Beyries, M. Otte & H. Plisson (eds.) *Traces et Fonction: les Gestes Retrouvés.* ed. E.R.A.U.L. 50, 97-104.
1994 Análisis funcional del utillaje en sílex en el yacimiento de Laminak II. *Kobie,* 21, 111-129.
in press The use of pebbles in some Upper Palaeolithic sites in the Basque country. In *Grinding and Pounding. Functional study of Groundstone Tools in Prehistory and Antiquity.* International Round Table at Clermond-Ferrand.

González Sainz, C.
1989 *El Magdaleniense superior final de la región cantábrica.* Tantín, Santander.
1992 Algunas reflexiones sobre las materias primas líticas y la variabilidad técnica y tipológica al término del Paleolítico superior de la región cantábrica. In R. Mora, X. Terradas, A. Parpal & C. Plana (eds.) *Tecnología y cadenas operativas líticas.* Universitat Autònoma de Barcelona, 57-72.

Hayden, B.
1989 From chopper to celt: the evolution of resharpening techniques. In R. Torrence (ed.) *Time, Energy and Stone Tools.* Cambridge University Press.

Hernández, F.
1994 La avifauna no passeriforme de Laminak II. *Kobie*, 21, 189-202.

Hoyos, M.
1995 Cronoestratigrafía del Tardiglaciar en la región cantábrica. In A. Moure & C. González (eds.) *El Final del Paleolítico Cantábrico*. Universidad de Cantabria, 15-76.

Ibáñez, J.J., J.E. González, R. Ruiz & E. Berganza
1993 Huellas de uso en sílex en el yacimiento de Santa Catalina. Consideraciones sobre la manufactura del utillaje óseo y la funcionalidad del asentamiento. In P.C. Anderson, S. Beyries, M. Otte & H. Plisson (eds.) *Traces et Fonction: les Gestes Retrouvés*. E.R.A.U.L. 50, 225-234.

Ibáñez, J.J. & J.E. González
1988 Tecnología de talla en el yacimiento de Laminak II (Berriatúa, Bizkaia). *Kobie* 17, 25-46.
1994 La fabricación del utillaje lítico tallado en Laminak II. *Kobie* 21, 85-110.
1996 *From Tool Use to Site Function. Use-wear analysis in some Final Upper Palaeolithic sites in the Basque Country*. BAR International Series 658.

Imaz, M.
1994 Los moluscos marinos de Laminak II. *Kobie*, 21, 221-224.

Jeske, R.J.
1989 Economies in raw material use by prehistoric hunter-gatherers. In R. Torrence (ed.) *Time, Energy and Stone Tools*, Cambridge University Press.

Keeley, L.H.
1982 Hafting and retooling: effects on the archaeological record. *American Antiquity*, 47, 4, 798-809.
1988 Lithic Economy, Style and Use: A Comparison of Three Late Magdalenian Sites. *Lithic Technology* 17, 1, 19-25.

López Quintana, J.C.
1996 Los yacimientos neolíticos de superficie de Sollube (Bizkaia): materias primas y territorio. In *I Congrés del neolític a la Península Ibérica*. Rubricatum, vol. 1, 177-182.

Mariezkurrena, C.
1990 Dataciones absolutas para la arqueologia vasca. *Munibe* 42, 287-304.

Merino, J.M.
1984 Industra lítica del yacimiento de Ekain. In J. Altuna & J. Merino *El yacimiento prehistórico de la cueva de Ekain (Deba, Guipúzcoa)*. Eusko Ikaskuntza B1, 65-175.

Myers, A.
1987 All shot to pieces? Interassemblage variability, lithic analysis and Mesolithic assemblage types; some preliminary observations. In A.G. Brown & M. Edmonds (eds.) *Lithic Analysis and Later British Prehistory. Some Problems and Approaches*. B.A.R. International Series, 162.

Nelson, R.K.
1986 *Hunters of the Northern Forests*. The University of Chicago Press.

Nelson, M.
1991 The Study of Technological Organization. In M. Schiffer (ed.) *Archaeological Method and Theory* vol. 3. University of Arizona Press, Tucson, 57-100.

Ortiz, L., J.J. Vivanco, A. Ferreira, P. Lobo, M.D. Muñoz, R. Pinillos, R.J.M. Tarriño & A. Tarriño
1990 *El hábitat en la prehistoria en el valle del río Rojo*. Eusko Ikaskuntza.

Peñalba, C.
1992 La vegetación y el clima en los montes vascos durante el Pleistoceno superior y el Holoceno según los análisis palinológicos. In A. Cearreta & F.M. Ugarte (eds.) *The Late Quaternary in the Western Pyrenean Region*. Universidad del País Vasco, 171-182.

Perlès, C.
1991 Economie des matières premieres et economie du débitage: deux conceptions opposees? In *25 Ans d'Etudes Technologiques en Préhistoire*. APDCA, 35-46.

Roselló, E., A. Morales & J.M. Cañas
1994 Los peces de Laminak II (Berriatua, Bizkaia). *Kobie* 21, 203-220.

Shott, M.J.
1986 *Settlement Mobility and Technological Organization among Great Lakes Paleoindian Foragers*. UMI Dissertation Services.

Straus, L.G.
1992 *Iberia before the Iberians. The Stone Age Prehistory of Cantabrian Spain*. University of New Mexico Press, Albuquerque.

Tarriño, A., M.I. Arriortua & J. Elorza
1989 Estudio geológico del área de Peñacerrada (Alava). Petrografía y cristalografía de las silicificaciones existentes. *Eusko Ikaskuntza, cuadernos de sección Ciencias Naturales* 5, 43-135.

Urtiaga, M.K., M.I. Arriortua & J. Elorza
1990 Estudio de las manifestaciones silíceas del noroeste de la provincia de Navarra. *Eusko Ikaskuntza, cuadernos de sección Ciencias Naturales* 7, 259-279.

Utrilla, P.
1995 El valle del Ebro durante el Tardiglaciar y comienzos del Holoceno. Las relaciones con el magdaleniense cantábrico. In A. Moure & C. González (eds.) *El Final del Paleolítico Cantábrico*. Universidad de Cantabria, 281-311.

Utrilla, P. & C. Mazo
1992 L'occupation de l'espace dans la grotte d'Abauntz, Navarre, Espagne. In *Le Peuplement Magdalénien*, Colloque de Chancelade, 365-376.

Van Noten, F., D. Cahen, L.H. Keeley & J. Moeyersons
1978 *Les Chasseurs de Meer*. Dissertationes Archaeologicae Gandenses 18. De Tempel.

Vaughan, P.C.
1985 *Use-wear Analysis of Flaked Stone Tools.* The University of Arizona Press.

THE PLEISTOCENE-HOLOCENE TRANSITION IN PORTUGUESE PREHISTORY: A TECHNOLOGICAL PERSPECTIVE

Nuno Ferreira Bicho

INTRODUCTION

Traditionally, Palaeolithic studies in Europe have been dominated by work in two core areas: the French Perigord and northern Spain. The cultural sequences derived from the French research were accepted as models for expected cultural variability in other areas around the Mediterranean region where the Palaeolithic occurred. This is a consequence of both historical precedence and the establishment of typological systems in France (Bordes 1950; Sonneville-Bordes & Perrot 1954, 1955, 1956a, 1956b).

Recently, it has become clear that the different cultural "phases" found in the French core area do not always describe accurately the variability in other areas of southern Europe (e.g. Italy, Portugal and Spain). While typological studies are still very important, the problems in the recognition and definition of cultural units seem to have increased in the last two decades due to new and more diverse data, methodologies, and paradigms regarding cultural diversity during the Palaeolithic. As a result, new research fields and methods were developed for the recognition and definition of cultural diversity.

One of these new research fields is the study of lithic technology. Different methodologies have been developed to study lithic reduction strategies. These can be divided into two areas, one the so-called Frison effect (Jelinek 1976) or the tool reduction processes, and the other core reduction sequences. This last one is the focus of this paper, and it is seen here as the long process starting with raw material procurement and ending with the abandonment of lithic artefacts. The most common methodology for reconstructing core reduction sequences is refitting. Although the results from this technique are, in fact, the most complete for the study of core reduction sequences, it is not always possible to undertake it. The method requires certain conditions to be able to be carried out and is time-consuming and, therefore, very expensive.

Another methodology has been developed which can be used to reconstruct the reduction sequences. This methodology, used in the analysis presented here, is based on various morphological characteristics of formal tools, blanks, cores, and core maintenance products. This kind of study has been rarer than refitting. It has been used in the Near East by Munday with Middle Palaeolithic assemblages from the Levantine Mousterian (Munday 1979), and by Ferring with Upper Palaeolithic assemblages (Ferring 1980, 1988). It has also been used in Europe by Sampson and Bradley (1986) to study Acheulian collections from Caddington, by Schild (1976) in Poland who worked with different Palaeolithic and Mesolithic assemblages, and in Portugal for Late Glacial lithic industries (Bicho 1992) from which some of the results are presented in this paper.

CULTURAL BACKGROUND

Portuguese Late Glacial and Early Holocene prehistory has been divided into 6 phases (Bicho 1994, 1996, in press a). This division was based on typology, technology, subsistence and settlement patterns. The six phases are Early Magdalenian (16,000-14,500 BP), Late Magdalenian (13,000-10,500 BP), Terminal Magdalenian (10,500-8500 BP), Transitional Phase (8500-7500 BP) and Mesolithic phase (7500-6000 BP). The first three are clearly Late Palaeolithic, while the phases dated to between 10,500 and 7500 BP have been labelled as Epipalaeolithic (Bicho in press b). At least two facies have been identified, the Rio Maior facies, covering the time span from 16,000 to at least 9000 BP, and the Carinated facies dated to between at least 11,000 and 8500 BP.

A note of caution is needed, though, when the term Magdalenian is used in Portugal. From a typological point of view, the Portuguese Magdalenian is within the range of variability known in the rest of Europe. From a technological and chronological point of view, the Portuguese Magdalenian is very distinctive. The Magdalenian assemblages in Portugal are mostly flake industries marked by a low frequency of bladelets, rather than being a blade industry. Blades are present but only as the tail of a normal distribution curve that clearly shows a domination of blanks with average dimensions of ca. 25 mm in length and 8 mm in width (Bicho 1992). The retouched tools are also mostly made on flakes, with a microlith index of less than 20% prior to 10,500 BP, after which the frequencies rise to ca. 35% (Bicho in press a). The term Magdalenian is used only because of the history of Portuguese research, where French archaeologists such as Henri Breuil and Jean Roche played a very important and fundamental role in the development of Palaeolithic archaeology.

Three different settlement patterns have been identified in Portugal, corresponding to the Palaeolithic, Epipalaeolithic and Mesolithic periods respectively. During the Palaeolithic period (Early, Middle and Late Magdalenian) there were two major areas of settlement on the flat plains of the Portuguese Estremadura, one near the modern coast line (Torres

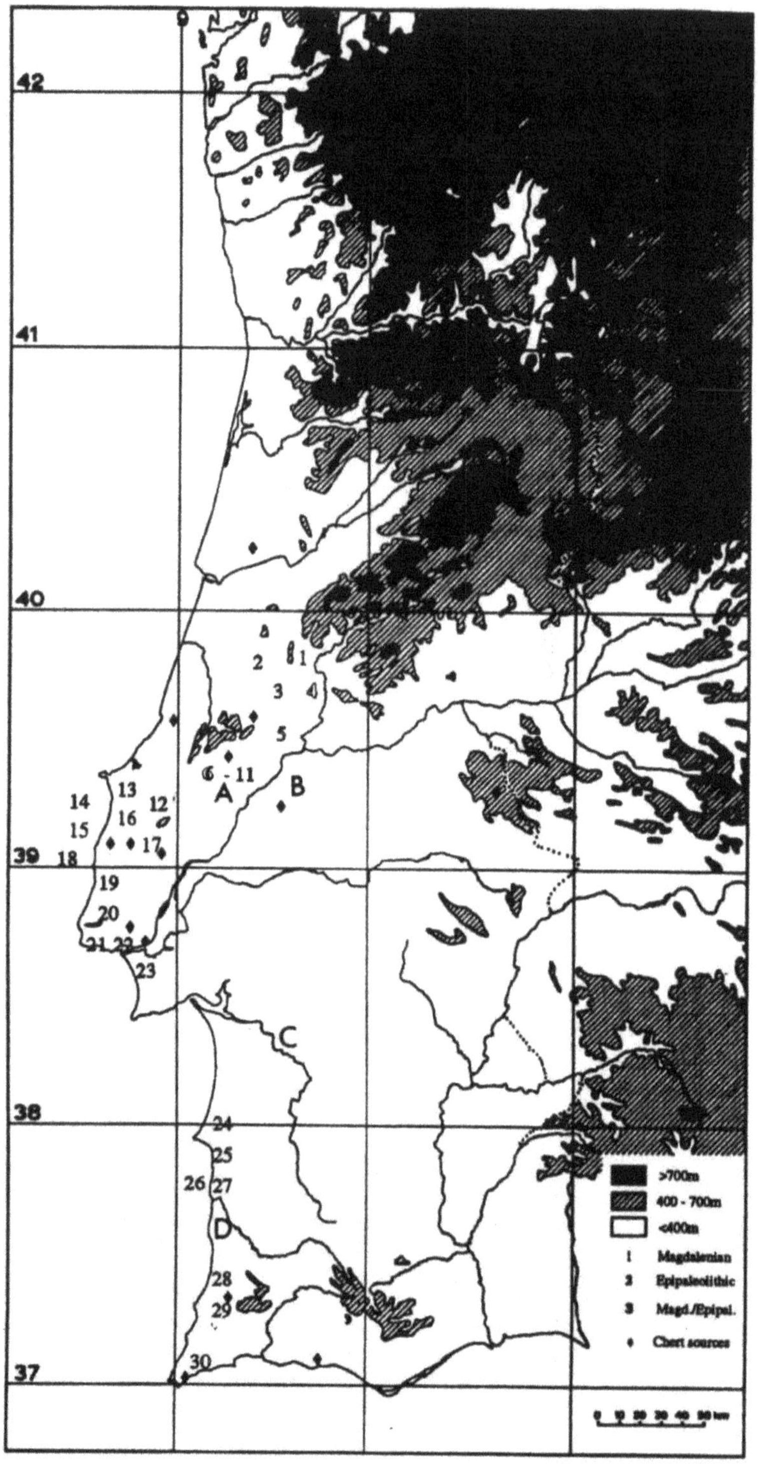

Figure 1: Sites from Central and southern Portugal. Magdalenian and Epipalaeolithic: 1 - Caldeirão cave; 2 - Casal Papagaio Cave; 3 - Pena de Mira rockshelter; 4 - Picareiro cave (2 levels); 5 - Santa Cita; 6 - Cabeço do Porto Marinho (CPM) (15 levels); 7 - Carneira (5 areas); 8 - Vascas; 9 - Areeiro I; 10 - Areeiro III (2 levels); 11 - Bocas rockshelter (3 levels); 12 - Suão cave; 13 - Toledo; 14 - Ponta da Vigia; 15 - Rosso do Cabo; 16 - Cerrado ovo; 17 - Vale da Mata; 18 - Cabeço do Curral Velho; 19 - S. Julião (2 areas); 20 - Magoito; 21 - Penha Verde; 22 - Quinta da Bicuda; 23 - Ponta do Cabeledo; 24 - Praia do Norte; 25 - Cabo de Sines; 26 - Milfontes Norte; 27 - Pedra do Patacho; 28 - Palheirões do Alegra; 29 - Zambujeira; 30 - Casteljo. Mesolithic: A - Fonte Pinheiro; B - Muge area in the Tagus River Valley; C - Sado River Valley; D - Mira River Valley

Vedras) and one inland (Rio Maior) (Figure 1). The inland limestone mountains were scarcely inhabited. During this period, both the open woodland and forested areas were exploited. The main prey species were red deer, wild boar, aurochs, horse and roe deer, while in the highlands the prey species were red deer and rabbit and less frequently ibex, and possibly chamois (Bicho 1996). Exploitation of marine fauna, both fish and shellfish, seems to increase drastically after 13,000 BP, as it is present in various inland sites.

After 10,500 BP, the prey species seem to have been the same as before. At this time there was a sharp increase in the number of sites and occupied areas, with an average of 3 sites/levels per millennia before 10,500 BP and 16 sites/levels per millennia after that. These changes were also matched by a diversification of site location. Caves and rock-shelters located at higher altitudes became more common, as well as Atlantic shore areas, from estuaries to high cliffs (Bicho 1996; Bicho and Haws 1996). These facts have been explained as the result of a demographic explosion which occurred sometime around 10,000 BP (Bicho 1994). It is also during this period that a shift towards a logistical mobility pattern is seen (Bicho in press a).

In the Portuguese Estremadura, the residential sites tended to concentrate around areas where chert was available, with specialised short-term camps for hunting, gathering and fishing marine resources, located respectively in the inland flat plains and limestone plateaux and on the coast. Unlike Central Portugal, the southern area was characterised by a very different geological context, where chert is not easily available. The residential sites seem to be located within 500 meters from the modern coast, while specialised sites for exploiting chert are likely to have been located inland, since the only raw materials available on the coast are greywacke, quartz and quartzite.

All these patterns changed during the Mesolithic, after 8500 BP (Arnaud 1993; Bicho 1994; Morales and Arnaud 1990). The sites tended to concentrate in the Estremadura and Alentejo regions, around the inland estuaries of the Tagus and Sado rivers. In the southern coast, the pattern was one of dispersion, with the site location near the mouth of small rivers running east-west in deeply cut limestone and schist valleys.

In the Estremadura, the diet included a strong marine component, and some of the sites were used seasonally and have been interpreted by Arnaud (1986, 1987, 1990) as year-round residential sites with specialised camps. The sites increased in size, sometimes reaching thousands of square meters, and were frequently used as burial grounds. In the southern coast, the shellmiddens seem to be much smaller and not as high, and characterised by a diet divided equally between marine resources and terrestrial ones, including large and small game and birds (Lubell and Jackes 1988). Occupation seems to have been mostly year-round, though some sites show evidence of seasonal occupation (Lubell and Jackes 1988). The Mesolithic settlement and subsistence patterns seem to represent a second demographic explosion, probably resulting from the new environmental and technological conditions during the Atlantic phase.

METHODOLOGY

Morphological analysis

The methodology used in the present study of Portuguese Late Glacial assemblages is a variation of Ferring's analysis (1980). The analyses for this study have been carried out since 1990. Samples were drawn from the different assemblages to cover the different excavation areas. It was believed that this would allow the examination of all lithic variability present in the assemblages. Consequently, whenever it was possible (meaning when there was a sufficiently large area excavated), different areas were sampled. In those cases where there was not the possibility for spatial division the sample was drawn in a way that eventually would allow a spatial separation for statistical work.

The analytical samples were drawn in most cases from very large assemblages. In general, each sample totals more than four hundred complete blanks. Although only unbroken pieces were the subject of extensive individual analysis, all the other lithic artifacts from most excavated units were counted and sorted by raw materials. Since this work is still under way, the complete counts for some sites or site levels are not yet complete. The artifacts were classified into the following categories: fragment; chip, flake proximal fragment, flake distal fragment, blade proximal fragment, blade distal fragment, bladelet proximal fragment, bladelet distal fragment, flake, blade, bladelet, burin spall, crested flake, crested blade/bladelet, core front, core trimming flake, core tablet, core and tool.

For the extensive individual analyses, debitage, cores and some of the formal tools were examined. The tool classes analysed are endscrapers, burins, notches, denticulate pieces, and backed microliths. These classes were chosen because they are the most numerous in all assemblages.

Only the complete pieces of debitage were analysed for morpho-technological attributes. These attributes were used to answer specific questions about those assemblages. The attributes chosen were different for each artifact class. The analysis of debitage was based on cortex, blank morphology, platform and dorsal scar patterns. The tools and core maintenance products were also analysed by these attributes, but additional attributes had to be included. The cortex characteristics were monitored as the percentage of cortex on the dorsal surface of the pieces. Since absolute precision was impossible in terms of the time required for such measurements, it was decided that general divisions measured intuitively would be sufficient. Thus, five categories were used: no cortex, less than 25%, between 25 and 74%, between 75 and 95%, and more than 95%. Another important characteristic of the cortex was its location. These are: distal, proximal, medial, lateral, lateral/distal, lateral/proximal, lateral/medial, lateral/lateral, distal proximal, distal/medial, proximal/medial and total cortex.

Other than the common statistics such as length, width and thickness, four other qualitative elements were monitored for blank morphology. The first one is shape. Blank shape is the plan view of the piece when viewed from the dorsal surface. Five basic shapes were recognised, plus a category to include unidentifiable or other rare shapes. The recognised categories are converging, rectangular, circular and semi-circular, "déjeté" and expanding toward the distal end. The second attribute for blanks was profile. Blank profile is the side view from proximal to distal ends. Again, the presence of four different categories were monitored: flat, curved, twisted and irregular. Blank tip was another attribute analysed. It was divided into feathered, pointed, hinged, plunged and bipointed. In the tools and core maintenance products, truncated and broken examples were added. Finally, cross-section was monitored. The cross-section is defined as the front view of the artifact, with triangular, trapezoid, flat, irregular and a fourth category for other possibilities.

Platform characteristics were monitored based on two different elements: the presence/absence of core edge abrasion, and platform types. These types on debitage and tools are unfacetted, dihedral facetted, multiple facetted, cortical, crushed, punctiform, lipped, Mesolithic, and a category for other possible types and unidentifiable. Again, some other types were added in the case of tools and core maintenance products: platform on truncation (for the burins); removed platform; and platform on flaking surface.

Dorsal scar characteristics were investigated through two different aspects: the number of scars and the dorsal scar patterns. The examination of dorsal scar patterns was based on different categories: parallel from proximal end, parallel from distal end, parallel from side, other unidirectional case, bi-directional, parallel to axis, bi-directional, perpendicular to axis, bi-directional, crossed, radial or partial radial, unidentifiable or other.

Other than these attributes, the tools and the core maintenance products were examined through such elements as the location of the tool on the blank (proximal, distal, lateral, bilateral, proximal/lateral, proximal/distal and distal/lateral); direction of scars in crested pieces (bifacial or unifacial); amount of scars on crested pieces (less than 1/3 of the length; between 1/3 and 2/3 of the length; more than 2/3 of the length); location of scars on core maintenance products (distal, proximal, medial, distal/proximal, lateral, all dorsal surface, and on platform).

The cores received a different treatment. While the attributes were different for the core analysis, there was a connection between the attributes used for the cores and those used for the debitage. The cores were classified into types, without distinction between the type of blank that they produced. Length, width, thickness and weight were measured. The percentage of the cortex was also monitored, with the same five groups used for the debitage.

Technological attributes were then examined on the cores. Cross-section and the differential treatment of the core faces were checked. Cross-section was simply divided into four categories; circular, triangular, quadrangular and irregular. The core face treatment was isolated in terms of the different faces. The core was analysed as if it was a rectangular polyhedron, such as a cube. Two of the faces are: (1) the most important flaking surface; and (2) the respective platform. The opposite face from the platform is called the distal face. The face opposite the flaking surface is called the back face and, finally, the two other faces are considered as the lateral faces and treated as a single observation. If the two lateral faces had different treatments, this was noted. The face treatment had eight categories that were either repeated or combined in the case of the lateral faces: these are cortical, flaking surface, platform, crest, keel (not prepared as a crest, but still a long edge formed by two faces of the core, with an angle around 45 degrees), crushing, random removal and, finally, a *varia* for other possibilities.

Platform characteristics were examined through the presence/absence of platform edge abrasion and platform type. These latter types are unfacetted, dihedral facetted, multifacetted, cortical, bipolar, crushed, and a combination when there is more than one platform.

Since the core typology was carried out without the distinction of the blank type produced, it was necessary to document this. Thus, the type of production was indicated separately from the core type. The four categories were flake, blade, bladelet, and a mixture when there was such evidence. In addition, it was noted if blank types were coming from the same platform or from different platforms. Finally, the reason for core abandonment was monitored and classified into six reasons: hinged, crushed platform, natural imperfection such as a geode, fracture, loss of angle (when the angle is more than 90 degrees) and no obvious reason.

Core typology is divided into different main groups, such as single platform, opposed platforms, complex multiple platforms, discoidal, chopper, and other types with little importance in these assemblages. The single platform cores include pyramidal type (after Brézillon 1983), prismatic type (after Brézillon 1983), flat, carinated (also counted as carinated endscraper or carinated burin in the typologies), other simple shaped cores with one platform, and cores with two platforms that are not related (i.e. two platforms using the same face of the core with the flaking surface located on parallel faces of the core). While this kind of core is not characterised by a single platform, technically there is no difference between the shaping of these two faces of the core and the shaping of the face for one single flaking surface.

Opposed platform cores include such types as one face opposed platform, opposed platform with twisted flaking surfaces (after Marks and Ferring 1976), and other cores like those with two opposed platforms and one or more platforms located on different core faces.

Complex multiple platform cores are those where the location of the platforms involved a complex preparation and shaping of the core, allowing the presence of either related platforms, related flaking surfaces, and/or related platforms and flaking surfaces. Such types are called "90 degree" after Marks and Ferring (1976), although the definition is not exactly the same. In this study, the cores typed as ninety degree are those where two working faces intercept each other by means of both flaking surface with flaking surface or platform with flaking surface. Another core type is the orthogonal one, with multiple flaking surfaces intersecting at about 90 degrees angles (after Brézillon 1983).

Other types are "chopper" (a common chopper but analysed here as a core and not as a tool), discoidal or partially discoidal, inform and bipolar (all types but "chopper" after Brézillon 1983). Other types were placed into a single group since the frequency was so low that no patterns were recognised in these types.

In general, the attributes selected for the study were chosen to clarify specific moments in the reduction sequence, to examine the differences among assemblages and the significance of technological changes. These attributes were not selected blindly or randomly, but were based on previous knowledge and preliminary examinations of the collections. Some attributes were included to elucidate specific questions that previous knowledge of the material suggested. Such attributes as blank shape, profile, tip and cross-section were chosen to give the general format of the blank. When those attributes are related to the amount of cortex present and to blank size, the results reflect the different moments in the reduction process.

Details such as platform type are used to monitor differences among the assemblages and the degree of control over knapping. Again, all of these in conjunction with the location of cortex shows the kind and amount of decortication and core shaping. Dorsal scar patterns on the blanks, when related to the amount of cortex and to the general blank shape, suggest when and if changes of flaking direction took place. This can also be confirmed by core typology, the type of removals, and the relative position of the different type of removals on the different core faces.

All these attributes were used primarily to reconstruct the different reduction sequences present in these assemblages. The examination of relationships also allowed, at different levels, a comparison and definition of the differences and similarities among these assemblages and, consequently, a clarification of the temporal and spatial diversity of the lithic assemblages during the Pleistocene/Holocene boundary in central Portugal. In addition, these analyses helped to define the "desired end-product", that would, ultimately, reveal the relationship between simple blanks and formal tools. This was possible since both the debitage and different classes of tools were analysed in a comparable manner.

Statistical Analysis

The descriptive analyses resulted in three large data sets. These were used in statistical analyses with very specific problems in mind. The first question of this study was how much technological variability existed among these assemblages. The second was the so-called "desired end-product" or, in other words, the blanks that were preferentially used to make the retouched tools.

To investigate the question of technological variability, different problems had to be resolved. The first was the presence of technological diversity within each assemblage. The second concerned the real significance of the different attributes used during the descriptive analyses, between and within assemblages. Finally, a third and more complex problem had to be addressed: the patterning of this diversity across both time and space. The answer to these problems will clarify the space-time patterns of the lithic technology and their relationship with the socio-economic organisation of the local prehistoric groups.

To approach these problems, two different kinds of data were available. These were the categorical variables such as blank shape, blank tip or platform type, and the metric variables such as length, width and thickness of the different artifact classes (i.e. flakes, bladelets, tools and cores).

The discrete variables were used first to identify intra-assemblage variability and inter-assemblage variability and, at the same time, to establish the attributes that were significantly different across the assemblages. Such results were obtained using a simple procedure previously introduced by Hietala (1984) in a very similar problem concerning technological variability within and between assemblages from the Levantine site of Boker Tachtit. This very useful procedure has two phases. The first phase was designated as "the log-linear model", and the second was an F statistic, using partitioned results of the log-linear model. The first part of this model summarised by Hietala states that:

"The general log-linear model is similar to a factorial analysis of variance model (Scheffé 1959: 121), where the natural logarithms of the cell frequencies are characterised by a linear sum of parameters associated with a general effect, main effects for the individual variables, two-way interaction effects linked with pairwise combinations of the variables, and so forth" (1984: 46).

Hietala also outlines that hypotheses associated with this model can be statistically tested through a G2 statistic.

"Fortunately, hypotheses associated with the log-linear model may be statistically tested through the G2 statistic (Fienberg 1977: 36). Interestingly, this statistic is a function of the likelihood ratio statistic (Rao 1965: 350), based on maximum likelihood estimates of the model parameters. If l denotes the likelihood ratio statistic, then G2 = -2 log λ and if, in addition, the alternative hypothesis is the simple negation of the null hypothesis (corresponding to the saturated model), then

$$G2 = 2\ \Sigma i\ \Sigma j\ n\ ij \log (nij / mij)$$

where *nij* is the observed frequency of the *(i,j)* cell and *mij* is the expected frequency of the *(i,j)* cell based on estimating the cell probabilities by the principal of maximum likelihood. If the null hypothesis is correct (i.e., if the fitted model is correct) and the sample size is not small, then the statistic has an approximate chi-squared distribution with degrees of freedom equal to the number of independent constraints, under the null hypothesis, placed on the parameters of the model" (1984:47).

The sample size should be considered inadequate whenever there are more than twenty percent of the expected frequencies with values smaller than five and if the number of empty cells does not exceed two. Otherwise, the sample size is adequate for this kind of procedure. For a hypothesis of independence, the degrees of freedom in a contingency table with *r* rows and *c* columns are (*r* - 1) (*c* -1). This procedure is based on the simple assumption that:

"if a partitioning theorem is used to separate technological variability (for specific variables observed on particular artifact classes) into within-level and between-level components, then (under the hypothesis of unpatterned total variability), the G2/degrees of freedom statistic for between-level variability divided by the G2/degrees of freedom for within-level variability will, asymptotically, be an F-statistic with numerator and denominator degrees of freedom given by the degrees of freedom for the statistics in the numerator and denominator of the ratio." (Hietala 1984: 49). [Consequently] "This F statistic can be used to determine if between-level variability is substantially larger than within-level variability. Needless to say, the same concept can be employed for studying within-area and between-area variability and, hence, can be used to study local and global patterning" (ibid.: 50).

In practical terms, the raw data were organised in counts per attribute, separated into flakes and bladelets. Each set of attributes (e.g. cross-section with triangular, trapezoidal and flat forms) was run for all sites simultaneously.

After the G2 between-sites was computed, the G2 within-sites was obtained for each assemblage, based on the two or three sub-samples that each assemblage had been divided into during the descriptive analysis. The F statistic was then computed, resulting in the clarification of which set of technological attributes was significant in terms of the inter-assemblage variability. Those cases had a low F statistic and a high P value where the within-site variability was more pronounced than between-site variability. Those sets of attributes where the between-site variability was more pronounced than within-site variability had high F statistic values and very low P values. With this procedure the most significant technological attributes were isolated and could be used for other purposes later. The results of these statistics allowed the elimination of those attributes that had more variability within a site than across the sites. These results also suggested that there were different technological procedures of lithic reduction in this area that may eventually pattern either in terms of space and time, or both. Another statistical procedure used in this study is the Cluster Analysis. This was used to address the question of how technological variability patterned across time and space. This procedure was carried out using the Euclidean distances as the proximity measure between assemblages, in an agglomerative hierarchical clustering procedure through complete-linkage.

Finally, One Way Analysis of Variance (ANOVA) was used to compare some of the metric variables. The significance level used was 5%.

The levels of significance of the statistical tests have to be interpreted in a relative manner. This procedure is a consequence of the type of data - archaeological material - being handled. One has to keep in mind that what may be statistically significant is not necessarily significant at the practical level of archaeological material research. In a practical example, this means that if a mean length of 35 millimetres of a flake assemblage is statistically different from another flake assemblage with 36 millimetres, it is clear that such difference lacks real meaning. Basically, those two assemblages have the same length, since such a difference would be meaningless for a prehistoric flint-knapper. Consequently, the levels of significance will be considered in a general way from "significantly different" to "not significantly different". Using Ferring's terminology (1980: 110), P Values between .000 and .01 are considered as "significantly different", those between 01 and 05 are viewed as "moderately different", and, finally, those above ca. 05 are considered as being "not significantly different".

REDUCTION SEQUENCES

The core reduction sequences presented here are based on the results of the analysis of morphological characteristics of the cores and debitage. After the analyses, it was clear that the reduction sequences are marked by a fairly wide range of differences among the assemblages. These differences, however, are present mostly at the detailed level, and likely resulted from the flintknappers idiosyncrasies as much as from raw material constraints, such as internal fractures and size of the initial core blanks, at least in the Magdalenian assemblages.

Most reduction sequences were used for more than one raw material. The reduction sequences for chert show some variability, while those used for quartz and quartzite are very similar across assemblages. There are a wide variety of reduction strategies: unidirectional, bi-directional, multidirectional and radial. Different core types resulted from different reduction strategies, and one reduction strategy produced different types of core. Differences in certain aspects of debitage morphology may have produced by different reduction sequences, though the relationship between these could not be identified without a refitting program.

Most reduction strategies (e.g. unidirectional flaking) produced three different blank types and their associated cores. In most cases these cores were intended to produce certain blanks with a specific desired shape. Thus, there were cores that produced flakes, those that produced bladelets, and those that produced both flakes and bladelets. In general, the difference between the last two was easily identifiable on those cores using either opposed or multidirectional flaking. On those cores where both types of blanks were produced and on unidirectional flaking this distinction was more difficult. The basic criteria used to distinguish between these core types was to consider bladelet cores as those without flake scars, and those cores for flake and bladelet production as those marked by both bladelet and flake scars.

Finally, before the actual description of these reduction sequences, it is necessary to keep in mind that though there is an incredible amount of morphological and technological data, the reconstruction of reduction sequences without refitting is the result of data interpretation. Thus, there may be more than one way to interpret the data. Consequently, this study is as much interpretative as it is speculative.

Quartz and quartzite

Quartz blanks were likely obtained near the sites, in gravel layers present in the local river cuts. Two sizes of core blanks were gathered: cobble and gravel sizes. The latter was used frequently and produced mostly prismatic and single platform cores though opposed, orthogonal and 90 degree cores were also produced. These cores weigh less than 100 grams and, frequently, less than 40 grams. They were not prepared extensively, and when shaping occurred it was carried out in a simple form, producing both flakes and bladelets. Sometimes, rarely, pyramidal cores were produced. No crested bladelets were made during the reduction sequence to produce bladelets. This was possibly a consequence of the type of fracture resulting from the available quartz. The quartz pebbles tend to create clean fractures with very regular and angular ridges that can be used as guide lines very much in the same manner as a crested ridge.

The cobbles were used to make chopper-like cores. These show evidence of a greater number of removals than the smaller cores. These flakes were never retouched unlike some of the flakes removed from the smaller cores. The two types of reduction sequence suggest the co-existence of an expedient technology for the large chopper-like cores with a different technology, more elaborate and less economic, for the smaller cores.

Quartzite use seems to follow those characteristics of quartz reduction. Contrarily to quartz, the most common core blanks were cobbles. These were fairly large blanks, weighing over 250 grams up to close to one kilogram. Preparation of flaking surfaces and of platforms was extremely rare. This fact is shown by the presence of cortex on all surfaces of the cores, including the platform, and by the lack of chips and fragments. One interesting aspect of quartzite is the high number of broken flakes. These fractures probably happened during the knapping process and resulted in a high number of Siret burins. Quartzite retouched tools are rare. When made, they tend to be of simple types, such as denticulates, notches and simple, large scrapers. Also made of quartzite are the heavy duty tools: hammer stone, anvils and grinding stones (Bicho in press a). Again, quartzite points to an expedient use of this raw material.

Chert

The relative frequency in the use of chert has been shown to change through time in the area of Rio Maior (Bicho in press a). This change is probably a consequence of cultural evolution stemming from subsistence and settlement as well as technological traits. Other changes seen in chert are the increase in bladelet tools and in microlithic weaponry.

Chert, unlike quartz and quartzite, was never located at the sites. These sites seem to be, in the case of residential areas such as Rio Maior and Torres Vedras, within a short distance from the chert sources, somewhere between 2 and 5 kilometres. The nodules were brought whole to the sites, and were prepared there. These nodules were, in general, of small size very rarely weighing more than 100 grams. They were used to exhaustion and were fully exploited. Multi-platform and opposed platform cores are very common, producing bladelets and flakes at different times in the reduction sequence. All types of flakes were used for the production of formal tools, suggesting a very economic way to deal with chert (Bicho 1993a).

The reduction sequences were fairly complex with careful core shaping, preparation and maintenance. All these aspects suggest that chert was an expensive resource, and thus the model stating that the residential areas were chosen near the chert sources (Bicho in press a).

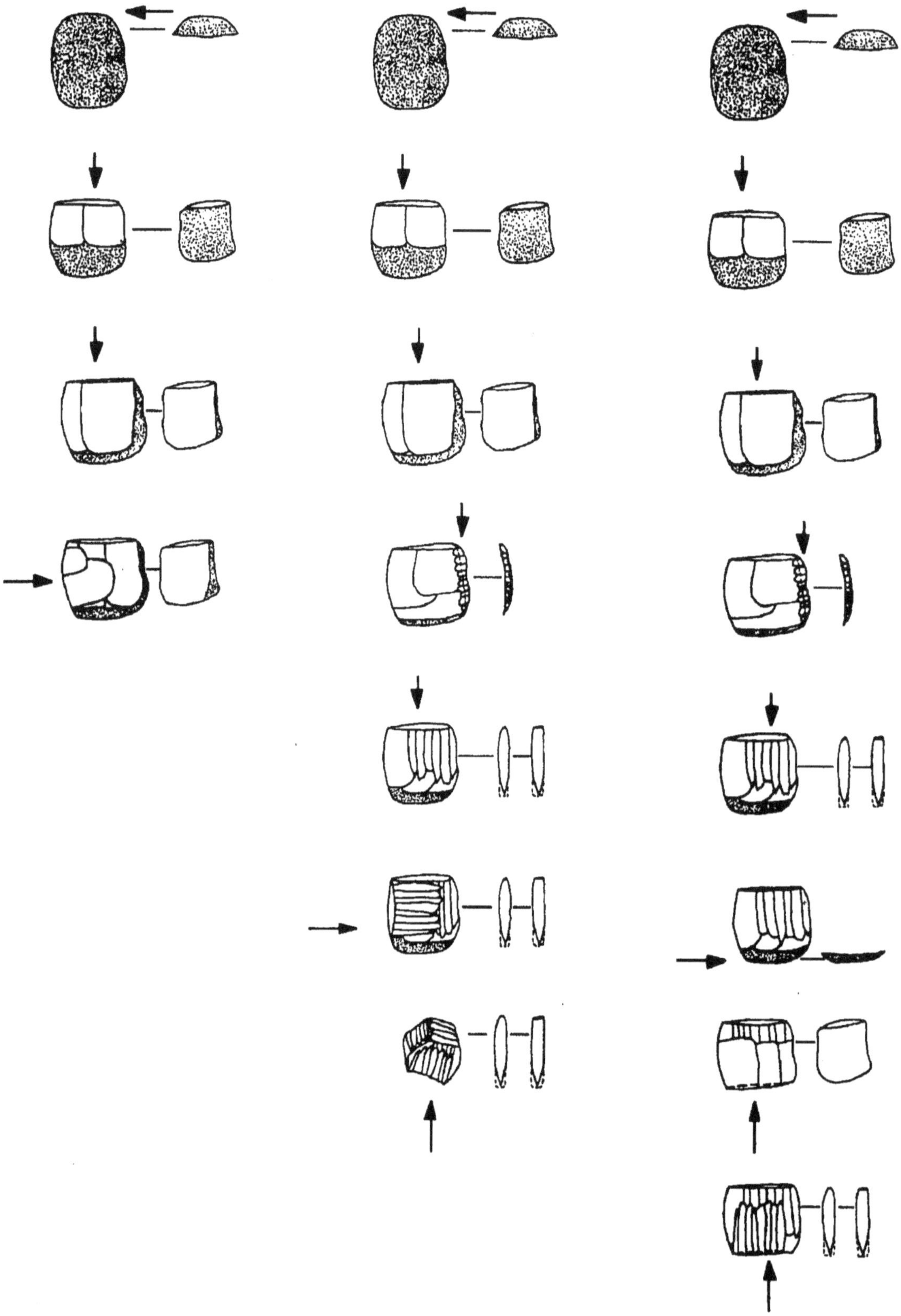

Figure 2: Chert reduction sequences from the Rio Maior area

Reduction sequences

All the reduction sequences, with the exception of radial and chopper cores, seem to have followed identical stages of reduction. The first step was the removal of a circular or sub-circular cortical flake, usually fairly thin (Figure 2). This flake was removed from the shorter axis of the pebble, using the long axis as the platform. The removal of this first flake made a clean, unfacetted scar that was used as the platform for core shaping. This was carried out through the removal of fully and partially cortical flakes from the front face of the core, creating a flaking surface. The flakes resulting from core shaping tended to be the largest flakes, characterised by more than 50% of the dorsal surface with cortex and a tendency to have curved profiles. These two facts suggest that this was the stage of the reduction sequence where the strongest blows were applied. Sometimes the flaking surface was extended to the lateral faces of the core, using the same platform to remove flakes. At this stage, the flakes produced had very little or no cortex and tended to be thin. At this point, different options had to be considered by the flintknapper:

a) the core was already fully exploited and the blanks were too small to be used, resulting in the abandonment of the core;
b) another platform was made due to hinge fractures, difficult platform angles or crashed platforms;
c) change to bladelet production.

In the case of b), platforms were made in a variety of manners. The most common form was to change the axis of the flaking surface using one of the lateral core surfaces. This process was likely to have been used often because of the size and shape of these little cores. Since the pebbles used as core blanks tended to be shaped roughly as a rectangular box, with the production of flakes from one surface, this would frequently create a natural platform on both sides of the flaking surface. From this new surface, blanks were produced until a new platform had to be found, usually following the same strategy. Frequently, the opposed platform was also used, and more rarely a new area of the core was used as platform and/or flaking surface.

The change in blank type was also very frequent, and was probably related to the length of the flaking surface. It was as common to go from flake to bladelet as it was from bladelet to flake. These changes took place in either direction during a single reduction sequence. In the case of the change to flakes, it was more common to happen simultaneously with a change in platform. The production of bladelets was carried out with the creation of a crested ridge, that was removed either as a flake or bladelet. Although crested pieces are very common in these assemblages, their use was likely to have been restricted to the preparation of the first set of bladelets. This is suggested by the fact that crested bladelets tend not to show opposed directional scar patterns. In the case of quartz, crests were not used to prepare the flaking surface for bladelet production and, in general, platforms tended to be unfacetted and cortical, unlike the chert core platforms.

Another type of reduction sequence used in these assemblages is known as the carinated technique. The first step was the removal of thick flakes from large cores. These flakes were then used as cores. Typologically, these are either carinated endscrapers or burins. If the flake did not have a keel, this was prepared by the removal of one or two flakes on each side of the blank. The original ventral surface of the thick flake was the platform used for the removal of these flakes and later for the bladelets. The result of this reduction sequence was very small, thin and slightly twisted bladelets, probably produced by the flintknapper with a hammerstone using the fingers as support for both the core and the bladelets.

The last reduction sequence used for chert is the Mesolithic type, using a unidirectional strategy for the production of bladelets. This type produced very standardised bladelets both in shape and in size. The cores present a very flat flaking surface, with the distal end showing a pronounced curvature, unlike the general convexity of the prismatic or pyramidal cores flaking surface. The reduction sequence started with larger and flatter core blanks. The cortex was removed almost completely if not totally through flake removal. This core shaping probably used a single platform. After that, a crest was prepared along the whole length of the core. After the removal of the crested bladelet, the core presented a fairly long flaking surface marked by two long and parallel ridges that were used as guide lines for bladelet removal. The core platform was multifacetted and the bladelets were likely to have been removed by pressure instead of by striking. These cores do not indicate any change in the flaking axis. The bladelets are characterised by a relatively large multifacetted platform, usually with a trapezoidal shape. The tip is very rarely pointed though it shows evidence of frequent plunging, probably as a result of the shape of the flaking surface.

Quartz and, more rarely, quartzite blanks were also produced by a radial reduction sequence. The blank for this reduction was not standardised, with both large and small sizes of pebbles and cobbles. The cortex was removed from the blank, in a more or less random manner. Only the cortex from the flaking surface was completely removed. Core platforms tend to be irregular with unfacetted, facetted and cortical areas. Hinge fractures seem to have been very common, and it was likely the main cause for core abandonment. The flakes resulting from this reduction were likely to have been fairly small and irregular.

Finally, the last type of reduction sequence is that using the chopper. This initial blank was, in general, very large, frequently quadrangular in shape. Cortex removal was very limited if existent at all. Flake production tended to eliminate gradually the cortex, resulting in flakes marked by one cortical edge, identical to a thick natural backed piece.

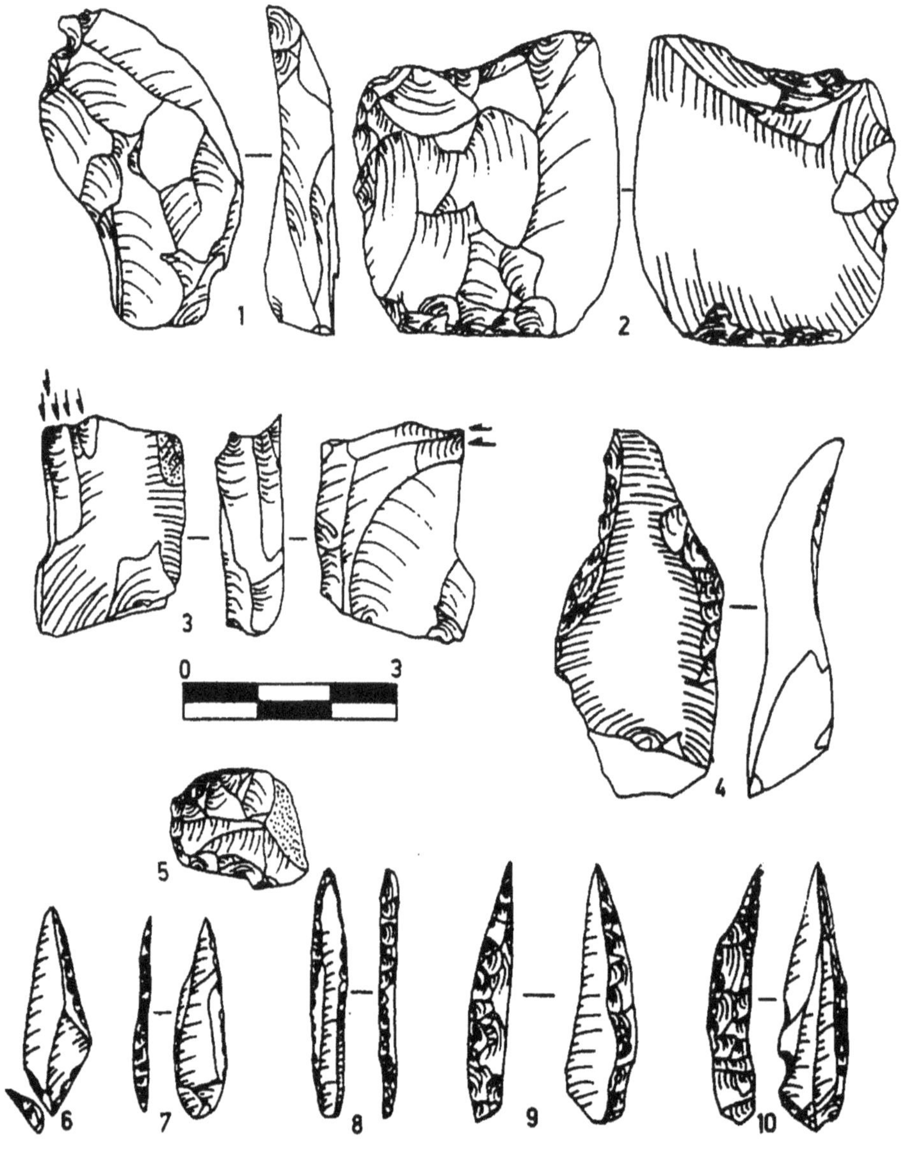

Figure 3: Lithic artefacts from Bocas rockshelter, level 0, dated to ca. 10,000 BP

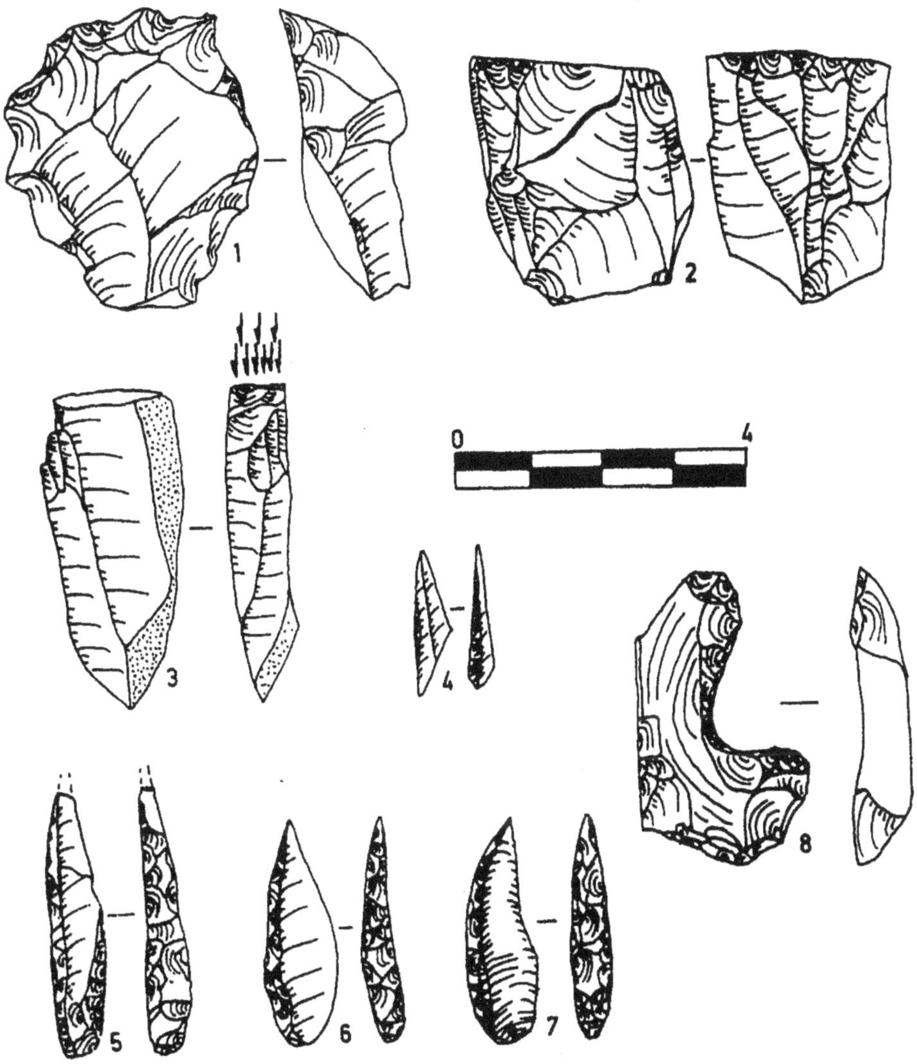

Figure 4: Lithic artefacts from Bocas rockshelter, level 1, dated to ca. 10,000 BP

In summary, there were 4 types of reduction strategies. These were unidirectional, opposed, crossed (both bi-directional and multidirectional) and radial. The first strategy used reduction sequences that resulted in various core types: prismatic, pyramidal, carinated, flat and chopper. The second strategy, opposed, also used different types of sequences. These were not, in general, alternating platforms, though this type of sequence was used for both flakes and bladelets. The crossed strategy resulted in two types of cores: 90 degree and orthogonal. These two types are not a result of core reduction stages since examples of both were found at all reduction stages. The first 3 strategies were used for both the production of flakes and bladelets, either isolated, one after the other or simultaneously in the case of alternating platforms.

Retouched tool blanks are characterised by a wide variability of sizes and shapes. Different types of flakes were used for different types of formal tools and some formal tools were produced on all types of blank. A pattern was found based on the Chi-square tests and on the differences in blank size from 12 assemblages from Rio Maior dated between 16,000 and 8500 BP (Bicho 1992). These show that there are both differences across tool classes within each assemblage and between assemblages.

It seems that blank profile was not a significant factor in the choice of blanks. Thin endscrapers, however, tended to use flakes with curved profiles. Blanks seem to have been carefully chosen to produce these scrapers.

The amount of cortex present on the dorsal surface dominates the choice of the blank for tool production. Non-cortical flakes are clearly preferred for tool production. Although this is in general true, there are a few exceptions in some sites and with the notches and thick endscrapers. These present a pattern similar to that of the debitage.

Another dominant factor in the choice of blanks seems to be blank cross-section. Although the triangular cross section seems to be the most common type in both tools and debitage, there is a change in the ratio from flat to trapezoidal in the formal tools. This suggests that blanks with simpler cross-sections (i.e. triangular and trapezoidal), were preferred to more complex forms such as the flat cross-section. This fact might be related to the number of ridges on the dorsal surface of the piece, since a flake with a flat cross-section has more ridges than a triangular or a trapezoidal piece. Of course, no apparent reason is known for this choice, which allows the possibility of some kind of cultural choice, including that of style.

In general, notches and thick endscrapers were made on randomly shaped flakes. Blank profile, cross-section and amount of cortex were important factors in the choice of the blanks for the production of thin endscrapers, while only the latter two were important for the production of both burins and denticulates.

The microlithic tools are characterised mostly by backed bladelets (Figures 3 and 4). Among these, there is a wide range of weapon types marked by, among others, Istres, arched pointed and Microgravette points. There is a tendency for the use of double backing, probably to ensure a more resistant piece. Geometrics are present in all assemblages, but they are rare in those dated prior to 8500 BP. Microlithic geometrics seem to have been made using the microburin technique. Bladelet blanks tend to look homogeneous because they have little or no cortex. It seems that the other morphological characteristics, such as cross-section or blank profile, were not important in the selection of blanks for microlithic tools. There was one exception however. This is the Dufour bladelet made on small, naturally pointed, twisted bladelets that were produced by a carinated reduction sequence.

The correlation between the tool class dimensions and the debitage seems to suggest a single general pattern across the assemblages. This pattern indicates that the backed tools as well as Dufour bladelets are always in the lower size range of the standard deviation of the debitage, for length, width and thickness. There is also some size diversity across tool flake class. Across assemblages the length, width, and thickness means are different, all assemblages have the same correlation between the dimensions of the debitage and the different tool classes. The size of thin endscrapers is identical to that of the debitage. These endscrapers are the smallest flake tools. Burin length and width are just slightly larger than that of the debitage. The thicker flakes were commonly chosen for burin production. The average length and width of thick endscrapers are, in general, larger than that of the debitage. The thicker flakes were necessarily used for the production of these types. Although the general pattern is similar to that of the burin class, the thick endscrapers are commonly larger than the burins.

The larger pieces of debitage were used to produce both notches and denticulated tools. No special criterion was used for the blank shapes chosen for producing these two tool classes. The blanks used to produce the thick endscrapers were smaller than those used to make the notches and the denticulated pieces, although still in the range of the flake blanks. Naturally, thick flakes characterise this tool class. The flakes that were used for these endscrapers were generally flat, with little cortex, and the cross-sections were preferably triangular and trapezoidal.

The blanks used to produce burins were mostly decorticated, the cross-sections were mostly triangular and often trapezoidal. These blanks were not very large, although the general tendency was to use thick flakes rather than thin ones. Thin endscrapers were produced on thin, mostly small, sometimes slightly curved flakes. Flakes produced in the first stages of reduction sequence were apparently rejected for the production of thin scrapers (and, in fact, for most of the tools except notches and denticulates) and preference was given to non-cortical flakes. These flakes were mostly of triangular

TABLE 1
FLAKE FREQUENCY BY AREA FOR SELECTED ASSEMBLAGES

Attributes	AR I		CPM 7		CPM 8		CPM 13		CPM 12		CPM 14		CPM 15		CPM 16		CR II		CPM 19		PC		AR III		
	1	2	1	2	1	2	1	2	1	2	1	2	1	2	1	2	1	2	1	2	1	2	1	2	6
Blank Shape																									
converging	29	37	31	21	18	18	43	56	87	87	42	52	36	38	37	11	44	46	34	39	46	52	89	88	70
rectangular	36	27	24	15	23	32	64	41	60	53	30	24	36	24	30	14	33	35	35	27	29	29	44	63	54
circular	34	37	36	20	30	29	46	36	65	69	50	45	19	28	44	27	51	46	27	47	45	61	126	152	102
déjeté	15	20	15	28	29	31	62	16	44	29	17	24	51	32	12	12	23	15	18	21	16	26	55	60	39
expanding	18	15	18	21	13	25	47	70	82	78	14	16	50	24	13	15	31	31	20	28	20	24	37	36	39
Blank Profile																									
flat	89	105	91	67	68	83	164	165	202	184	18	119	111	88	90	53	117	107	86	103	100	118	277	271	195
incurvate	45	25	39	32	38	54	91	46	123	118	43	42	65	31	47	23	58	58	44	47	43	73	70	110	97
twisted	3	14	5	12	13	16	28	12	25	27	13	10	20	25	10	9	21	17	14	20	18	16	28	32	24
Blank Tip																									
feathered	108	117	95	76	101	119	208	142	250	248	121	132	137	106	116	70	146	135	111	124	123	154	178	293	255
hinged	16	16	24	23	8	19	41	42	56	59	26	25	39	28	12	5	32	13	23	26	21	25	61	74	36
overpassed	13	9	8	10	6	13	28	27	33	18	7	10	17	8	13	4	13	1	4	14	7	17	20	33	14
Cross-section																									
triangular	98	96	75	63	62	79	157	103	209	212	99	114	124	83	73	36	123	102	75	103	92	97	236	254	194
trapezoidal	14	23	27	24	26	38	55	44	106	88	19	24	37	27	19	16	22	29	15	16	20	35	53	45	52
flat	26	25	33	25	30	34	70	60	33	29	46	33	35	33	55	32	51	51	54	51	49	75	84	115	70
Platform type																									
unfacetted	99	95	88	59	63	89	188	119	184	176	112	133	117	77	92	51	125	123	82	117	106	140	267	294	244
dihedral fac.	7	12	6	18	11	19	38	26	31	50	18	10	27	26	13	5	16	14	11	4	10	24	14	11	8
multifactted	2	1	2	1	1	2	3	7	18	5	3	1	6	5	1	1	2	1	3	3	3	3	4	2	2
cortical	23	19	21	29	28	21	28	39	47	46	18	10	24	24	13	14	20	25	24	23	15	20	73	92	39
crushed	7	12	14	5	9	18	20	20	29	21	8	16	18	9	24	10	19	15	20	15	22	13	16	15	17
# of Scars																									
0	6	18	10	10	7	9	29	22	24	24	11	9	13	12	7	10	16	18	11	15	8	13	28	32	25
1	31	26	27	28	29	35	80	66	73	74	32	34	41	31	32	18	53	30	29	32	55	67	109	106	75
2	59	60	52	40	51	60	103	72	111	110	68	74	80	49	54	25	85	75	58	74	60	60	148	153	129
3	29	37	31	27	27	35	52	49	106	75	27	40	49	34	32	22	32	51	35	38	28	47	62	90	62
4 and more	13	3	15	7	5	14	19	15	36	47	26	14	13	17	22	10	10	8	11	11	10	20	29	25	25
% of cortex																									
no cortex	99	93	76	66	68	90	156	39	194	168	112	114	111	94	84	45	118	106	93	108	93	120	232	244	194
<25%	21	21	26	24	23	32	67	33	72	77	22	32	46	22	35	20	34	38	26	31	27	45	75	85	59
25-75%	9	9	19	7	16	17	24	25	41	39	14	12	20	12	14	11	16	16	10	9	24	19	27	38	26
76-95%	3	2	4	5	5	5	7	6	17	19	5	4	6	4	6	1	13	4	3	7	8	9	11	13	11
>85%	7	18	10	10	7	9	28	21	26	27	11	9	13	12	8	9	15	18	12	15	8	14	30	34	26
Loc. or cortex																									
distal	8	14	10	13	12	20	38	15	48	35	13	15	22	10	18	11	20	20	15	19	16	12	45	42	38
lateral	20	11	28	14	18	19	39	26	44	54	24	30	24	14	25	8	33	27	15	22	24	37	54	73	46
lateral/distal	2	6	6	3	9	8	11	9	25	24	1	2	10	10	7	8	7	5	4	2	11	14	6	9	7
complete	7	18	10	10	7	9	26	20	26	24	10	9	13	12	8	9	15	18	12	15	8	14	29	32	26
Trimming																									
absent	133	140	125	110	118	145	280	218	338	325	155	165	191	138	134	78	175	165	130	161	156	177	369	406	309
present	5	4	10	2	2	8	10	6	12	5	9	6	5	6	13	7	21	17	14	9	5	30	6	8	7
Sample size	138	144	135	112	120	153	290	224	340	330	164	171	196	144	147	85	196	182	144	170	161	207	375	414	316

TABLE 2

F-STATISTICS FOR COMPARING WITHIN- AND BETWEEN-ASSEMBLAGE TECHNOLOGICAL VARIABILITY FOR FLAKES

	Between-assemblages		Within-assemblages			
Variable	G^2	d.f.	G^2	d.f.	F-statistic	*p*-value
% of cortex	65.53	44	45.63	52	1.69	.05>p>.025
Location of cortex	77.79	33	39.46	39	2.33	.01>p>.005
Blank shape	276.35	44	98.97	52	3.31	p<.001
Blank profile	57.63	22	64.02	26	1.06	p>.25
Blank tip	51.16	22	51.65	26	1.17	p>.25
Cross-section	239.44	22	29.25	26	9.63	p<.001
Platform type	269.51	44	102.72	52	3.09	p<.001
# of dorsal scars	106.78	44	66.52	52	1.90	.025>p>.01
Trimming	85.46	11	29.65	13	3.41	.025>p>.01

TABLE 4

F-STATISTICS FOR COMPARING WITHIN- AND BETWEEN-ASSEMBLAGE TECHNOLOGICAL VARIABILITY FOR BLADELETS

	Between-assemblages		Within-assemblages			
Variable	G^2	d.f.	G^2	d.f.	F-statistic	*p*-value
Blank shape	52.27	11	8.52	13	7.20	p<.001
Blank profile	39.00	22	54.68	26	.84	p>.25
Blank tip	129.59	22	37.53	26	4.09	p<.001
Cross-section	23.77	11	30.36	13	.92	p>.25
Platform type	90.45	22	26.49	26	4.03	.005>p>.001
# of dorsal scars	48.46	33	66.64	39	.86	.1>p>.05
Trimming	76.11	11	26.88	13	3.34	.025>p>.01

TABLE 3

BLADELET FREQUENCY BY AREA FOR SELECTED ASSEMBLAGES

Attributes	AR I		CPM 7		CPM 8		CPM 13		CPM 12		CPM 14		CPM 15		CPM 16		CR II		CPM 19		PC		AR III		
	1	2	1	2	1	2	1	2	1	2	1	2	1	2	1	2	1	2	1	2	1	2	1	2	3
Blank Shape																									
converging	29	15	29	15	24	27	68	27	71	60	25	27	23	24	28	11	22	20	19	22	27	48	18	53	67
rectangular	10-	15	11	5	25	27	55	28	27	20	20	14	14	15	34	18	25	28	15	13	12	19	10	28	33
Blank Profile																									
flat	19	19	10	7	15	21	47	25	48	47	21	14	15	23	20	10	17	18	16	10	15	11	10	35	27
incurvate	15	6	19	9	19	22	54	10	37	42	19	11	13	18	27	13	15	22	5	16	14	26	4	20	37
twisted	12	6	11	5	12	22	44	23	32	17	5	16	10	10	17	9	16	13	14	9	11	32	14	26	36
Blank Tip																									
feathered	32	18	21	12	24	40	76	30	70	69	27	22	19	35	39	23	32	41	17	17	14	35	11	40	53
pointed	12	7	17	8	17	16	38	12	16	18	9	15	8	3	22	4	9	4	12	13	22	31	16	36	37
hinge+overp.	2	6	2	1	5	9	31	16	30	19	9	4	11	13	3	5	7	8	6	5	4	3	1	5	10
Cross-section																									
triangular	38	21	25	17	26	41	101	36	77	86	36	33	29	41	51	22	35	45	25	24	28	37	22	59	61
trapezoidal	7	7	14	3	14	23	34	16	38	17	6	6	9	6	20	6	12	8	5	10	11	22	4	7	29
Platform type																									
unfacetted	22	15	16	9	24	33	75	30	40	49	26	25	23	25	22	15	17	30	16	11	15	32	14	39	51
crushed	6	4	7	3	11	12	15	3	21	11	3	5	4	10	16	6	7	8	5	6	11	18	4	7	10
punctiform	12	6	15	4	8	8	11	1	26	22	10	5	5	5	30	7	16	8	10	13	10	17	8	27	14
# of Scars																									
1	7	1	2	3	6	8	10	10	16	15	1	4	3	6	3	3	4	5	4	5	4	6	4	8	3
2	18	11	15	10	19	26	50	23	28	40	20	17	14	22	28	6	23	18	14	9	13	11	9	33	32
3	15	10	14	4	12	13	56	13	36	25	12	8	14	14	17	16	16	19	10	8	14	34	10	25	30
4 and more	5	8	9	3	7	16	24	12	36	25	12	11	7	8	16	7	5	10	6	13	9	18	4	14	35
Trimming																									
absent	38	25	27	21	44	61	139	57	107	104	37	35	33	48	44	28	34	46	27	31	31	53	23	70	90
present	8	6	13	1	2	4	6	1	10	2	8	6	5	3	20	4	14	7	8	4	9	16	5	11	10
Sample size	46	31	40	22	46	65	145	58	117	106	45	41	38	51	64	32	48	53	35	35	40	69	28	81	100

and trapezoidal cross-section, although some flakes with flat cross-sections were also chosen.

There was some relationship between tool classes and the different stages of the reduction sequence. The initial stages of the reduction sequence produced thicker, larger flakes that were largely cortical. These flakes were used for the production of notches and denticulates. During the preparation and shaping of the flaking surface, relatively large flakes were produced, probably some of which were thick. These flakes were used for the production of thick endscrapers and some of the thicker burins. Finally, the last stages of core reduction produced flakes which were completely decorticated, sometimes slightly curved and, in general, very thin. These flakes were used for the production of the thin endscrapers and the thinner burins (Bicho 1992, 1993a).

INTRA- AND INTER-SITE VARIABILITY

This section is divided into two parts. The first part focuses on the identification of those attributes that are technologically significant. The second section focuses on the patterning of assemblages based on their technological characteristics. The results presented here were investigated using the statistical procedures described before. This research was conducted in two phases. The first phase took place in 1990-1992 as part of a doctorate research. After the first phase, the methodology to analyse the lithic materials was changed according to the results obtained. Thus, two different data sets are presented here, corresponding to those two working phases.

Intra- and inter-assemblage variability: significant technological attributes

The first step in examining patterns of technological variability was to separate those attributes that showed random variability from those which showed patterned variability. It was assumed that attributes showing more variability within assemblages than between assemblages are not important in terms of technology and core reduction, and may result from individual aspects of core blank or personal style of the flintknapper. On the other hand, those attributes that showed more variability between assemblages than within assemblages were assumed to pattern, either temporally or spatially, and are important for the study of lithic technology. The two part statistical method used to isolate the random technological attributes from those that are assumed to pattern is the "log-linear model" followed by the F Statistic described previously. This procedure was applied only to flakes and bladelets since there was insufficient information for cores due to sample size.

Flakes

The morphological analysis of the flake debitage was based on 10 different attribute classes (e.g. location of cortex, blank shape, blank profile, cross-section). Dorsal scar pattern was not used for this statistical procedure due to inadequate sample size. The results from the statistical tests show that variability in "blank profile" and "blank tip" is no larger between sites than within sites (Table 1). The P values for the F Statistic are very large, over .25, and consequently these technological attributes are considered to be not significantly different across assemblages. "Trimming", "percent of cortex" and "number of dorsal scars" are considered to be moderately different (Table 2). The other attribute classes, "location of cortex", "blank shape", "cross-section" and "platform type" show significant variability among assemblages.

Bladelets

The statistical results from the data set on bladelets are different from the flake results. The first aspect to be noted is that, since the bladelet samples were smaller, attribute classes such as "location of cortex", "percent of cortex" and "dorsal scar patterns" were not used for this statistical procedure (Table 3). Also, the fact that some of the single attributes were numerically insignificant caused a smaller number of single attributes in each class. The statistical tests showed that "blank shape", "blank tip" and "platform type" were significantly different across assemblages, while "trimming" was only "moderately different" (Table 4). The other attribute classes (e.g. cross-section, blank tip) were proven to be homogeneous across assemblages, presenting a large variability within sites. This variability is probably random, and not significant for accessing patterns in lithic technological variability among these assemblages.

Summary

The significant flake attribute classes are "blank shape", "platform type", "location of cortex" and "cross-section". The latter two attribute classes are probably related to the extension of core decortication, while blank shape was probably related to different stages of the core reduction sequence. Platform type variability, although clearly a technological attribute, was not necessarily a consequence of different stages of the reduction sequence (i.e. the frequent use of cortical platforms instead of unfacetted ones). Other types, such as dihedral facetted and multifacetted platforms, can serve the same technological purpose and, consequently, their frequency should be similar (random). This is not the case, however. There are differences between the frequencies of the two platform types, suggesting that some of the technological procedures (i.e. platform type) were intended during the reduction process. This difference means that while other attributes such as "blank shape", "cross-section" and "location of cortex" are dependent upon the type of reduction sequence, platform type is independent of the reduction sequence that was used and independent of the final result - the type of flake. Platform type may be related to the technique of removal (e.g. soft hammer versus hard hammer), resulting in a characteristic platform type such as lipped. While this type was numerically insignificant in flakes, there were clear differences between unfacetted, cortical and dihedral platforms. Multifacetted platforms were

TABLE 5
Dates and Selected Variables for Clustering Final Upper Paleolithic, Epipaleolithic and Mesolithic Assemblages from Portugal

Assemblages	Chert %	Non-Chert %	Carination index %	Geometrics %	Dufour bladelets %	Backed Points %	Distance to chert sources Km	Dates C14-BP	
AR I	79.1	20.9	19.4	.5	.5	.5	3	-	-
CPM 7	78.6	21.5	15.4	1.5	1.0	0.0	5	16,340±420	SMU-2015
								15,820±400	ICEN-542
CPM 8	75.1	24.9	12.3	0.0	1.7	1.7	5	15,410±195	SMU-2476
CPM 10	72.8	27.2	16.7	0.0	2.3	0.0	5	14,050±850	SMU-2668
CPM 11	77.0	23.0	27.8	1.3	1.3	1.9	5	-	-
LP.G	79.7	20.3	0.0	0.0	0.0	22.2	22	12,310±90	OxA-5527
CPM 12	89.1	10.9	26.0	1.1	.8	5.6	5	11,810±110	ICEN-689
CPM 13	94.9	5.1	16.0	.4	1.0	1.2	5	12,220±110	ICEN-687
								11,680±60	SMU-2011
CPM 14	89.6	10.4	6.5	1.9	.8	3.8	5	11,160±280	ICEN-545
CPM 15	90.4	9.6	19.2	1.1	.8	5.6	5	11,110±130	SMU-2637
LP.F	65.4	34.6	0.0	0.0	0.0	6.4	22	11,780±90	Wk-4219
CPM 16	88.3	11.7	24.2	0.0	1.1	4.4	3	10,940±210	ICEN-690
PC	93.2	6.8	53.2	0.0	7.0	5.0	22	10,880±90	SMU-2635
LP.E	71.8	28.2	50.0	0.0	0.0	12.5	5	10,070±80	Wk-4217
CPM 18	92.0	8.0	7.0	0.0	0.0	12.2	5	9,270±170	SMU-2666
CPM 19	86.4	13.6	6.8	3.2	0.0	1.9	5	9,100±160	ICEN-688
CR II	91.5	8.5	6.3	8.2	.6	5.3	3	-	-
CR III	96.7	3.3	7.7	6.8	0.0	5.1	3	-	-
OCR	98.5	21.1	21.1	5.5	.4	2.2	3	-	-
AR II	94.1	48.4	48.4	.2	8.6	.7	3	8,680±80	ICEN-547
								8,850±50	ICEN-494
								8,570±130	ICEN-546
								8,380±90	ICEN-488
FP	88.5	11.5	9.5	44.7	.5	0.0	2	8,450±190	ICEN-973

*TL date obtained by the Laboratory of th eBritish Museum averaging 11,200-9,700 (Zilhão 1995)

**CR III is in the same stratigraphical position and it is located within 10 meters of CR II, thus, the date should be identical

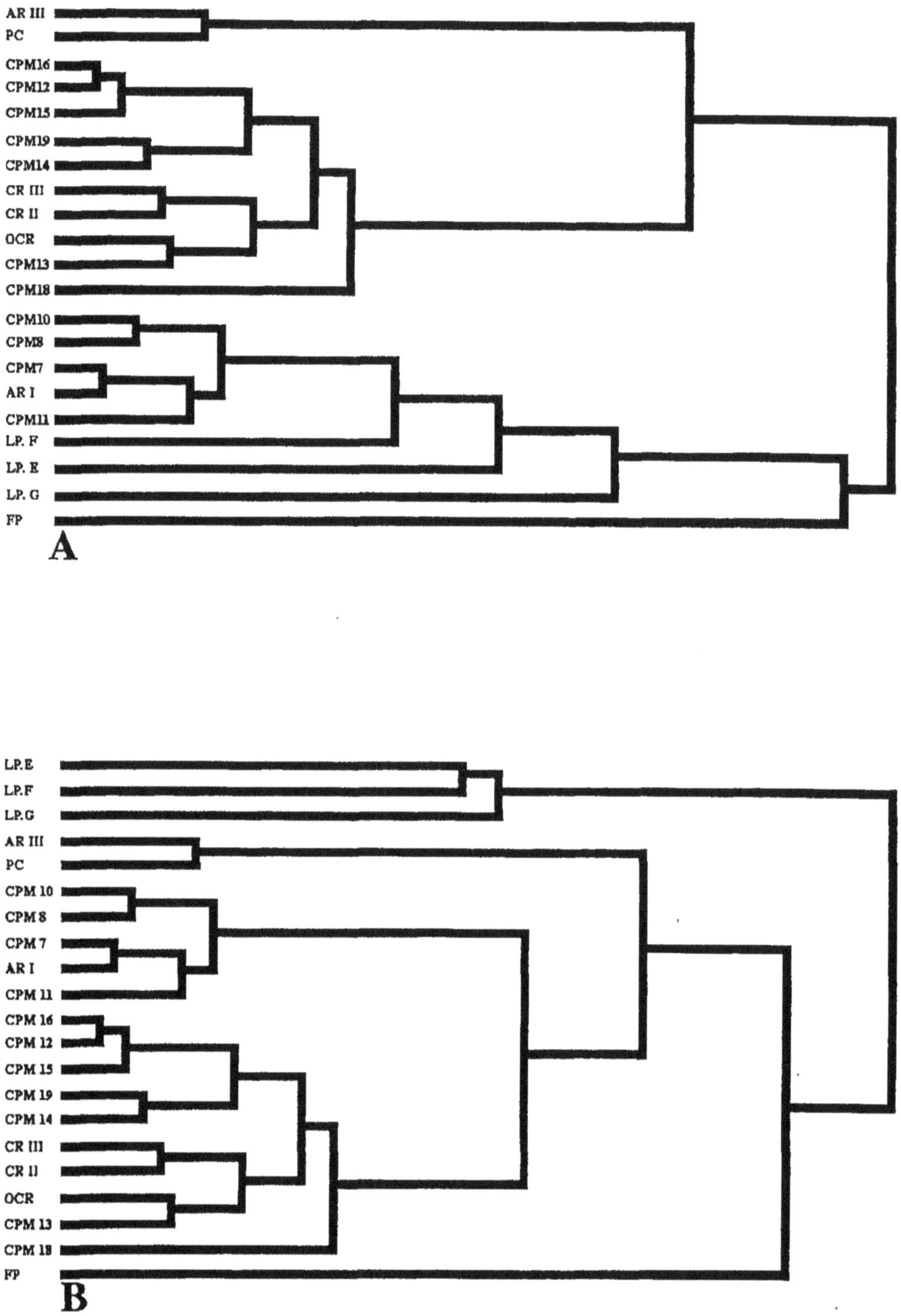

Figure 5: Dendrogram of the cluster analysis. A - excluding distance to chert sources; B - including distance to chert sources

numerically insignificant for this specific problem, but crushed platforms were fairly common. In the case of the latter type, this was probably related to technological factors (i.e. this type of damaged platform was independent of the conscious choice of the maker, independent of the reduction sequence, and probably dependent on the flint-knapping tools and on the technological ability of the flintknapper).

The bladelet attributes that show significant variability between assemblages suggest two things: first, technologically there are no major differences among Magdalenian bladelet assemblages; and, second, some of the attribute classes that are significantly different between sites (i.e. "blank shape" and "blank tip") are related to the shape of the bladelet, and consequently to the "desired end-product". The other bladelet attribute class that shows significant variability between sites is platform type. Contrary to the case of flake production, in bladelet production platform type is dependent upon the final result - the type of bladelet - and vice-versa (i.e. the type of bladelet is dependent on the platform type). Consequently, the platform type was not an option. The option was the type of bladelet to be produced and, necessarily, this option was chosen before the reduction sequence began.

According to the technological study, the two attribute classes that distinguish and characterise the two kinds of bladelets produced are "blank shape" and "blank tip". These two classes point to the same difference, a dichotomy between pointed bladelets and non-pointed bladelets. All the other attributes within these two classes are either numerically insignificant (e.g, déjeté or expanding shapes), or independent of reduction sequences and dependent upon flint-knapping tools or technological ability (e.g. hinged and plunged tips). Other attribute classes such as "blank profile" and "cross-section", that *a priori* seemed to be significantly different between sites, are not. This was the case of the twisted, flat pointed bladelets so common at such sites as Areeiro III and Pinhal da Carneira. In fact, from those attributes only the fact that the bladelets are pointed is statistically significant.

Inter-assemblage variability: patterns and clusters

The cluster analysis was carried out using two sets of variables (Table 5 and Figure 5). These variables are either technological, reflecting technological aspects of the reductions sequences or having strong impact on the lithic technology. The first set includes the frequencies of: raw materials, divided into chert and non-chert; carinated cores (endscrapers and burins) among all cores; geometrics (including geometrics and microburins) among all retouched tools; Dufour bladelets among all retouched tools; and backed points among all retouched tools. The second data set includes, in addition to those variables, the distance from the site to the nearest chert source. The results are very similar, except in the case of Picareiro cave, with all three levels clustering separately from all other sites when distance is considered.

The results of these cluster analyses confirm the results obtained in 1992 that used only 12 assemblages with different variables and statistical procedures. Clearly, there are 3 clusters, possibly 4 clusters and one isolated case. In general terms, these clusters identify both phases and facies of the Portuguese Late Glacial. Cluster A corresponds to the Early and Middle Magdalenian. Clusters B and C mark the time span between 12,500 and 8500 BP (Late and Terminal Magdalenian), separating the Rio Maior and Carinated facies. Cluster D, corresponding to the Picareiro assemblages, is isolated only when distance to chert sources is computed. Otherwise, these assemblages tend to connect in a late clustering stage to Cluster A. Finally, Fonte Pinheiro is completely isolated from the other assemblages. Although, a reasonable amount of work has been carried out in the Portuguese Mesolithic, no lithic materials have been published in a systematic manner. Thus, only one Mesolithic assemblage was used in this procedure, clearly showing the lack of similarity with the Palaeolithic and Epipalaeolithic assemblages.

TRENDS AND PATTERNS IN LITHIC TECHNOLOGY: FROM THE MAGDALENIAN TO THE MESOLITHIC IN CENTRAL PORTUGAL

Based on the technological information presented above, it is possible to define four clear groups in the Final Late Palaeolithic and Mesolithic of Central Portugal. Each group of assemblages is marked by at least one aspect of its reduction sequence.

The first phase of the Magdalenian is marked by the common use of high frequencies of quartz and quartzite. This characteristic seems to be due to the cultural preferences of these hunter-gatherer groups, since the locations of chert sources were within 5 km of the sites (Bicho 1992, 1993b). This characteristic changed after 14,000 BP, with a sharp increase in the use of chert to ca. 90% of the assemblages around 12,000 BP. This frequency seems to have increased to over 95% after 10,000 BP. Though the frequency of use of quartz and quartzite seems to have been the result of cultural choices before 14,000 BP, this fact does not seem to hold for some assemblages with a low frequency of chert dated to after the Prebölling. These are lithic assemblages from cave sites, Caldeirão and Picareiro, both located at 20 or more kilometres from the nearest chert source.

All three assemblages from Picareiro are very small samples, and this could help to explain why these assemblages are marked by high frequencies of quartz and quartzite. There is, however, another possible explanation: exactly because of the distance to the chert sources, the reduction sequences are different than those carried out in sites near chert sources. Instead of chert, quartz and quartzite were used frequently at Picareiro cave, reserving the few pieces of chert for specific tasks. This explains the link in the cluster analysis of

TABLE 6

Bladelet Technological Attributes - CPM 18 and Fonte Pinheiro

Atributes	CPM 19	FP - Total	FP - Paleolithic	FP- Mesolithic
% of Cortex				
no cortex	72.6	94,6	94.1	95.6
<25%	14.3	4.6	4.7	4.4
25-75%	13.1	.8	1.2	0
Blank Shape				
converging	59.5	74.6	72.9	77.8
rectangular	38.1	23.9	24.7	22.2
expanding	2.4	1.5	2.4	
Platform Type				
mesolithic		36.0		100.0
unfacetted	29.8	17.7	27.5	
dihedral facetted	10.7			
multi-facetted	11.9	.8	1.2	
cortical	1.1	.8	1.2	
crushed	34.5	5.4	9.2	
punctiform	6.0	17.7	27.5	
lipped	6.0	20.6	32.2	
other		.8	1.2	
Dorsal Scar Pattern				
undirectional from butt	75	100.0	100.0	100.0
uniderctional from distal end	1.2			
uniderctional from side	1.2			
bidirectional opposed	10.7			
crossed	10.7			
bidirectional perpendicular	1.2			
Blank Profile				
twisted	29.8	36.9	36.5	37.8
flat/incurvated	70.2	63.1	63.5	62.2
Blank Tip				
feathered	40.5	51.5	45.9	62.2
pointed	45.2	33.9	40	22.2
hinged	8.3	4.6	5.9	2.2
overpassed	6	10	8.2	13.4

TABLE 7

Bladelet Measurements from CPM 18 and Fonte Pinheiro

	CPM 19	FP - Total	FP - Paleolithic	FP- Mesolithic
Length				
mean	35.2	23.6	22.8	25.8
standard deviation	12.6	6.3	5.6	7.9
range	13-80	13.4-48	13.4-35.6	18-48
Width				
mean	12.1	8.6	8.6	8.7
standard deviation	4.9	1.8	1.9	1.8
range	4-24	3.7-14.9	3.7-13.1	5.5-14.9
Thickness				
mean	4.0	2.3	2.4	2.2
standard deviation	2.3	.9	1	.5
range	1-10.1	1-5.1	1-5.1	1.3-3.6

TABLE 8

Analysis Of Variance of paleolithic and mesolithic Bladelets from Fonte Pinheiro

Length

Source	DF	Sum Squares	Mean Squares	F-test
Between assemblages	1	107.822	107.822	2.789
Within assemblages	64	2474.663	38.667	p=.0998
Total	65	2582.484		

Width

Source	DF	Sum Squares	Mean Squares	F-test
Between assemblages	1	.069	.069	.02
Within assemblages	134	449.361	3.353	p=.8864
Total	135	449.429		

Thickness

Source	DF	Sum Squares	Mean Squares	F-test
Between assemblages	1	1.672	1.672	2.234
Within assemblages	134	100.321	.749	p=.1374
Total	135	101.993		

Picareiro assemblages with the Early and Middle Magdalenian assemblages. The same pattern is found in Caldeirão cave (Zilhão 1995, 1996). Quartz and quartzite were frequently used, though most retouched tools were made on chert. Thus, and unlike the open air sites of the flat Estremadura plains, the cave sites were located at some distance from the chert sources. As a result, chert was probably highly curated, used only for certain tasks and certain reduction sequences. Though curation is not a characteristic of raw material economy in the open air sites, it has been shown that distance to chert sources, though short, had an important impact on the use of quartz and quartzite and on the size of blanks in the Epipalaeolithic assemblages of Rio Maior (Bicho 1993b, in press a).

The second phase, dated to between 12,000 and 8500 BP, is marked by two technological groups corresponding with two facies. These groups are clearly distinguished in the cluster analysis and by the technological aspects of the reduction sequences. In one group, the majority of the core reductions used a carinated technology, that ultimately was used to produce Dufour bladelets. In the other group, the majority of the core reduction used single and prismatic cores to produce "normal" bladelets that were later backed. It is clear that these two facies can be separated not only on the basis of technological traits, but also based on typology.

Though this group can be separated into two different facies, the technological traits do not distinguish any phases. These, apparently, can only be separated based on other aspects, such as typology, settlement system, mobility patterns and, possibly, subsistence (Bicho 1994, in press a, in press b). Based on these, two phases can be defined, the Late (12,000-10,500 BP) and Terminal (10,500-8500 BP) Magdalenian.

Finally, the fourth group is composed of the Mesolithic assemblages, dated to after 8500 BP. Unfortunately, no significant data have been published on Mesolithic lithic assemblages. Thus, only the assemblage from Fonte Pinheiro was used in this study. This site, however, could be considered problematic for different reasons. The first is that only a very small area of the site was *in situ,* due to deep ploughing. The second is that the radiocarbon date obtained is 1000 years earlier than any other dated assemblage characterised by a high frequency of geometrics. Finally, the typology and technology show some Palaeolithic characteristics.

With this set of problems, the lithic assemblage could very easily be considered as a mix between a Palaeolithic and a Mesolithic assemblage. This, however, does not seem to be true after detailed scanning of the site conditions and morphology of the lithic assemblage. The only area where the site was *in situ* showed a clean stratigraphy with a single archaeological horizon from which the charcoal sample was obtained. This stratigraphy was identical to that of Areeiro III, located about 1 km north-east, and also radiocarbon dated to ca. 8500 BP.

The analysis of the lithic material clearly points to two different reduction sequences, one basically following the Palaeolithic "traditional" manner for bladelet production and another, with a different technology, probably using pressure flaking, resulting in flat flaking surfaces on the cores and perfectly homogeneous paralleled sided bladelets (Tables 6 and 7). The bladelets produced in both cases are virtually identical, except for their platforms and tips. Naturally, the Palaeolithic reduction sequence produced more pointed bladelets, while the Mesolithic one resulted in feathered tips, frequently plunged. Also, the Mesolithic bladelets may be characterised by a highly standardised platform, while the Palaeolithic types could have any type of platform, including cortical. All the dimensions are identical when the two groups are compared through ANOVA (Table 8).

Thus, based on the conclusion that the assemblage is not a result of mixing, one interpretation is possible: that this assemblage is marked by two different reduction strategies, of which one was used to produce highly standardised bladelets that were subsequently transformed into geometrics. The other reduction strategy, following the traditional Palaeolithic technology, produced bladelets to be used for backing, retouching or notching. This assemblage represents the first stage of a technical shift towards the almost exclusive use of pressure flaking for the production of bladelets that, ultimately, were transformed, with the microburin technique, into geometrics. This reduction sequence became the most common one after 7500 BP. This technological shift was certainly one of the aspects of the complex process, marked by changes in settlement system, land use, mobility, demography and subsistence that marked the local Mesolithic spanning from ca. 8000 to 6000 BP.

REFERENCES

Arnaud, J.

1986 *Post-Glacial Adaptations in Southern Portugal.* World Archaeological Congress, Southampton, 1986 (pre-print).

1987 Os concheiros mesolíticos dos vales to Tejo e Sado: semelhanças e diferenças. *Arqueologia* 15, 53-64.

1990 The Mesolithic communities of the Sado valley, Portugal in their Ecological Setting. In C. Bonsall (ed.) *The Mesolithic in Europe.* John Donald Publishers, Edinburgh, 614-631.

1993 O mesolítico e a neolitização: balanços e perspectivas. In *O Quaternário em Portugal.* Colibri, Lisbon, 173-184.

Bicho, N. F

1992 *Technological change in the Final Upper Palaeolithic of Rio Maior, Portuguese Estremadura.* Ph.D. Dissertation. Southern Methodist University, Ann Arbor, University Microfilms.

1993a O Paleolítico Superior Final de Rio Maior: perspectiva tecnológica. *Trabalhos de Antropologia e Etnologia* 33 (3-4), 15-36.

1993b *How two Kilometers can make a difference: A raw material study of Epipaleolithic assemblages from Central Portugal.* Paper presented at the 58th Annual Meeting of the Society for American Archaeology, St. Louis.

1994 The end of the Paleolithic and the Mesolithic in Portugal. *Current Anthropology* 35, 664-674.

1996a *Caves, Rock Shelters and Open-air sites: land use during the end of the Paleolithic in Central Portugal.* Paper presented at the XIII Congress of the UISPP. Forli, Italy.

1996b The role of quartz and quartzite in the Magdalenian of Cabeço de Porto Marinho, Rio Maior, Portugal. In N. Moloney, L. Raposo & M. Santonja (eds.) *The role of quartzite and other non-flint raw materials in the Iberian Palaeolithic.* British Archaeological Reports International Series, Oxford, 175-181.

in press a Lithic Raw Material Economy and Hunter-Gatherer Mobility in the Late Glacial and Early Postglacial in Portuguese Prehistory. In L. Fisher & B. Eriksen (eds.) *Lithic Raw Material Economy in Late Glacial and Early Postglacial Western Europe.*

in press b Spatial, Technological, and Economic Organization after the Last Glacial Maximum in Portuguese Prehistory. In J. Fullola i Pericot & N. Soler i Masferrer (eds.) *El Món Mediterrani després del Pleniglacial.* Girona.

Bicho, N. F. & J. Haws

1996 *What to eat, where to go - subsistence and settlement pattern in the Portuguese Tardiglacial.* Paper presented at the 61st Annual Meeting of the Society for American Archaeology, New Orleans.

Bordes, F.

1950 Principes d'une méthode d'étude des techniques de débitage et de la typologie du Paléolithique ancien et moyen. *L'Anthropologie* 54, 19-34.

Brézillon, M.

1983 *La Dénomination des Objects de Pierre Taillée.* IV Supplément à Gallia Préhistorique, 2nd Edition. CNRS, Bordeaux.

Ferring, C. R.

1980 *Technological Variability and Change in the Late Paleolithic of the Negev.* Ph.D. dissertation, Southern Methodist University. Ann Arbor, University Microfilms.

1988 Technological change in the Upper Paleolithic of the Negev. In H. Dibble & A. Montet-White (eds.) *Upper Pleistocene Prehistory of Western Eurasia.* The University Museum, University of Pennsylvania, Philadelphia, 333-348.

Fienberg, S. E.

1977 *The analysis of Cross-Classified Categorical data.* Massachusetts Institute of Technology Press, Cambridge.

Hietala, H.

1984 Variations on a categorical data theme: local and global considerations with Near-Eastern paleolithic applications. In H. Hietala (ed.) *Intrasite Spatial Analysis in Archaeology.* Cambridge University Press, Cambridge, 44-53.

Jelinek, A.

1976 Form, Function, and Style in Lithic Analysis. In C.E. Cleland (ed.) *Cultural Change and Continuity: Essays in Honor of James Bennett Griffin.* Academic Press, New York, 279-316.

Marks, A. E., N. Bicho, J. Zilhão & R. Ferring

1994 Upper Paleolithic Prehistory in Portuguese Estremadura: Preliminary results. *Journal of Field Archaeology* 21, 53-68.

Marks, A. E. & C. R. Ferring

1976 Upper paleolithic Sites near Ein Avdat. In A.E. Marks (ed.) *Prehistory and Paleo-environments in the Central Negev, Israel.* SMU Press, Dallas, 141-198.

Morales, M. G. & J. Arnaud

1990 Recent Research on the Mesolithic in the Iberian Peninsula: Problems and Perspectives. In P. M. Vermeersch & P. Van Peer (eds.) *Contributions to the Mesolithic in Europe.* Leuven University Press, Leuven, 451-462.

Munday, F.

1979 Levantine Mousterian technological Variability: A perspective from the Negev. *Paléorient* 5, 87-104.

Rao, C. R.

1965 *Linear Statistical Inferences and its Applications.* John Wiley and Sons, New York.

Sampson, G. & B. Bradley

1986 Analysis by replication of two Acheulian artifact Assemblages from Caddington, England. In G.N. Bailey & P. Callow (eds.) *Stone Age Prehistory.* Cambridge University Press, Cambridge, 29-45.

Scheffé, H.

1959 *The Analysis of Variance.* John Wiley and Sons, New York.

Schild, R.

1976 Flint Mining and Trade in Polish Prehistory as seen from the perspective of the perspective of the Chocolate Flint of Central Poland: second approach. *Acta Archeologica Carpathica* 16, 147-177.

de Sonneville-Bordes, D. & J. Perrot

1954 Lexique typologique du paléolithique supérieur: Outillage lithique. *Bulletin de la Société Préhistorique Française* 51 (7), 327-333.

1955 Lexique typologique du paléolithique supérieur: Outillage lithique. *Bulletin de la Société Préhistorique Française* 52 (2), 76-79.

1956a Lexique typologique du paléolithique supérieur: Outillage lithique. *Bulletin de la Société Préhistorique Française* 53 (7), 408-412.

1956b Lexique typologique du paléolithique supérieur: Outillage lithique. *Bulletin de la Société Préhistorique Française* 53 (9), 547-559.

Zilhão, J.
1995 *O Paleolítico Superior da Estremadura Portuguesa.* Unpublished Ph.D Dissertation. Faculdade de Letras de Lisboa, Lisbon.
1996 The use of Non-Flint Raw Materials in a Late Palaeolithic Assemblage from Gruta do Cladeirão (Tomar, Portugal). In N. Moloney, L. Raposo & M. Santonja (eds.) *The role of Quartzite and other non-flint raw materials in the Iberian Palaeolithic.* British Archaeological Reports International Series, Oxford, 167-174.

THE ROLE OF RAW MATERIAL AVAILABILITY IN TECHNOLOGICAL ORGANIZATION: A CASE STUDY FROM THE SOUTH-EAST ITALIAN LATE PALAEOLITHIC

Sarah Milliken

INTRODUCTION

In the last decade, archaeologists have become increasingly interested in the relative importance of lithic raw material availability in the organization of technology (e.g. Bamforth 1986; Johnson 1989; Odell 1989; Kuhn 1991; Marks *et al.* 1991; Nelson 1991; Andrefsky 1994; Newman 1994). According to Nelson (1991), if hunter-gatherers have settled in a non-source area then it is social, economic and technological decisions which primarily affect tool and core design, and any resultant scarcity of raw material is only secondary in conditioning tool maintenance and recycling. This view contrasts with that of Bamforth (1986), who places availability of raw material ahead of other conditioners of technological organization. According to Bamforth, maintenance and recycling are closely related to raw material availability, and are not directly or solely related to settlement organization, or to time limits on the activities for which the tools were used. Therefore tools will be maintained and recycled only when raw material is in short supply, in other words, when it costs more to replace a tool than to rejuvenate it (Bamforth 1986). This paper discusses the results of a study carried out on two Late Palaeolithic assemblages in south-eastern Italy, one of which is located in a non-source area while the other is located in the proximity of abundant supplies of good quality local raw material. This study shows that while both of the arguments outlined above are theoretically tenable, multiple factors should be kept in mind when trying to understand the organization of prehistoric technology, rather than prime mover models.

THE GEOLOGICAL CONTEXT

South-east Italy (the region of Puglia) is composed almost exclusively of exogeneous rocks, of which karstic limestone is the dominant type. However, there is great variation in altitude, structure and hardness, and according to this variation the region can be neatly divided into four distinct geographic areas (Figure 1):

The *Salento peninsula* has a low relief (maximum height 200 metres) and a varied surface geology. The rocks, which are mainly calcareous, consist of four principal types: compact Cretaceous limestones (*Calcareniti del Salento*) which form the substratum of the whole peninsula and outcrops in the hilly areas; Miocene marly limestone (*Pietra leccese*) in scattered lowland areas to the east; a wide distribution of Pliocene calcareous tufa (*facies astiana*); and unconsolidated Pliocene sands occurring in beds resting on the tufa in the south. In addition to these four major rock types, deposits of Pleistocene *terra rossa* (red earth) up to five metres in depth cover parts of the troughs between the *serre*, asymmetrical ridges which were originally caused by fracturing associated with folding of the limestone, but have since been deepened by erosion. There are no flint-bearing deposits.

The *Murge* are tablelands consisting of homogeneous blocks of gently folded stratified limestone of Cretaceous age and, as in the Salento peninsula, there are no flint-bearing deposits. The land rises in a series of broad steps with north-east facing scarps, from a narrow coastal plain to an undulating stony zone with a maximum height at Torre Disperata of 686 metres.

The *Tavoliere* is composed of nearly horizontal beds of comparatively soft Pliocene sands and Pleistocene clays over a Mesozoic carbonate substrate. These beds are covered by a calcareous soil (calcrete), over which lie thick deposits of Holocene alluvium of both marine and continental origin. The highest point of the Tavoliere is 400 metres above sea-level, where the tablelands meet the Capitanata Apennines in the west.

The *Gargano promontory*, which rises steeply from the Tavoliere lowland up to a broad plateau with heights of more than 900 metres, is composed of secondary limestones with dolomitic outcrops in the centre. On the north-western and southern flanks there are small areas of Pliocene deposits. The Gargano promontory is the only part of the region where there are outcrops of cryptocrystalline rocks. Chert is found in the form of both nodules and tablets in Upper Cretaceous limestone and marly limestone formations and in Cretaceous/Upper Jurassic micritic limestones and clayey micrites (*Formazione di Rodi Garganico, Formazione di Cagnano Varano, Calcari dolomitici di Monte Jacotenente, Calcari di Mattinata*). In south-central Italy there are outcrops of chert in the Jurassic limestones and marls and Upper Triassic limestones in the adjacent region of Basilicata.

THE PALAEOENVIRONMENTAL CONTEXT

In south-east Italy the end of the last glacial maximum, between 18,000 BP and 15,000 BP, was a cold arid phase with an open steppe vegetation composed of *Artemisia*, Chenopodiaceae and *Ephedra*. The sea-level was at a low

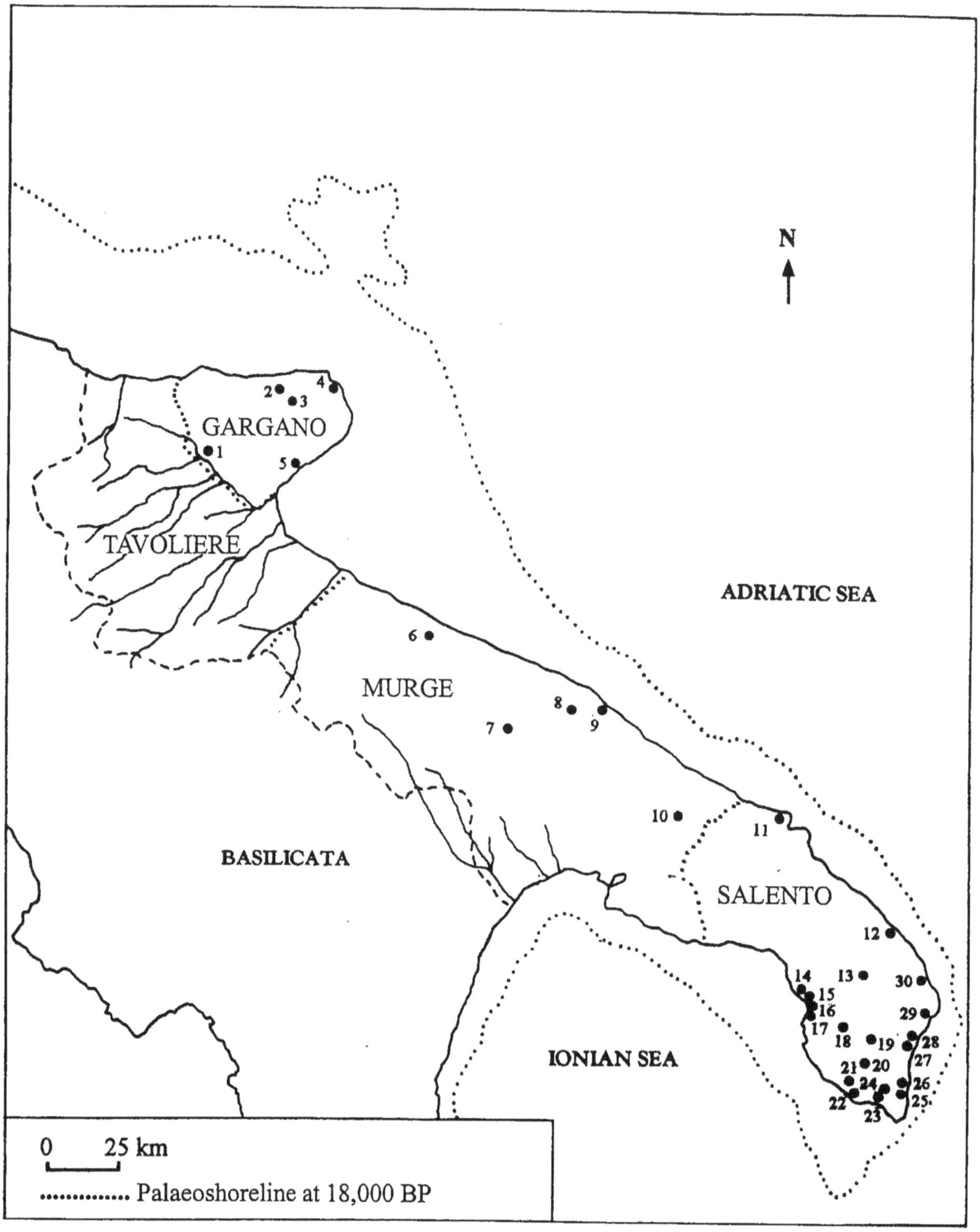

Figure 1: Figure 1: Distribution of Late Palaeolithic sites in south-east Italy: (1) Grotta Paglicci; (2) Grotta Scaloria; (3) Macelli; (4) Foresta Umbra; (5) Macchione; (6) Grotta Santa Croce; (7) Grotta di Corto Martino; (8) Grotta Simone; (9) Grotta delle Mura; (10) Monticello; (11) Torre Testa; (12) San Foca; (13) Soleto; (14) Torre dell'Alto; (15) Grotta di Uluzzo; (16) Grotta Carlo Cosma; (17) Grotta del Cavallo; (18) Grotta delle Venere; (19) Supersano; (20) Taurisano; (21) Ugento; (22) Punta Macchione; (23) Torre Vado, (24) Monti Russi; (25) Grotta delle Prazziche; (26) Grotta delle Cipolliane; (27) Grotta Zinzulusa; (28) Grotta Romanelli; (29) Porto Badisco; (30) Laghi Alimini

point of -118 metres and a plain varying between 60 and 10 kilometres in width fringed the south-east Italian coast.

From 15-13,000 BP (Oldest Dryas), an open, treeless vegetation of steppic type indicates an arid climate; the period from 13-11,000 BP (Bölling-Alleröd) was characterized by higher temperatures and higher precipitation rates, as indicated by the increase in arboreal pollen values. 15,000 BP marks the beginning of the deglaciation, and the sea-level rose rapidly at a rate of 16m/kyr to a level of -54 metres at 11,000 BP. The coastal plain would therefore have been reduced by at least 50%, and in some places totally submerged. During this period, a progressive decline of the steppe vegetation and the appearance of mixed deciduous woodland elements such as oak and poplar suggests an increasingly mild and humid climate.

The Younger Dryas event, between 11,000 BP and 10,000 BP, saw a halt in the sea-level rise. The return of a steppe vegetation suggests that the climate was cold and arid, though the presence of some trees indicates that the climate was less severe than during the last glacial maximum. 10,000 BP saw the return of mild and humid conditions. A rapid rise of sea-level at a rate of 17m/kyr brought the sea-level up to -37 metres, and the coastal plain would have been more or less submerged throughout the region. The postglacial climatic amelioration triggered an increase and expansion of thermophilous vegetation characterized by a mixed deciduous woodland of oak, beech, and hazel.

The effects of rising sea-levels would certainly have been noticeable on a human time-scale. Firstly, rising sea-levels would have caused a corresponding upward movement of the inland water table, which would have affected the distribution of fresh-water springs. This might have been particularly important in the Salento peninsula, which lacks surface water in the form of rivers. Secondly, the rising sea-levels would have influenced the spatial and temporal availability of subsistence resources. The Adriatic coastal plain before 15,000 BP may have been sufficiently productive to support semi-sedentary hunting bands throughout the year, without necessitating summer forays into high country as is often assumed (e.g. Higgs and Vita Finzi 1972; Barker 1981).

Between 15,000 BP and 9,000 BP, the rising sea-levels would have resulted in the progressive segmentation of these once continuous plains into smaller fragments that would have been no longer capable of supporting large migratory herds. One could also postulate that a much closer seashore would have allowed a more substantial marine contribution to the diet. Accordingly, one would expect to see a proportional change in the subsistence base, and a change in the technology adapted to the exploitation of these new resources.

THE SOUTH-EAST ITALIAN LATE PALAEOLITHIC

The first detailed study and classification of Italian Late Palaeolithic assemblages was published by G. Laplace in 1964, in which the 'Tardigravettian' was subdivided into three phases, Ancient, Evolved and Final, which were designed to correspond neatly with the French Upper Palaeolithic cultural divisions of Solutrean, Magdalenian and Azilian. The Laplace chronostratigraphic scheme is a classic example of the culture historic method, based exclusively on the typological classification of the tools. The south-east Italian assemblages were divided into two cultural facies which retain their identities throughout the Tardigravettian: the 'Garganian facies' in the north and the 'Salentine facies' in the centre and south. The distinction between these facies was based on the presence/absence of certain tool types (circular endscrapers, truncated backed bladelets and geometric microliths) and fluctuations in the burin and endscraper indices.

This scheme was subsequently revised by Bartolomei *et al.* (1979), in which the 'Tardigravettian' was renamed the 'Epigravettian', and most recently by Palma di Cesnola *et al.* (1983). Again, the supposedly diagnostic indices and tool types provided the basis for the classificatory scheme. While no distinction was made between the assemblages from northern and southern south-east Italy for the Ancient Epigravettian, the Evolved and Final Epigravettian assemblages were again divided into two sub-regions on the basis of the fluctuations in the burin and endscraper indices and the relative percentages of circular endscrapers, truncated backed bladelets and geometrics.

In the Salento peninsula the Final Epigravettian is often divided into three subphases: pre-Romanellian, Romanellian and Epiromanellian (eg. Palma di Cesnola *et al.* 1983). The term 'Romanellian', whose name derived from the type site of Grotta Romanelli in the Salento peninsula, was originally introduced by A.C. Blanc (1939a, 1939b) and was defined as a microlithic industry characterised by small circular and subcircular endscrapers, truncated and pointed backed bladelets, polyhedral and nucleiform burins, geometric microlithis (semilunates and triangles) and microburins. The 'Epiromanellian' is characterised by a relative increase in the quantity of small circular endscrapers and geometric microliths.

The traditional approach to the Late Palaeolithic in south-east Italy has focused on the description and classification of assemblages within a chronostratigraphic framework, and the seriation of typological sequences of artifacts to establish inter-site cultural correlations. The process of periodization using abstract typological indices as temporal markers and ethnic identifiers was grounded on the assumption that the stone tool record conformed to what was essentially an organic model of diachronic change, with time and culture as parallel causal variables. Although time is continuously connected, the tendency to divide and classify has resulted in a synchronic view of time as a sequence of homogeneous periods, rather than a diachronic view of time as continuous change. The naming of these periods on binomial principles, such as Ancient, Evolved and Final Epigravettian, affords the illusion of classed order, but the traditional chronostratigraphic schemes neither adequately describe the range of assemblage variability, nor do they serve to explain that variability. The aim of my own research (Milliken 1991) was therefore to seek explanation for diachronic and synchronic variability in the Epigravettian assemblages in south-east Italy in terms of social and economic strategies.

THE MODEL

The only chert sources in south-east Italy are located in the Gargano promontory, and in the adjacent region of Basilicata, and consequently all lithic raw materials had to be imported at least 250 kilometres to the sites in the Salento peninsula. Since technological strategies can be envisaged as being determined by the need to maximise efficiency and economy as defined by their relative costs, hunter-gatherers in the Salento peninsula can therefore be expected to have employed economising strategies in lithic procurement, tool production and use. In contrast, the model does not expect to find evidence for such economising behaviour at sites in the Gargano promontory where good quality chert was locally available. In this study the following indicators were used to measure raw material economy:

- *Laminarity*. A dominance of laminar forms in an assemblage has been interpreted by some as an indication of a technological strategy to economise on raw material (Leroi-Gourhan 1964: 190-197). Laminar blanks have a longer cutting edge for a given volume of raw material than flakes; they permit more frequent rejuvenations than flake blanks and are more flexible and can be transformed more easily into a wide variety of tools. Standardized preforms such as bladelets can be removed from cores with less waste than amorphous flakes (Johnson 1987). Blade and bladelet cores maximise or conserve raw material because a large quantity of tool edge can be produced in relation to the amount of raw material used.

- *Size*. It is logical to assume that the smaller the tools are, the more economical is the use of raw material. A distance-decay relationship is generally recognized in situations in which prehistoric populations inhabited areas distant from raw material sources (cf. Newman 1994). Johnson's (1989) distance-decay model predicts decreasing size of artifacts with distance from the source, whereby the progressive decline in the size of artifacts with the increase in distance is essentially viewed as an economising strategy by which a greater number of implements could be made from scarce raw material.

- *Maintenance*. Rejuvenation and resharpening are two alternative maintenance strategies designed to prolong the use-life of tools and therefore minimize the costs associated with lithic procurement (Bamforth 1986; Shott 1986). Resharpening involves the retouching of a blunted tool edge, while rejuvenation involves the removal of an edge to provide a new edge. Resharpening is therefore a more conservative strategy in terms of raw material than rejuvenation.

- *Breakage*. Assemblages in non-source areas are likely to be characterised by higher rates of tool breakage than those in source areas, if breakage indicates an intense utilization of the toolkit.

- *Curation*. High rates of breakage and maintenance are features characteristic of a curated technology (sensu Binford 1979), whereby tools are useful in a variety of tasks, manufactured in anticipation of use, maintained through a number of uses, transported between sites, and recycled into other functional modes. Therefore tools were designed to stay in the system for as long as possible. A relative importance of retouched tools in an assemblage could be used as an indicator of the degree of curation, regardless of whether the retouch pertains to investment in design, as in the case of backed bladelets, or to maintenance.

- *Cores*. In non-source areas cores are likely to have been exploited to the absolute maximum possible, and therefore they will tend to be both small sized and amorphous in shape. Bipolar cores are also likely to be present, since this method permits further blanks to be produced once a core has become too small for freehand reduction.

- *The import of prefabricated tools*. Production efficiency would have been increased by advancing the core reduction process closer to the finished tool while still in the source area, as in this way the cost of portaging raw material destined to be lost in production failures would have been reduced or eliminated.

- *The use of non-cryptocrystalline raw materials*. Where a strategy of raw material economy was in operation one might expect to find non-cryptocrystalline raw materials used to make part of the toolkit. An example of this is found in the Iberomaurusian assemblages in northern Africa where the use of limestone and other non-siliceous rocks in addition to flint can be accounted for by the difficulty of obtaining adequate supplies of flint along the Maghreb littoral. There, the non-siliceous rocks were used for the chopping and scraping component of the toolkit (Lubell *et al.* 1984: 149).

- *Recycling of older artifacts*. This is a practice that has been observed among the Nakada of Australia, who gleaned out old implements from archaeological sites and reworked them when they moved to areas where they were unable to find mining places for new stone (Tindale 1974: 87).

LITHIC ASSEMBLAGE VARIABILITY IN THE SOUTH-EAST ITALIAN LATE PALAEOLITHIC

Grotta delle Cipolliane

Grotta delle Cipolliane lies on the Adriatic coast of the Salento peninsula at 30 metres above sea-level, cut into a rocky coastal cliff of Plio-Pleistocene tufaceous limestone. Above the cliff the land rises steeply up to a plateau at 120 metres above sea-level, and then rises more gradually up to a plateau at 190 metres. At 15,000 BP there would have been a 10 kilometre wide coastal plain in front of the cave, watered by springs emerging at the foot of the cliff; by 11,000 BP, rising sea-levels would have reduced this plain to 0.5 kilometres, and by 9,000 BP it was completely submerged. The nearest chert sources lie 250 kilometres to the north.

The cave was excavated in the 1960s (Palma di Cesnola 1962; Gambassini 1970). The sequence from this site spans the Ancient (level 4), Evolved (level 3) and Final Epigravettian, with a possible hiatus between the earlier (level 2) and later stages of the Final Epigravettian (level 1). A series of three radiocarbon dates was obtained on burnt bone only for level 3: level 3 I-II 15,200±100 (R-353), level 3 III 15,270±300 (R-355) and level 3 VI-VII 15,200±100 (R-356) (Alessio *et al.* 1976). The faunal remains consist of *Equus caballus*, *Equus asinus hydruntinus*, *Bos primigenius* and *Cervus elaphus* with, in level 1, the addition of *Capreolus capreolus* and *Sus scrofa*. Marine molluscs were also found only in level 1, consisting of *Monodonta turbinata*, *Patella* sp., and *Glycimeris* sp. On the basis of the ungulate faunal remains and the radiocarbon dates, Palma di Cesnola *et al.* (1983) assigned the temperate humid phase of level 3 to the Angles interstadial, and the following continental arid phase of level 2 to Dryas I. Between level 2 and level 1, a phase of erosion suggests a hiatus and level 1 was therefore assigned to the Preboreal. However, a reconstruction of the palaeoshorelines (Milliken 1991) suggested that level 1 predates the Preboreal, and therefore it was tentatively reassigned to the Bölling or Alleröd.

A technological study was made on 849 of the tools recovered during Gambassini's 1964 excavation:

Backed bladelets (n.324, 38%): of these, 47% are fragments. Level 1 has a considerably higher rate of breakage (67%) than any of the other spits. Of the unbroken bladelets, the majority (84%) are smaller than 2.5 centimetres in length, while the remaining 16% are smaller than 5 cm. There are no backed blades.

Linear retouched tools (n.146, 17%): while blades are rare and sporadic (4%), flakes and bladelets are present in roughly equal quantities (50% and 46% respectively), with no significant temporal fluctuations in the relative proportions of each type. With regard to the size of the tools, there is only a slight predominance of medium sized blanks (between 2.5 and 5 centimetres) over small ones (54% and 41% respectively), while large blanks (longer than 5 centimetres) are rare and sporadic. The cross-sections of the blanks are predominantly flat throughout the stratigraphy, with carinated examples present only in spits 4III-I, 3II and level 1. 52% of all linear retouched tools are broken. Comparisons between the levels are hampered by the small sample sizes, but there seems to be a significantly higher rate of breakage in spits 4IV-III (73%) than in spits 4II-I (35%). In level 1 the rate of breakage is 53%.

Burins (n.123, 15%; Figure 2: 1-8): throughout the stratigraphy the burins are predominantly made on flakes (76%). The majority of the bladelet burins are clustered in spits 3VII-II (86%), with only one example in level 4 (spit 4IV) and three in level 1. Blade burins are entirely absent. Blank size is predominantly smaller than 2.5 cm (78%), and no burins are larger than 5 cm in length. There is no significant variation between levels 1, 2, 3 and 4 on the basis of mean length. There is only one fragment, in spit 3V.

Gambassini (1970: 163) observed that many burin spalls amongst the debitage are longer than the facets of the burins, and in some cases are longer than the burins themselves. This suggests that the burins were being rejuvenated to a considerable extent. As the distinction between bladelets and flakes is based on the value of the length:width ratio, the constant rejuvenation of burins by the removal of more spalls might therefore account for the observed predominance of flake forms. If so, the relative abundance of bladelet forms in spits 3IV-II might suggest less intense rejuvenation compared with the other spits.

Endscrapers (n.105, 13%; Figure 3: 1-12): the relative proportion of flake and bladelet forms fluctuates through the stratigraphy, and considering the assemblage as a whole there is only a marginal dominance of flake types (51%). The blanks are predominantly smaller than 2.5 cm (87%), accompanied by sporadic occurrences of medium sized types (2.5-5 cm); none is larger than 5 cm in length. Long endscrapers, whose length is greater than twice their width, are present in small quantities from spits 4II-2VI, but do not reappear in level 1 after the hiatus. The majority of the blanks are flat (90%), that is their thickness is less than half their width, with carinated types present in spits 4I-3III and again in level 1.

According to the traditional chronostratigraphic schemes, certain characteristics of the endscraper class are considered to be temporally diagnostic. The Salentine facies of both the Evolved and the Final Epigravettian is characterized by a long/short endscraper index (Gfl/Gfc index) of less than one, and the Final Epigravettian sees the appearance and subsequent relative increase during the Romanellian and Epiromanellian of circular endscrapers (Laplace 1964; Bartolomei *et al.* 1979; Palma di Cesnola *et al.* 1983). In addition, carinated endscrapers are thought to be a characteristic type of the Ancient Epigravettian (Palma di Cesnola and Bietti 1983). At Grotta delle Cipolliane, the

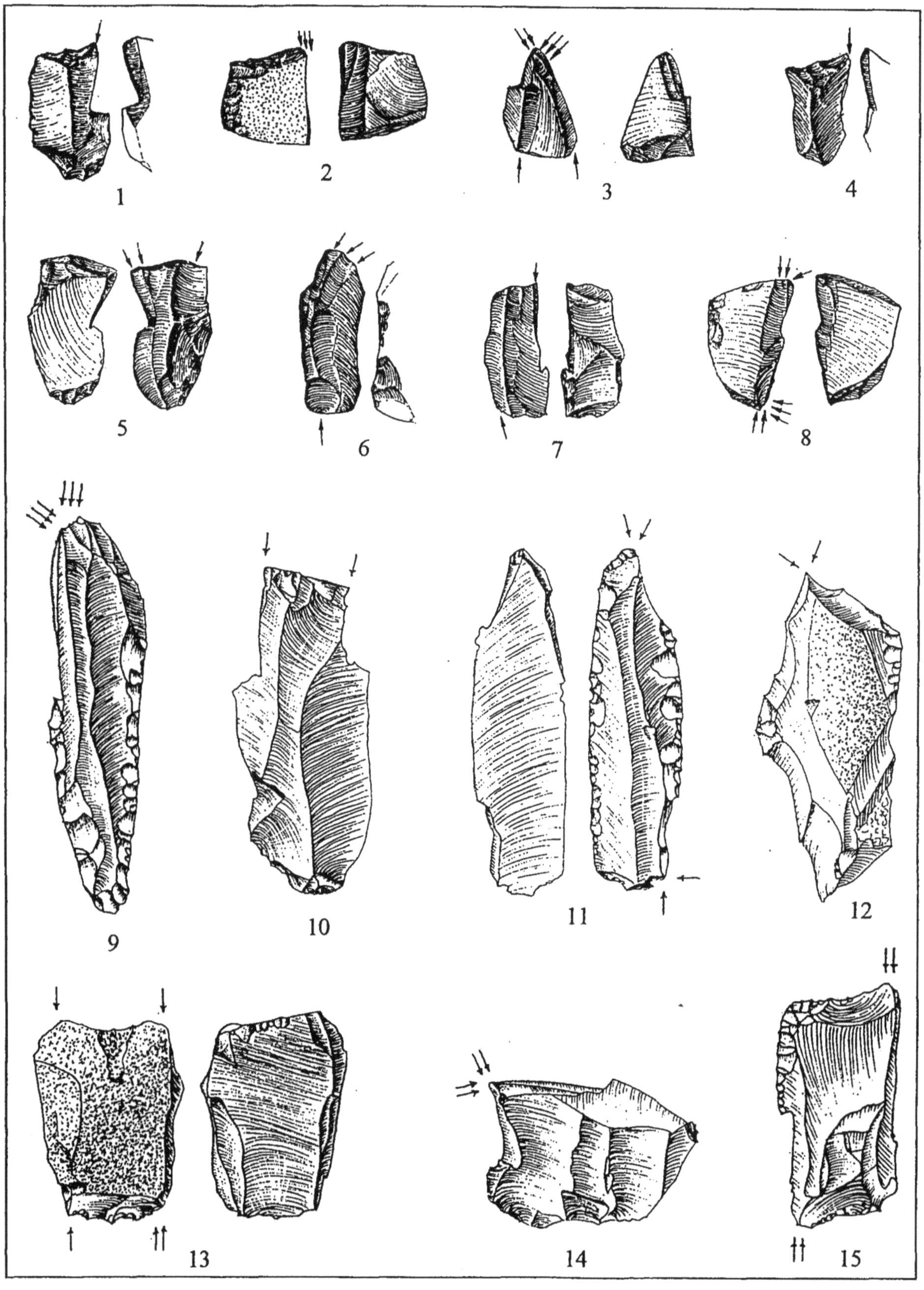

Figure 2: Burins from Grotta delle Cipolliane (nos. 1-8) and Grotta Paglicci levels 7-4 (nos. 9-15) (after Gambassini 1970; Mezzena and Palma di Cesnola 1967; full size)

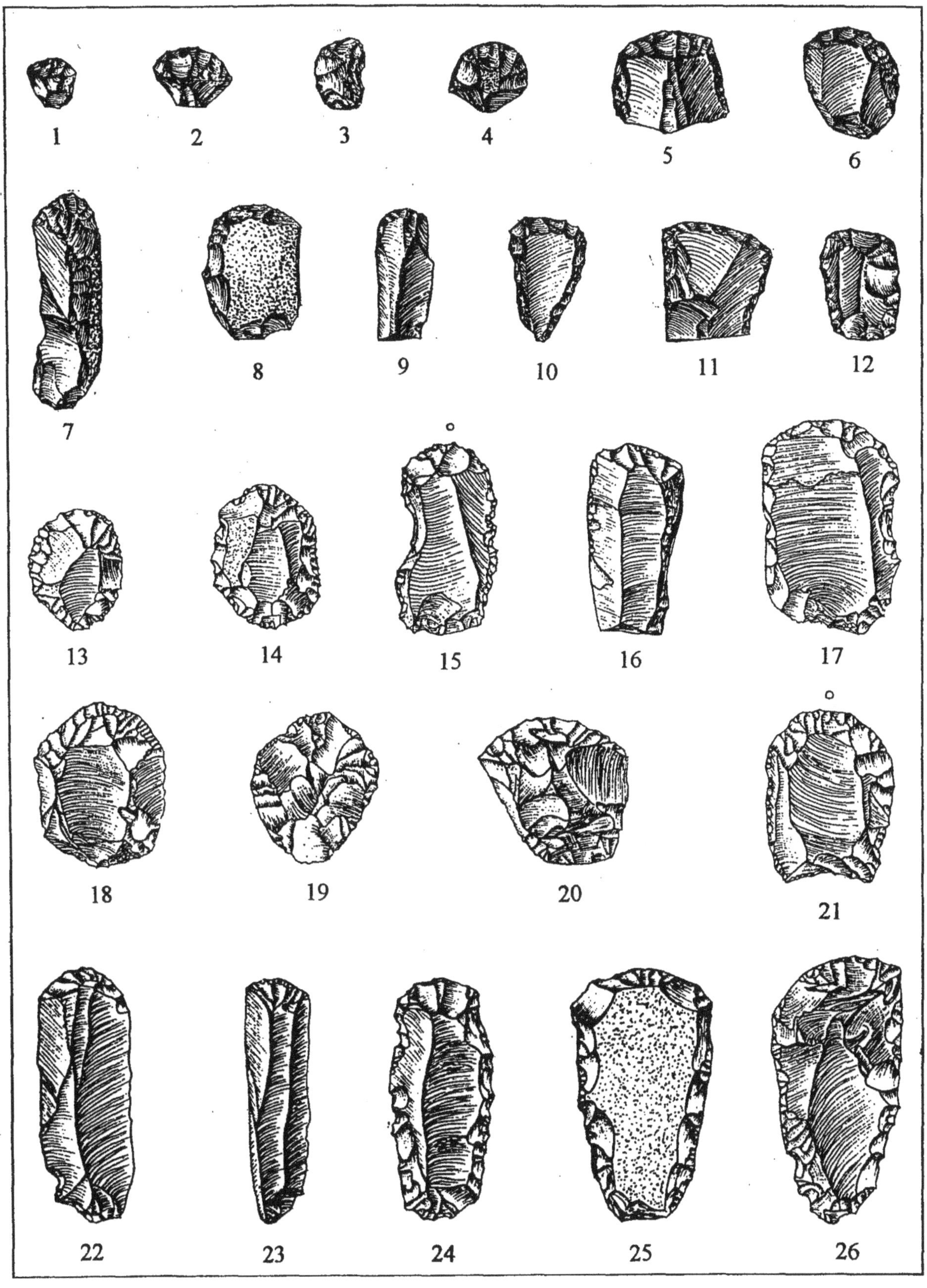

Figure 3: Endscrapers from Grotta delle Cipolliane (nos. 1-12) and Grotta Paglicci levels 7-4 (nos. 13-26) (after Gambassini 1970; Mezzena and Palma di Cesnola 1967; full size)

endscraper assemblage conforms to the Gfl/Gfc index of less than one, and circular endscrapers appear for the first time in level 1. According to Bietti and Burani (1985: 23), circular endscrapers represent a typical example of the influence of raw material scarcity on lithic typology. Repeated resharpening of endscrapers made on small flakes may have resulted in the creation of a circular endscraper, with a continuous retouched edge. With regard to the carinated endscrapers, they are not stratigraphically restricted to the Ancient Epigravettian (level 4), and indeed there are proportionately more in level 1 than in any other level. Carinated endscrapers are also likely to result from repeated resharpening of the tool, which shortens its length in relation to its width and increases the steepness of the scraper edge angle (Wilmsen 1968), and this is supported by the fact that all five of the carinated endscrapers in level 1 are smaller than 1.2 centimetres, while those in levels 3 and 4 all fall in the range of 1.2 to 2.5 centimetres with the exception of one.

Only nine (5%) of the endscrapers are broken, five of which come from level 1 (24% of the endscrapers in level 1), the others from spits 3I, 3V and 3VII. There would therefore appear to be a higher rate of breakage in level 1, which might suggest that the endscrapers were being used and resharpened more intensively before being discarded, and therefore that a strategy of raw material economy was in operation. This is supported by the relatively high proportion of carinated endscrapers in level 1, and the presence of circular forms.

Notches (n.51, 6%): these are predominantly made on flakes (73%), and the majority are made on blanks smaller than 2.5 cm (73%). Nine (18%) are broken.

Denticulates (n.30, 4%): of these, nine (30%) are broken. They are predominantly made on flakes (70%), and in contrast to the notches the majority are larger than 2.5 cm (73%).

Points (n.21, 3%): these are made predominantly on bladelets (62%), and the majority are larger than 2.5 cm (71%). Four (19%) are broken, in spits 4II, 4I, 2VI and level 1. Carinated forms are rare (19%), and stratigraphically dispersed.

Truncations (n.15, 2%): the blanks are predominantly bladelets (60%), with one blade in spit 4II, and flakes are present from spit 3IV onwards. The size of the blanks is either small or medium, with large blanks present only in spits 4II and 3IV. Five (33%) of the truncations are fragments, and these occur in spits 4I, 3IV, 3III and 3II (n.2).

Perforators (n.17, 2%): the blanks are predominantly flakes (71%) and there are marginally more medium sized (53%) than small ones (41%), with one large perforator in spit 2VIII. Only in spits 3III-II are the perforators predominantly small. Four (24%) of the perforators are fragments, occurring in spits 3VII, 3III, 3II and level 1.

Composite tools (n.17, 2%): none of the composite tools are broken. The class is composed of combinations of burins, endscrapers, denticulates, truncations, notches and perforators. The composite tools are predominantly small (76%) and are made on either bladelets or flakes in equal proportions. Those made on flakes are restricted to a tight stratigraphic cluster from spits 3VI-II.

In summary, the tools are predominantly laminar (61%) except in spits 3V and 2VI, though only in spit 3V is this significant. Backed bladelets are the only tools which are exclusively, and indeed by definition, made on laminar blanks. When the backed bladelets are excluded from the calculation, the assemblage is for the most part made on flake blanks (60%) except in spits 4IV and 3VIII, though only in spit 3VIII is this significant. The only tool classes which are predominantly laminar are the points (62%) and the truncations (60%). All the other tool classes have some laminar forms, ranging from 24% (burins) to 50% (linear retouched tools and composite tools). The majority (99%) of these laminar blanks are bladelets, with blades only occurring in spits 4II (n.2), 3VIII (n.1), 2VIII (n.1) and level 1.

Considering the assemblage as a whole, the majority (68%) of the tools at Grotta delle Cipolliane are smaller than 2.5 centimetres, except in spits 4VI, 3VIII and 2VII. When the backed bladelets are excluded from the calculation, the majority (59%) are still smaller than 2.5 cm, except in spits 4IX, 4VI, 4IV, 3VIII and 2VIII-VII, though only in spits 3VIII and 2VIII is this significant.

Finally, 33% of the tools are fragments. Spits 3VI-IV have a significantly lower rate of breakage than any other part of the stratigraphy, while level 1 has a higher rate of breakage than spits 3VII-I. The postulate that the high rate of breakage in level 1 may indicate a relatively intense utilization of the toolkit is supported by the high proportion of carinated and circular endscrapers which suggests heavy maintenance.

The percentage of retouched tools from the 1962 excavation (Palma di Cesnola 1962) ranges between 4% and 25%, while the percentage of cores ranges from 0% to 4%. While all of the cores from the 1964 excavation are bipolar and grade morphologically into *pièces ésquillées*, in the 1962 excavation prismatic bladelet cores were the predominant type found in level 3, with a maximum dimension of 3 centimetres. The second most common type are semiprismatic bladelet cores, while subdiscoidal cores are rare. Level 1 sees a reversal, with a predominance of discoidal flake cores which have a maximum dimension of 2 centimetres, and then semiprismatic and rare prismatic bladelet types. However, the scars on a core are only an indication of the last blank removals, and in level 1 there are marginally more bladelet tools than in level 3 (44% compared with 41%), while among the debitage there are marginally fewer bladelets than in level 3 (5% compared with 9%). This suggests that there was no change in the strategy of blank reduction, but rather that the cores in level 1 were reduced more intensively by the continued removal of small flakes when it was no longer possible to remove bladelets, and this is supported by the smaller size of the residual cores.

In the absence of evidence from core refitting, the reconstruction of the core reduction strategies at Grotta delle Cipolliane has to remain hypothetical. However, the fact that the backed bladelets are made on blanks with a fairly specific and standardized morphology, while other tool classes are made on blanks that vary markedly in form suggests the use of a single reduction strategy throughout the stratigraphy, oriented towards the production of bladelets, and the by-products of this operational sequence, such as the blanks produced during the initial shaping of the core, preliminary blanks, blanks of inferior quality and blanks broken during the flaking process were used for the manufacture of the other tool classes. A single reduction strategy is the most conservative in terms of raw material consumption.

The stratigraphy at Grotta delle Cipolliane presents a fairly continuous Epigravettian sequence through which to monitor eventual diachronic variation. A marked continuity was observed in the technological attributes of the retouched tools from the 1964 excavation, which suggests that there was no change in technical tradition. However, the combined data from the 1962 and the 1964 excavations indicate an increase in raw material economy between levels 3 and 1, characterised by more intense core reduction and tool utilization.

Grotta Paglicci

Grotta Paglicci is situated on the north side of the Valle Settepenne, on the southern side of the Gargano promontory. The cave, which is cut into Jurassic limestone, lies at 150 metres above sea-level and faces southwest, overlooking the valley floor and the Tavoliere plain. Grotta Paglicci is therefore located at the interface of two distinct environments: the barren, rugged mountains of the Gargano promontory which rise up to 1000 metres; and the low, flat Tavoliere plain, which probably hosted either prairie or steppe vegetation communities during the late glacial, depending on the temperature and humidity (Sala 1983a, 1983b). The nearest chert sources lie 24 kilometres to the east of the cave.

The site has been excavated since the 1960s (Mezzena and Palma di Cesnola 1967; Galiberti 1979). The Epigravettian sequence was divided into two main complexes: a lower one, comprising levels 18-10; and an upper one comprising levels 9-1. The Final Epigravettian levels, with which we are concerned here, have been dated to between 14,820±210 (F-65) for level 7 and 11,950±190 (F-95) for level 4 (Azzi *et al.* 1977). The faunal remains consist of *Equus caballus*, *Equus asinus hydruntinus*, *Sus scrofa*, *Capra ibex*, *Cervus elaphus* and *Rupicapra rupicapra*. The mammalian and micromammalian fauna from levels 7-4 indicate a temperate climatic phase (Prebölling-Bölling).

A technological analysis was made of 870 retouched tools from the Final Epigravettian levels 7-4 at Grotta Paglicci:

Backed bladelets (n.274, 31%): of these, 84% are fragments. The rate of breakage ranges between 75% and 95%, but there is no linear temporal trend. 75% of the bladelets are smaller than 2.5 centimetres in length, and 24% are smaller than 5 cm. Two of the backed bladelets are longer than 5 centimetres, in spits 7c and 5b, and therefore these should be classified as blades.

Linear retouched tools (n.314, 36%): flake and bladelet blanks were used in roughly equal quantities (40% and 44% respectively). The majority of the tools are medium sized (52%), then small (30%) and large (18%). The maximum length is 9 cm. The cross-section of the blank is predominantly flat (91%). There are no significant temporal variations in the relative proportions of blank types and sizes.

Burins (n.30, 3%; Figure 2: 9-15): the burins are predominantly made on flakes (60%), then bladelets (23%) and blades (17%), and there is no temporal variation from this norm. The majority are medium sized (60%), with small and large burins equally abundant (20% each). Medium sized burins are predominant throughout the stratigraphy with the exception of spit 5b, where the majority are smaller than 2.5 cm. None of the burins is broken.

Endscrapers (n.57, 7%; Figure 3: 13-26): the endscrapers are predominantly made on flakes (82%), then bladelets (16%), with only one endscraper on a blade in spit 7a. The majority are medium sized (60%), then small (35%) and large (5%). Carinated endscrapers are present only in spits 5a-4b. The majority of the endscrapers are short (88%), with long endscrapers present in low quantities throughout. None is circular. Thirty (53%) of the endscrapers are fragments, but the sample sizes are too small to be able to monitor changes in the rate of breakage through the spits.

Notches (n.35, 4%): the notches are predominantly made on flakes (89%), then bladelets (8%), with one on a blade in spit 6a. The size of the blanks is for the most part small (54%). Fifteen (43%) of the notches are fragments.

Denticulates (n.19, 2%): the denticulates are predominantly made on flakes (63%), and blade forms are absent. The majority (58%) are smaller than 2.5 centimetres. Seven (37%) of the denticulates are fragments.

Points (n.21, 2%): the points are predominantly made on laminar blanks (85%), with bladelets and blades accounting for 52% and 33% respectively. The size of the blanks is for the most part medium (43%), then large (33%) and small (24%). Carinated points are present only in spit 6d. Twelve (57%) of the points are fragments.

Truncations (n.78, 9%): the truncations are predominantly made on bladelets (50%) and flakes (44%), while blades (6%) are restricted to spits 6a-5b. There is only a marginal predominance of small blanks (50%) over medium (44%), and few large blanks (6%). Twenty-four (31%) of the truncations are fragments.

Perforators (n.11, 1%): perforators are made exclusively on laminar blanks, which are predominantly bladelets (73%). The majority are small (45%), with medium and large ones amounting to 27% each. Five (45%) of the perforators are fragments.

Composite tools (n.12, 1%): these are significantly absent from level 4. The composite tools are made on flakes and bladelets in equal quantities (42% each), and on blades only in spits 7b and 5a. They are predominantly medium (75%) or large. None are smaller than 2.5 centimetres. The tool class combinations are endscraper-truncation (42%), burin-endscraper (42%), endscraper-point (8%) and truncation-notch (8%). Only one (8%) composite tool is a fragment.

Geometrics (n.19, 2%): the geometrics are all made on bladelets, with the exception of one (a rhomboid) in spit 5b which is made on a flake. All are small, apart from one medium sized isosceles trapeze in spit 5a. Eight (47%) of the geometrics are fragments.

In summary, the industry from levels 7-4 is predominantly laminar (68%), with the percentage of laminar blanks ranging between 61% and 83%, and with no linear temporal trend. When the backed bladelets are excluded from the calculation, the industry is still predominantly laminar (53%), except in spits 7c, 7a-6c and 4b. Again, there is no linear temporal trend, although the percentage of laminar tools in spit 4a is considerably higher than in any other level. Blade tools (9%) are absent from spits 6c and 4a, but only in spit 4a is this significant. The size of the industry is predominantly small when the backed bladelets are included (49%), except in spits 5b-5a, and predominantly medium sized (between 2.5 and 5 centimetres; 49%) when they are excluded, except in spits 7b-6d and 4a. There is no linear temporal trend according to either calculation. Finally, 57% of all the retouched tools are fragments. The rate of breakage within the individual levels ranges from 46% to 90%, with no linear temporal trend.

The debitage from level 4a comprises six bladelet cores (0.2%), 2002 unretouched flakes and bladelets smaller than 2 centimetres (70%), and 325 unretouched flakes and bladelets larger than 2 centimetres (11%), while retouched tools account for the remaining 18% (Donahue 1986). On the basis of the abundance of tertiary retouched flakes, the presence of bladelet cores, comparatively few primary (cortical) flakes and the presence of many large and well-formed blades among the tools, Donahue (1986) suggests that during the occupation of level 4a the raw material was imported onto the site in the form of prepared bladelet cores, prepared blanks, and finished tools.

Grotta delle Cipolliane and Grotta Paglicci compared

The assemblage from Grotta Paglicci levels 7-4 is broadly contemporary with level 2 at Grotta delle Cipolliane. The assemblage from the 1964 excavation of level 2 is too small to be able to make meaningful comparisons with that from Grotta Paglicci, but as both of the samples analysed showed strong continuity in their technological attributes, it is possible to compare the entire sequence from Grotta delle Cipolliane (n.849) with the assemblage from Grotta Paglicci levels 7-4 (n.870), and still expect to get meaningful results.

There are more laminar tools at Grotta Paglicci, in both the total assemblage (68% compared with 62%) and in the assemblage excluding backed bladelets (53% compared with 40%). A large part of this difference can be accounted for by the fact that the burins, points and perforators are all predominantly made on laminar blanks, compared with a predominance of flake blanks at Grotta delle Cipolliane. The tools are larger at Grotta Paglicci, with a predominance of tools larger than 2.5 centimetres in both the total assemblage (51% compared with 32%) and in the assemblage excluding backed bladelets (64% compared with 41%). The difference in the size of the tools is largely accounted for by the fact that the burins, endscraper and composite tools at Grotta Paglicci are all predominantly larger than 2.5 centimetres, while at Grotta delle Cipolliane they are predominantly smaller than 2.5 centimetres.

If these differences in size and laminarity result from the effect of raw material availability on strategies of tool production, we would expect these differences to characterize the whole assemblage from each site. Instead, however, the differences are restricted to certain tool classes, and this therefore requires an alternative explanation. The fact that the burins, endscrapers and composite tools at Grotta Paglicci are the only tools which are significantly larger than those at Grotta delle Cipolliane might result from a more intense rejuvenation of these tools at the latter site, which would suggest different strategies of tool use and maintenance rather than of tool production. On the other hand, the fact that the proportion of blade tools at Grotta Paglicci is significantly higher than at Grotta delle Cipolliane, both in the assemblage as a whole (9% compared with 1%), and in the assemblage excluding the backed bladelets (12% compared with 1%), suggests that the differential availability of raw material did have some influence on tool production strategies at the two sites. The fact that there is a higher rate of breakage at Grotta Paglicci (57% compared with 33%) is surprising, as one might expect that because of the comparatively local availability of raw material the tools at this site would have been used more expediently than those at Grotta delle Cipolliane. However, the higher rate of breakage at Grotta Paglicci is caused primarily by a significantly higher rate of breakage among the backed bladelets (84% compared with 47%), while the other tool classes show a lower rate of breakage compared with those at Grotta delle Cipolliane.

Enlarging the database

Apart from Grotta delle Cipolliane, there is a series of other cave and rockshelter sites in the Salento peninsula with assemblages that have been attributed to the Late Palaeolithic: Grotta Romanelli (Blanc 1928; Laplace 1966); Ugento (Cardini 1965; Laplace 1966); Grotta del Cavallo (Palma di Cesnola 1963; Laplace 1966); Grotta di Uluzzo (Borzatti von Löwenstern 1963, 1964); Grotta Carlo Cosma (Borzatti von Löwenstern 1965); Grotta delle Prazziche (Borzatti von Löwenstern 1969); Taurisano (Laplace 1966; Bietti 1979); and Porto Badisco (Guerri 1984, 1986). There are also some open sites where large assemblages have been found on the surface which can be dated on the grounds of their typology to the Epiromanellian: Torre Testa (Cremonesi 1967, 1978; Punzi 1967, 1968); San Foca (Ingravallo 1980); and Laghi Alimini (Piccinno and Piccinno 1974, 1978; Milliken and Skeates 1990, in press). The sites are all located within an area of approximately 1500 square kilometres in the southern part of the peninsula, and all but seven of them are situated along the present day coastline (Figure 1).

One characteristic of this sub-region is that there appears to be little significant temporal or spatial variability in the typological composition of the assemblages. In terms of tool classes, the majority of the assemblages are dominated by backed bladelets and linear retouched tools, except at Grotta Romanelli and where the predominant combination is endscrapers and backed bladelets, and at Grotta di Uluzzo where the assemblage composition fluctuates through a variety of predominant classes. On the other hand, the relative increase in the abundance of circular, carinated and nosed endscraper forms is a feature which does seem to have some temporal significance. The available data concerning the assemblages with debitage are summarized in Table 1, with the sites arranged in approximate chronological order. Considering the large absolute distances involved, the significance of differences in the relative proximity of the Salento peninsula sites to the raw material sources is likely to be negligible. If the relative degree of curation of an assemblage can be measured using the percentage of retouched tools and rejuvenation flakes, Table 1 shows that the assemblages from Taurisano and Grotta delle Cipolliane seem to have been more heavily curated than those from Grotta delle Prazziche, Grotta di Uluzzo and Grotta Carlo Cosma. Looking at the chronology of these Salento peninsula sites, there would seem to be some temporal significance in the relative degree of curation and expediency, although it must be taken into account that the sample provided by the sites studied here is very small, slightly more than a quarter of the sites in the sub-region, and therefore it cannot be assumed that these data are necessarily representative.

Table 1: Debitage data for sites in the Salento peninsula

	Retouched tools (%)	Unretouched blanks (%)	Rejuvenation flakes (%)	Cores (%)	Period
Uluzzo	0-13	84-100	n.d.	1-3	Final Epigravettian
Carlo Cosma	15	74	n.d.	1	
Prazziche	15	79	n.d.	6	
Cipolliane 1	16	73	9	2	
Cipolliane 2 upper	16	72	8	4	
Cipolliane 2 lower	25	60	11	4	
Cipolliane 3	11	77	10	2	Evolved Epigravettian
Taurisano 1	20	72	7	1	
Cipolliane 4 upper	9	78	12	1	Ancient Epigravettian
Cipolliane 4 lower	4	69	7	0	
Taurisano 14-9	0-18	68-86	10-13	2-5	

Sources: Grotta di Uluzzo (Borzatti von Löwenstern 1963, 1964); Grotta Carlo Cosma (Borzatti von Löwenstern 1965); Grotta delle Prazziche (Borzatti von Löwenstern 1969); Grotta delle Cipolliane (Palma di Cesnola 1962); Taurisano (Milliken 1991).

A study of the cores from various sites in the Salento peninsula which are stored at the Museum of Maglie showed that they form a very homogeneous group from the point of view of size and type (Milliken 1991). At Grotta Romanelli the cores are of the prismatic bladelet type and measure between 1.5 and 3.5 centimetres; at Grotta Uluzzo and Ugento Pozzo Zecca they are again prismatic bladelet cores, measuring between 2 and 2.5 centimetres in length at the first site and between 1.9 and 3.8 centimetres at the second. In spits 9-14 at Taurisano the cores are either prismatic bladelet cores or bipolar cores, and measure between 2 and 3 centimetres in length, while the single core from level 1 is a discoidal flake core which measures 2 centimetres in its largest dimension. One prismatic bladelet core was found at Grotta Carlo Cosma, while 25 small irregular flake cores were found at Grotta delle Prazziche.

Other features suggestive of raw material economy, and which seem to have some chronological significance, are the use of silicified limestone for the manufacture of macrolithic tools in the Epiromanellian levels at Grotta del Cavallo, Grotta delle Prazziche, Porto Badisco and Grotta di Uluzzo, and the cache of Middle Palaeolithic artifacts found at Torre Testa which had been used as raw material by the Late Palaeolithic occupants at the site. Finally, the presence of various blades measuring between 6 and 9 centimetres among the microlithic industry at Grotta Romanelli suggests that these were imported prefabricated tools.

Unfortunately the data available for the five Late Palaeolithic sites in the Murge region (Figure 1), which are situated midway between those in the Gargano promontory and those in the Salento peninsula, are too scarce to be able to assess the relative degree of raw material economy. However, it was noted that both the Ancient and Final Epigravettian assemblages at the site of Grotta delle Mura are made on chert from the Gargano promontory, which lies 150 kilometres to the north (Palma di Cesnola 1987), even though the chert outcrops in the adjacent region of Basilicata would have been closer (circa 100 kilometres).

DISCUSSION: THE ROLE OF RAW MATERIAL AVAILABILITY IN TECHNOLOGICAL ORGANIZATION

Johnson's (1989) distance-decay model predicts decreasing size of artifacts with distance from the source, whereby the progressive decline in the size of artifacts with the increase in distance is essentially viewed as an economising strategy by which a greater number of implements could be made from scarce raw material. The comparison of the assemblages from Grotta delle Cipolliane and Grotta Paglicci failed to support the expected distance-decay trend with regard to size. The difference in tool size was found not to characterize the assemblages as a whole, but rather it was restricted to specific tool classes, namely the burins, endscrapers and composite tools, and I suggest that the smaller size of these tools at Grotta delle Cipolliane might result from their curation, while at Grotta Paglicci they may have been used in a relatively more expedient fashion, and therefore less intensively maintained. Analogous situations where the assemblage characteristics failed to support a distance-decay model with regard to size have been found in industries from Natufian sites in Jordan and Late Prehistoric sites in Oklahoma (Henry 1989). The reduction of tools and the relative degree of curation and expediency are therefore likely to be closely linked to the role of a site within a larger land use system and the kinds of activities carried out at the site, rather than with distance from the source.

The distance between the raw material source and a site can be expected to have influenced strategies of raw material transportation. Units of raw material can enter a site in various different stages of reduction, and the weight of the transported raw material, and therefore the cost in terms of energy, will vary accordingly. As the transport costs of cores are high, one would expect the degree of core reduction to be closely linked to distance from the source. A rough measure of the degree of core reduction can be ascertained from the size of the residual cores, or alternatively from their relative frequency on the premise that intense reduction to the point of exhaustion will lead to the loss of their archaeological visibility. Unfortunately, there are no data concerning the size of the cores from the sites in the Gargano promontory. With regard to their relative frequency, the percentage of cores at Grotta Paglicci level 4a (0.2%) is considerably *lower* than at the sites in the Salento peninsula for which data are available, and where the range is between 0.2% (at Grotta Carlo Cosma) and 6% (at Grotta delle Prazziche). This would seem to suggest either that cores were being transported to Grotta Paglicci to a lesser degree than they were to the sites in the Salento peninsula, or that they were being more intensively reduced there, neither of which conforms to the expectations of the model.

One possible explanation for this apparent ambiguity resides in the fact that if raw material was transported in the form of cores, the spectrum of possible uses of that raw material would have been considerably wider than if the raw material was transported in the form of tool blanks and finished tools, where the exact uses to which the raw material could be put would have been more circumscribed. This difference in the relative flexibilty of transported raw material is likely to have been related to considerations such as the kinds of activities to be performed at and around the sites, and also to the relative availability of opportunities for replacing tools.

Tools are the means by which subsistence choices are carried out, and by their presence or absence during a foraging trip they may make some courses of action possible and remove others. Therefore, while the transportation of tool blanks and finished tools to the Salento peninsula would have been least costly in terms of energy, this may have limited the types of prey which could be exploited and therefore increased the possibility of failing to meet dietary requirements. The greater flexibility offered by the transport of cores could thus be seen as part of a strategy of risk reduction. On the other hand, production efficiency would have been increased by advancing the core reduction process closer to the finished tool while still in the source area, as in this way the cost of portaging raw material destined to be lost in production failures would have been reduced or eliminated. The import of a small component of prepared blanks or tools is suggested by the presence of rare blades in the assemblages at Grotta Romanelli and Grotta delle Cipolliane. However, prepared blanks and tools were also imported in level 4a at Grotta Paglicci.

If the costs of raw material procurement were simply a function of the distance from the source to the site, one would expect to see no temporal variability in raw material economy. However, a change in raw material economy *was* observed at Grotta delle Cipolliane, indicated by a more intense reduction of the cores to the point of exhaustion, and a more intense utilization and maintenance of the toolkit. At Taurisano, the relative increase in retouched tools and the decrease in rejuvenation flakes suggests a change in the strategy of toolkit maintenance to one focusing on resharpening, which is more economical in terms of raw material conservation than maintenance by rejuvenation (Hayden 1989). In addition, the change in the size and type of cores suggested, as at Grotta delle Cipolliane, more intense reduction. At both of these sites the changes can be interpreted in terms of an *increase* in raw material economy, while the trend towards relative expediency that was observed in the Epiromanellian assemblages from the Salento peninsula can be seen as a *decrease*.

The analysis of the debitage from the five sites in the Salento peninsula suggests that there was a change in raw material economy which appears to have some temporal significance. This trend towards relative expediency in the Epiromanellian may be related to changing economic and social strategies, such as a change in the procurement strategy which resulted in increased access to raw material sources. Assemblage composition may also be affected by ecological shifts which alter either the regional pattern of the man-land relationship, or the role of a particular site in terms of the seasonal round or in relationship to local resources. Alternatively, or in addition, raw material availability may be related to social constraints on access, or may have been at least partially achieved through exchange, even over long distances.

Another feature suggestive of raw material economy which seems to have some temporal significance is the use of limestone for the manufacture of macrolithic tools in the Epiromanellian assemblages at Grotta delle Prazziche, Grotta del Cavallo, Grotta di Uluzzo and Porto Badisco. This phenomenon marks a reversal in a trend which saw, from the Lower Palaeolithic to the Upper Palaeolithic, a progressive *decrease* in the use of non-cryptocrystalline raw material. In the Salento peninsula in the lower Palaeolithic, all the handaxes are made from local limestone, in contrast to those from the Gargano promontory which are made from local chert. In the Middle Palaeolithic, the industries at Grotta delle Prazziche, Grotta del Cavallo, Grotta Carlo Cosma and Grotta Romanelli are again made from limestone, in contrast to those in the Gargano promontory, such as at Irchio (Mancini and Palma di Cesnola 1958), where the industry is made from chert. In the other Middle Palaeolithic industries of the Salento peninsula, the majority of the tools are made from a combination of local limestone, shell *(Meretrix chione)*, and a very small component of chert. At Grotta del Cavallo there is a change in raw material economy through the Early Upper Palaeolithic, from a predominance of limestone accompanied by rare chert in the archaic Uluzzian, to a predominance of chert accompanied by rare limestone in the late Uluzzian.

An interesting analogy with the assemblages of the Salento peninsula is provided by Bacho Kiro cave in Bulgaria, where the Upper Palaeolithic industry is made exclusively from imported flint, from sources 100-200 kilometres away in both an eastern and a western direction from the site. Unmodified blocks of flint were absent from all layers, and the low proportion of cores (1% to 4%) in the assemblage indicates the import of ready-made blanks, and also suggests a very intensive utilization of all the cores available (Koslowski 1982: 159). This pattern of activity was dictated by the lack of flint deposits or other similar rocks within a 100 km radius from Bacho Kiro. The Upper Palaeolithic raw material strategy contrasts with that of the Middle Palaeolithic levels at the site, where local volcanic rocks make up between 67% and 80% of the chipped stone assemblage (Koslowski 1982: 82).

The return to a limestone component in the Epiromanellian industries of the Salento peninsula marks a reversal in a long-term trend which requires explanation. Possible reasons for the change in raw material economy that was observed in the assemblages from Taurisano and Grotta delle Cipolliane, and for the relative increase in expediency at the Epiromanellian sites in the Salento peninsula, also need to be explored. One factor which might promote a temporal change in strategies of raw material procurement is that of restricted access to raw material sources, which may have been imposed by both the physical environment and the social environment. With regard to the physical environment, raw material procurement may be hampered by seasonal variation in ground water levels and snow cover (Rolland 1981; Hayden and Nelson 1981).

The chert sources in the Gargano promontory are located at between 850 and 1000 metres above sea-level, while those in the Basilicata Apennines are located at 1500 metres above sea-level. The lowering of the snow line during continental arid phases, and in particular during the Younger Dryas event (11-10,000 BP) which saw the return to almost full glacial conditions, may have induced a change in the strategy of raw material procurement at sites in the Salento peninsula, such as a shift from exploiting the sources in Basilicata which were nearer but which would have become inaccessible under such conditions, to those in the Gargano promontory. This may have been accompanied by a more intense utilization and maintenance of the toolkit, in order to minimize procurement costs which would have risen with the increase in the distance to a raw material source. However, the increase in raw material economy at both Taurisano and Grotta delle Cipolliane occurs with the passage from a continental arid phase to a more temperate humid one, and snow is therefore unlikely to have restricted access to the raw material sources.

Access to raw material may have changed through time as a result of changes in group mobility and settlement patterns, which may have been prompted indirectly by palaeoclimatic change. Vegetation shifts and animal biomass fluctuations would undoubtedly have affected hunter-gatherer land use strategies, and these in turn are likely to have affected strategies of raw material procurement, tool use and maintenance. In milder episodes, higher proportions of arboreal vegetation would have resulted in a lower herbivore biomass. Greater mobility may have been necessary in response to these changes, and this in turn may have resulted in more opportunities for raw material procurement, thereby allowing for a more profligate use of lithic raw material. Colder episodes, on the other hand, would have led to an expansion of more open landscape which would have been favourable to an increase in the size of game herds. More intensive settlements would thus have been possible, and the reduction in mobility would have necessitated a more parsimonious use of lithic raw material.

Again, the fact that the increase in raw material economy at Grotta delle Cipolliane and Taurisano coincides with a temperate climatic phase, which therefore should have seen an increase in mobility, does not conform to the expectations of this model. One reason for this may be that the model assumes that an *embedded procurement strategy* (Binford 1979) was in operation, whereby raw material procurement was done in conjunction with subsistence activities, and that the ensemble of economic activities dictated the mobility of the whole group. In a strategy of *direct procurement*, on the other hand, the scheduling of raw material procurement would have been independent of other economic activities, and would have been performed by a specific task-group. Binford (1979) argues on theoretical grounds that hunter-gatherers will rarely make a trip specifically to secure raw material for tool manufacture, although this strategy has been observed in the Central and Western Deserts of Australia (Gould, Koster and Sontz 1971; Gould 1978: 831-832; Gould and Saggers 1985: 121).

One way of minimizing the effort incurred by inconvenient procurement trips would have been to cache raw material at regularly used campsites, but the only evidence for a raw material cache in the Late Palaeolithic of south-east Italy is the cache of older artifacts that was found at the Epiromanellian site of Torre Testa. It is possible that stock-piling of raw material did take place elsewhere, and the absence of caches from the archaeological record of this region can easily be accounted for by the fact that they would have been exhausted. However, stock-piling can only provide a short-term solution to the problem of raw material availability. Absolute limits on the amount of lithic material that can feasibly be transported, combined with its steady consumption, mean that raw material would have been in regular demand and the caches would have needed to be constantly replenished. Gould (1977) estimated that among the Aborigines of the Western Desert, twenty kilograms of lithic raw material would have been used up per person per year, or approximately forty tools a year. However, it has to be taken into account that each situation is different, and the number of tools that were needed at any of the sites in the region will have depended in part on the type and duration of the activities being performed, and the extent to which these activities consumed raw material.

The question of whether raw material procurement strategies in the Salento peninsula were embedded or not depends on the nature of the occupations at these sites, and on their role within a larger settlement system. If the sites in the Salento peninsula represent parts of a settlement system that was restricted in its seasonal mobility to just that part of the region, then direct procurement would have been necessary. If, on the other hand, the settlement system incorporated either the mountains of the Gargano promontory or Basilicata in its seasonal round, then embedded procurement would have been possible. General patterns of residential mobility, scheduling of subsistence resources, and the spatial concordance between raw material sources and subsistence resources are therefore likely to have determined how and when it was practical to procure raw material.

In south-east Italy, there has been no comprehensive or systematic regional survey to look for Late Palaeolithic sites. Research has tended to be localized in certain areas, and in particular has focused on the excavation of caves and rockshelters. As a result, one cannot know what percentage of a total settlement system is represented by the sites discussed here and, more importantly, how representative they are from the point of view of their role within that settlement system. In addition, there is the problem of establishing to what extent any of them can be said to be contemporaneous, either broadly or precisely, and therefore of whether they can be regarded as part of the same settlement strategy. These factors have to be borne in mind in the course of making a hypothetical reconstruction of settlement patterns and social relations (Milliken 1997).

Critical resources in the form of food, water, shelter and raw materials are rarely aggregated and clustered in such a way as to enable the procurement of all these things in one spot. *Mobility* is a settlement strategy which solves the problems of the spatial and temporal differences between resources, by redistributing people in the environment (Hayden 1978;

Kelly 1983). Binford's (1978a, 1978b, 1979, 1980, 1982) model of hunter-gatherer mobility recognizes two distinct forms of mobility: *residential mobility*, whereby small groups of producers and consumers move as a unit throughout the seasonal round; and *logistical mobility*, in which consumers remain at a more or less permanent base camp for several seasons, while small groups of producers procure distant resources and bring them back to the consumers. For example, in Late Palaeolithic central Italy the hunters developed mobile and specialized economies focused on two seasonally migrant species, red deer and steppe horse, and followed the herds in their seasonal migrations up into highland areas in the summer, and down into lowland areas in winter (Barker 1981). This kind of settlement strategy would therefore be described as residentially mobile.

A residential mobility settlement pattern is composed of a series of residential camps occupied by a microband for a short period of time only, and for the express purpose of exploiting the vicinity. A logistical mobility settlement pattern, on the other hand, is composed of a base camp as well as field camps which are often associated with special activities. *Base camps*, or residential sites, are the locus of mixed sex maintenance and habitation activities, and should therefore contain large and heterogeneous assemblages. *Field camps*, on the other hand, are special purpose, short-term, single sex occupations, and therefore one should expect smaller and more homogeneous assemblages. The differences in settlement strategy are also likely to result in different tool production and maintenance strategies. According to Binford's model, most tool production takes place in base camps of prolonged occupation, or at factory sites related to quarries. In field camps such as hunting stands, on the other hand, lithic production activity is limited to the maintenance and repair of armatures.

Mobility also places constraints upon technology by imposing carrying costs (Torrence 1983). As the gross number of tools which can be transported is ultimately limited, the degree of specialization of the toolkit is also limited, and therefore highly mobile groups can be expected to have small toolkits composed of versatile and flexible tools which are not function-specific. Thus assemblage diversity, or the number of different tools, will decline as mobility increases either in terms of frequency or magnitude (Gibson 1984; Shott 1986). Mobility may also impose other design criteria on the tools, such as size and weight (Ebert 1979; Keeley 1982). This relationship between mobility, tool design and the size of technological inventories has been documented ethnographically among the !Kung San of southern Africa, where tools are few in number, and are designed to be multipurpose and lightweight (Lee 1979: 119). Therefore, residentially mobile groups would be expected to have a smaller, lighter, and more homogeneous and versatile toolkit than a logistically mobile group.

Problems are encountered when moving from behavioural, or ethnographic, contexts to archaeological contexts. Unknown variables, such as the possibility that sites were reoccupied for a different purpose, and the stratigraphic mixing of more than one discrete behavioural assemblage into a single archaeological assemblage, raise the question of whether it is realistic to try to identify site type at all. Because of the sample size-effect, larger assemblages will contain greater diversity, while smaller assemblages will appear more homogeneous (Thomas 1983). In addition, there is no *a priori* reason to assume the presence of activity areas and functionally specfic tool kits in a site. The assumption that the functional and spatial dimensions of tool technology overlap, such that artifacts which are found together must have been used together, is unfounded. Artifact co-occurrence may have as much to do with stochastic processes, such as dropping, breakage, loss, as by the performance of specific assemblages (Ammerman and Feldman 1974). In fact, a multitude of factors and processes may affect artifact class co-occurrence as it is revealed in the archaeological record, and specific human activities need not be among them. According to Binford's (1981) "Pompeii Premise", assemblages formed by the sudden loss or abandonment of tools are representative of the systemic frequency of various tool classes. However, such assemblages are rare. Most assemblages, even those from brief occupations, are the product of several formation processes rather than a single one. Assemblages do not represent a single technological strategy, but are likely to be composed of tools representing several strategies. These may have been deposited at one time by a single individual or group, or may be the result of several separate but indistinguishable episodes. Yellen's (1977) description of !Kung artifact production, whereby the production of tools is not limited in time and place, but carried out at all base camps, has important implications for interpreting site type.

Level 4a at Grotta Paglicci has been interpreted as a winter hunting stand (Donahue 1985, 1986, 1988). The season of occupation was suggested on the basis of the microwear analysis, which revealed the presence of hide working and the absence of antler or plant processing, while the interpretation as a hunting stand was made on the basis of the site's location, the small size of the occupation area, the composition of the lithic assemblage, and the scarsity of long bones, which are richest in meat. Although this supports Barker's (1975, 1981) idea of seasonal transhumance following migratory herds between the highlands in summer and the lowlands in winter, the implication is that Grotta Paglicci is a specialized site rather than a residential site, and therefore that the settlement system of which it forms a part was logistically rather than residentially mobile (Donahue 1985, 1986, 1988).

Beyond this, the identification of site type becomes more difficult. The faunal data is of too poor quality to be able to identify factors such as seasonality, or even to allow for the distinction to be made between, for example, kill and butchery sites. In terms of tool class composition, the south-east Italian assemblages are fairly homogeneous, both in time and in space. At Grotta delle Cipolliane and Grotta Romanelli, the toolkits are dominated by endscrapers and backed bladelets, while at the vast majority of the other sites in the Salento peninsula the toolkits are dominated by backed bladelets and linear retouched tools. The assemblage composition at Grotta Paglicci is similar to that of the majority of the Salento sites, with backed bladelets and linear retouched tools dominant. This homogeneity would seem to suggest that there is no great difference between the sites in terms of the activities that were being performed.

According to Binford's model, if the sites are part of residentially mobile settlement strategies, then there should be no base camp or limited activity site distinction, while if the sites are part of logistically mobile settlement strategies, then base camps and limited activity sites should be recognizable by their distinctive faunal and lithic components. It is not possible on the basis of the available data to be confident which, if any, of the two models applies here, but the theoretical likelihood of each can be discussed. Firstly, the duration of the occupations of these sites is difficult to estimate precisely. While the range of activities apparently represented in the assemblages in the Salento peninsula argues for more than a fleeting visit to the majority of these sites, the lack of both raw material and water is likely to have precluded year-round occupation. The size of the assemblages also suggests that none of these are 'residential sites'. The Adriatic coastal plain before 15,000 BP may have been sufficiently productive to support semi-sedentary hunting bands throughout the year, without necessitating summer forays into high country. If this was so, then the sites in the Salento peninsula may represent the field camps of residentially mobile settlement strategies, where the residential sites located on the coastal plain would have been submerged by the sea-level rise. In such a settlement system, direct procurement strategies of raw material provisioning would have been necessary. The drowning of the residential sites on the Adriatic coastal plain would not have necessitated a change in the settlement system, but simply a spatial shift, and the large surface assemblages at San Foca and Torre Testa, and the abundance of sites around the Alimini lakes, may represent the new residential sites.

The residential site of the settlement system of which Grotta Paglicci forms a part may well have been located on the Tavoliere, which is now covered by thick deposits of Holocene alluvium. The Tavoliere would have been well watered by springs and rivers, and it is likely to have been able to support dense populations of both herbivores and humans during the late glacial period. In such a settlement system, raw material procurement could easily have been embedded in subsistence forays into the mountains of the Gargano promontory. If the raw material procurement strategies of the sites in the Salento peninsula were direct rather than embedded, according to the model outlined above, then this would have necessitated the maintenance of long-distance social relations with groups living in the source area(s). In the analysis of spatial and temporal variability in lithic artifacts, it is commonly assumed or implied that prehistoric populations of hunter-gatherers lived in spatially discrete and isolated units, forming tightly bounded social, economic and cultural entities. The analysis of the lithic assemblages failed to find much evidence either for spatial or temporal discreteness. It is more likely that the occupation residues at these sites represent a network of loosely defined social groups and communities linked by a variety of social relationships, and that these extended over vast areas. While some of these social relationships would have been clearly economic in nature, such as those conditioning raw material procurement, others are likely to have been organized in accordance with less overtly economic factors, such as the acquisition of mating partners. By focusing on strategies of settlement organization oriented around economic activities, Binford's approach to hunter-gatherer settlement systems overlooks the importance of considering the social environment (Milliken 1997). Although there is certainly a strong association between resource procurement strategies and settlement mobility, the redistribution of personnel may also have been a social strategy in order to facilitate communication, and therefore help to avoid risks and reduce stress. Although ecological relations may be *determinant*, the social relations must be viewed as ultimately *dominant* (Ingold 1980, 1981) in specifying how an environment was exploited.

Organizational variation in the Late Palaeolithic of south-east Italy should thus be seen to be the result of both organization around resources *and* organization around other persons in social relations of production. Long-distance social relations are well documented in the ethnographic record. For example, the circulation of both goods and personnel in Australia created 'chains of connection' (Mulvaney 1976) stretching over distances of more than 800 kilometres, which Gould (1980) interprets as indicators of regional insurance policies, where the exchange of items promotes interaction and facilitates the movement of personnel in times of crisis. In this context, it is interesting to mention the obsidian bladelet from Grotta Romanelli, which probably comes a distance of 340 kilometres from the island of Lipari (Milliken 1991).

CONCLUDING REMARKS

This paper has tried to stress the importance of considering the ways in which local conditions mediate the effects of global aspects of human adaptations. Different strategies of tool production, use and maintenance were recognized in the south-east Italian Late Palaeolithic assemblages, and three important factors operating at a regional scale have been identified as contributing to the structure of the observed assemblage variability: the availability of raw material, and the alternative strategies by which it was procured; the particular activities in which the tools were made and used; and the role of the sites within a settlement system. It has also been demonstrated how assemblage variability is likely to arise from the interaction between *multiple* causal factors as conditioners of technology, and consequently it was not always possible to attribute a specific phenomenon to a specific cause. The common focus on single conditioners of technology, such as either raw material availability (Bamforth 1986), or time stress (Torrence 1983), or risk reduction (Bleed 1986), or mobility (Shott 1986; Parry and Kelly 1987), therefore artificially reduces the complexity of human behaviour, and glosses the problem of translating globally defined adaptive patterns into a concrete set of local conditions under which tools were made and used.

Acknowledgements

This research was partially funded by the following awards: British Academy Major State Studentship; British School at Rome Grant in Aid of Research; Meyerstein Award for Archaeological Research; Centre National de la Recherche Scientifique Research Fellowship.

REFERENCES

Alessio, M., F. Bella, S. Improta, G. Belluomini, C. Cortesi & B. Turi

1976 University of Rome Carbon-14 dates XIV. *Radiocarbon* 18: 321-349.

Ammerman, A.J. & M.W. Feldman

1974 On the "making" of an assemblage of stone tools. *American Antiquity* 39: 610-616.

Andrefsky, W.

1994 Raw-material availability and the organization of technology. *American Antiquity* 59: 21-34.

Azzi, C.M., L. Bigliocca & F. Gulisano

1977 Florence Radiocarbon Dates III. *Radiocarbon* 19: 165-169.

Bamforth, D.B.

1986 Technological efficiency and tool curation. *American Antiquity* 51: 38-50.

Barker, G.W.

1975 Prehistoric territories and economies in central Italy. In E.S. Higgs (ed.) *Palaeoeconomy*: 111-175. Cambridge: Cambridge University Press.

1981 *Landscape and Society: Prehistoric Central Italy.* London: Academic Press.

Bartolomei, G., A. Broglio & A. Palma di Cesnola

1979 Chronostratigraphie et écologie de l'Epigravettien en Italie. *Actes du Colloque International 'La Fin des Temps Glaciaires en Europe', Talence 1977*: 297-324. Paris: CNRS.

Bietti, A.

1979 Le gisement paléolithique supérieur de Taurisano (Lecce, Italie) et sa position chronologique et culturelle dans l'Epigravettien. *Actes du Colloque International 'La Fin des Temps Glaciaires en Europe', Talence 1977*: 333-344. Paris: CNRS.

Bietti, A. & A. Burani

1985 The late Upper Palaeolithic in continental Italy: old classifications, new data and new perspectives. In C. Malone & S. Stoddart (eds.) *Papers in Italian Archaeology IV. The Cambridge Conference. Part ii: Prehistory*: 7-27. Oxford: British Archaeological Reports International series 244.

Binford, L.R.

1978a Dimensional analysis of behavior and site structure: learning from an Eskimo hunting stand. *American Antiquity* 43: 330-361.

1978b *Nunamiut Ethnoarchaeology*. London: Academic Press.

1979 Organization and formation processes: looking at curated technologies. *Journal of Anthropological Research* 35: 255-273.

1980 Willow smoke and dogs' tails: hunter-gatherer settlement systems and archaeological site formation. *American Antiquity* 45: 4-20.

1981 Behavioural archaeology and the "Pompeii Premise". *Journal of Anthropological Research* 37: 195-208.

1982 The archaeology of place. *Journal of Anthropological Archaeology* 1: 5-31.

Blanc, A.C.

1939a Dei "microbulini" e della precoce comparsa del Mesolitico in Italia. *Rivista di Antropologia* 32: 3-38.

1939b Les 'microburins' dans les niveaux à faune glaciaire de la Grotte Romanelli, en Terre d'Otrante (Italie). *Bulletin de la Société Préhistorique Française* 36: 115-131.

Blanc, G.A.

1928 Grotta Romanelli II. Dati ecologici e paletnologici. *Archivio per l'Antropologia e l'Etnologia* 58: 365-411.

Bleed, P.

1986 The optimal design of hunting weapons: maintainability and reliability. *American Antiquity* 51: 737-747.

Borzatti von Löwenstern, E.

1963 La Grotta di Uluzzo (campagna di scavi 1963). *Rivista di Scienze Preistoriche* 18: 75-89.

1964 La Grotta di Uluzzo (campagna di scavi 1964). *Rivista di Scienze Preistoriche* 19: 41-52.

1965 La grotta-riparo Uluzzo C (campagna di scavi 1964). *Rivista di Scienze Preistoriche* 20: 1-31.

1969 Industrie romanelliane e neolitiche nella Grotta delle Prazziche (Novaglie-Lecce). *Rivista di Scienze Preistoriche* 24: 91-143.

Cardini, L.

1965 Sui giacimenti di recenti scoperti nella penisola Salentina. *Atti del VI Congresso Internazionale delle Scienze Preistoriche e Protostoriche (U.I.S.P.P.), Roma 1962*: 144-148.

Cremonesi, G.
1967 Industria litica di tradizione paleolitico superiore rinvenuta a Torre Testa (Brindisi). *Rivista di Scienze Preistoriche* 22: 251-280.
1978 Nuovi rinvenimenti del Paleolitico superiore e Mesolitico a Torre Testa (Brindisi). *Rivista di Scienze Preistoriche* 33: 109-159.

Donahue, R.E.
1985 Landscape and site function: a preliminary report on Paglicci cave level 4a. In C. Malone & S. Stoddart (eds.) *Papers in Italian Archaeology IV. The Cambridge Conference. Part ii: Prehistory*: 28-49. Oxford: British Archaeological Reports International series 244.
1986 *Technomorphology, Tool Use and Site Function in the Italian Upper Palaeolithic.* Unpublished Ph.D dissertation, Michigan State University.
1988 Microwear analysis and site function of Paglicci Cave level 4a. *World Archaeology* 19: 357-375.

Ebert, J.I.
1979 An ethnoarchaeological approach to reassessing the meaning of variability in stone tool assemblages. In C. Kramer (ed.) *Ethnoarchaeology: Implications of Ethnography for Archaeology*: 59-74. New York: Columbia University Press.

Galiberti, A.
1979 Grotta Paglicci - strati 4-7. *Archivio di Tipologia Analitica* 7: 67-119

Gambassini, P.
1970 Risultati della campagna di scavi 1964 nel Riparo C delle Cipolliane (Lecce). *Rivista di Scienze Preistoriche* 25: 127-181.

Gibson, E.
1984 Reconstructing Corbiac: the context of manufacturing at an Upper Palaeolithic quarry. In J.E. Ericson & B.A. Purdy (eds.) *Prehistoric Quarries and Lithic Production*: 139-146. Cambridge: Cambrdge University Press.

Gould, R.A.
1977 Ethno-archaeology; or where do models come from? A closer look at Australian Aboriginal lithic technology. In R.V.S. Wright (ed.) *Stone Tools as Cultural Markers: Change, Evolution, Complexity*: 162-168. Canberra: Australian Institute of Aboriginal Studies.
1978 The anthropology of human residues. *American Anthropologist* 80: 815-835.
1980 *Living Archaeology.* Cambridge: Cambridge University Press.

Gould, R.A. & S. Saggers
1985 Lithic procurement in central Australia: a closer look at Binford's idea of embeddedness in archaeology. *American Antiquity* 50: 117-136.

Gould, R.A., D. Koster & A. Sontz
1971 The lithic assemblages of the Western Desert Aborigines of Australia. *American Antiquity* 36: 149-169.

Guerri, M.
1984 Porto Badisco (Otranto, Prov. di Lecce). *Rivista di Scienze Preistoriche* 39: 349-351.
1986 Porto Badisco (Otranto, Prov. di Lecce). *Rivista di Scienze Preistoriche* 40: 385-386.

Hayden, B.
1978 Snarks in archaeology; or, inter-assemblage variability in lithics (a view from the Antipodes). In D.D. Davis (ed.) *Lithics and Subsistence: The Analysis of Stone Tool Use in Prehistoric Economies*: 179-198. Nashville: Vanderbilt University Publications in Anthropology 20.
1989 From chopper to celt: the evolution of resharpening techniques. In R. Torrence (ed.) *Time, Energy and Stone Tools*: 7-16. Cambridge: Cambridge University Press.

Hayden, B. & M. Nelson
1981 The use of chipped lithic material in the contemporary Maya Highlands. *American Antiquity* 46: 885-898.

Henry, D.O.
1989 Correlations between reduction strategies and settlement patterns. In D.O. Henry & G.H. Odell (eds.) *Alternative Approaches to Lithic Analysis*: 139-155. Archaeological Papers of the American Anthropological Association Number 1.

Higgs, E.S. & C. Vita Finzi
1972 Prehistoric economies: a territorial approach. In E.S. Higgs (ed.) *Papers in Economic Prehistory*: 27-36. Cambridge: Cambridge University Press.

Ingold, T.
1980 *Hunters, Pastoralists and Ranchers.* Cambridge: Cambridge University Press.
1981 The hunter and the spear: notes on the cultural mediation of social and ecological systems. In A. Sheridan & G.N. Bailey (eds.) *Economic Archaeology*: 119-130. Oxford: British Archaeological Reports International Series 96.

Ingravallo, E.
1980 La stazione Mesolitica di S. Foca. *Studi di Antichità* 2: 59-77.

Johnson, J.K.
1987 Cahokia core technology in Mississippi: the view from the south. In J.K. Johnson & C.A. Morrow (eds.) *The Organization of Core Technology*: 187-206. Boulder: Westview Press.
1989 The utility of production trajectory modeling as a framework for regional analysis. In D.O. Henry & G.H. Odell (eds.) *Alternative Approaches to Lithic Analysis*: 119-138. Archaeological Papers of the American Anthropological Association Number 1.

Keeley, L.H.
1982 Hafting and retooling: effects on the archaeological record. *American Antiquity* 47: 798-809.

Kelly, R.L.
1983 Hunter-gatherer mobility strategies. *Journal of Anthropological Research* 39: 277-306.

Kozlowski, J.K.
1982 *Excavation in the Bacho Kiro Cave, Bulgaria (Final Report).* Warsaw: Panstwowe Wydawnictwo Naukowe.

Kuhn, S.L.
1991 "Unpacking" reduction: lithic raw material economy in the Mousterian of west-central Italy. *Journal of Anthropological Archaeology* 10: 76-106.

Laplace, G.
1964 Les subdivisions du Leptolithique Italien. Etude de typologie analytique. *Bullettino di Paletnologia Italiana* 73: 25-63.
1966 *Recherches sur l'Origine et l'Evolution des Complexes Leptolithiques.* Ecole Française de Rome Mélanges d'Archéologie et d'Histoire Supplement 4. Paris: Boccard.

Lee, R.B.
1972 !Kung spatial organization: an ecological and historical perspective. *Human Ecology* 1: 125-147.
1979 *The !Kung San: Men, Women and Work in a Foraging Society.* Cambridge: Cambridge University Press.

Leroi-Gourhan, A.
1964 *La Geste et al Parole. Technique et Language.* Paris: Michel.

Lubell, D., P. Sheppard & M. Jackes
1984 Continuity in the Epipalaleolithic of Northern Africa with emphasis on the Magrheb. In F. Wendorf & A.E. Close (eds.) *Advances in World Archaeology* 3: 143-191. London: Academic Press.

Mancini, F. & A. Palma di Cesnola
1958 Il giacimento musteriano delle sorgenti di Irchio nel Gargano (geologia e paletnologia). *Rivista di Scienze Preistoriche* 13: 1-46.

Marks, A.E., J. Shockler & J. Zilhão
1991 Raw material usage in the Palaeolithic. The effects of local availability on selection and economy. In A. Montet-White & S. Holen (eds.) *Raw Material Economies among Prehistoric Hunter-Gatherers*: 127-139. University of Kansas Publications in Anthropology 19.

Mezzena, F. & A. Palma di Cesnola
1967 L'Epigravettiano della Grotta Paglicci nel Gargano (scavi F. Zorzi 1961-63). *Rivista di Scienze Preistoriche* 22: 23-156.

Milliken, S.
1991 *Aspects of Lithic Assemblage Variability in the Late Palaeolithic of South-East Italy.* Unpublished D.Phil thesis, University of Oxford.
1997 *Regional approaches to hunter-gatherer land use in early Holocene Europe.* Paper presented at the 62nd Annual Meeting of the Society for American Archaeology, Nashville.

Milliken, S. & R. Skeates
1990 The Alimini survey: the Mesolithic-Neolithic transition in the Salento peninsula (S.E. Italy). *Institute of Archaeology Bulletin* 26: 77-98.
1998 I laghi Alimini: contributo allo studio del passaggio Mesolitico-Neolitico nella Penisola Salentina (Puglia). *Edizioni Scientifiche del Museo Comunale di Paleontologia di Maglie,* Quaderni 5: 43-67.

Mulvaney, D.J.
1976 'The chain of connection': the material evidence. In N. Peterson (ed.) *Tribes and Boundaries in Australia*: 72-94. Canberra: Australian Institute of Aboriginal Studies.

Nelson, M.C.
1991 The study of technological organization. In M.B. Schiffer (ed.) *Archaeological Method and Theory* 3: 57-100. Tucson: University of Arizona Press.

Newman, J.R.
1994 The effects of distance on lithic material reduction technology. *Journal of Field Archaeology* 21: 491-501.

Odell, G.H.
1989 Fitting analytical techniques to prehistoric problems with lithic data. In D.O. Henry & G.H. Odell (eds.) *Alternative Approaches to Lithic Analysis*: 159-182. Archaeological Papers of the American Anthropological Association Number 1.

Palma di Cesnola, A.
1962 Contributi alla conoscenza delle industrie epigravettiane nell'Italia centro-meridionale. *Rivista di Scienze Preistoriche* 17: 1-75.
1963 Prima campagna di scavi nella Grotta del Cavallo presso Santa Caterina (Lecce). *Rivista di Scienze Preistoriche* 18: 41-74.
1987 Il Paleolitico della Puglia Centrale. *Atti della XXV Riunione Scientifica dell'Istituto Italiano di Preistoria e Protostoria 'Preistoria e Protostoria della Puglia Centrale', Monopoli 1984*: 35-53.

Palma di Cesnola, A. & A. Bietti
1983 Le Gravettien et l'Epigravettien ancien en Italie. *Actes du Colloque International 'La position taxonomique et chronologique des industries à pointes à dos autour de la Mediterranée européenne', Siena 1983. Rivista di Scienze Preistoriche* 38: 181-228.

Palma di Cesnola, A., A. Bietti & A. Galiberti
1983 L'Epigravettien évolué et final dans les Pouilles. *Actes du Colloque International 'La position taxonomique et chronologique des industries à pointes à dos autour de la Mediterranée européenne', Siena 1983. Rivista di Scienze Preistoriche* 38: 267-300.

Parry, W.J. & R.L. Kelly
1987 Expedient core technology and sedentism. In J.K. Johnson & C.A. Morrow (eds.) *The Organization of Core Technology*: 285-304. Boulder: Westview Press.

Piccinno, A. & F. Piccinno
1974 Estremo Salento. Nota preliminare su alcune scoperte preistoriche. *Ricerche e Studi* 7: 61-65.
1978 Otranto. Laghi Alimini. Stazione Preistoriche. *Ricerche e Studi* 11: 122-132.

Punzi, Q.
1967 Torre Testa. Stazione preistoriche costiere del Brindisino. *Ricerche e Studi* 3: 13-33.
1968 Le stazione preistoriche costiere del Brindisino. *Rivista di Scienze Preistoriche* 23: 205-221.

Rolland, N.
1981 The interpretation of Middle Palaeolithic variability. *Man* 16: 15-42.

Sala, B.
1983a Le associazione faunistiche a grossi mammiferi nel Pleistocene superiore in Italia. *Geografia Fisica Dinamica Quaternaria* 6: 182-184.
1983b Varaiations climatiques et séquences chronologiques sur la base des variations des associations fauniques à grands mammifères. *Actes du Colloque International 'La position taxonomique et chronologique des industries à pointes à dos autour de la Mediterranée européenne', Siena 1983. Rivista di Scienze Preistoriche* 38: 161-180.

Shott, M.J.
1986 Technological organization and settlement mobility: an ethnographic examination. *Journal of Anthropological Research* 42: 15-52.

Thomas, D.H.
1983 *The Archaeology of Monitor Valley 2: Gatecliff Shelter.* New York: Anthropological Papers of the American Museum of Natural History 59.

Tindale, N.B.
1974 *Aboriginal Tribes of Australia.* Los Angeles: University of California Press.

Torrence, R.
1983 Time budgeting and hunter-gatherer technology. In G. Bailey (ed.) *Hunter-Gatherer Economy in Prehistory: A European Perspective*: 11-22. Cambridge: Cambridge University Press.

Wilmsen, E.N.
1968 Functional analysis of flaked stone artifacts. *American Antiquity* 33: 156-161.

Yellen, J.
1977 *Archaeological Approaches to the Present. Models for Reconstructing the Past.* New York: Academic Press.

LATE AHRENSBURGIAN BLADE PRODUCTION AT JERXEN-ORBKE, DETMOLD (GERMANY)

Christina Kempcke-Richter

INTRODUCTION

The open-air site of Jerxen-Orbke (Stadt Detmold) was discovered in 1987 by an employee of the Lippisches Landesmuseum of the department of Ostwestfalen-Lippe (Western Germany). She found the site while carrying out surveys in this area, because of the construction of a drainage system. During this survey, a lot of Late Palaeolithic and Mesolithic stone artefacts had been collected, which called for careful archaeological research. This included an extensive survey to locate the several artefact concentrations, as well as excavations, both of which were carried out by a team from the University of Cologne from 1990 to 1992. The surveys located three concentrations, one with Mesolithic artefacts, one with Late Palaeolithic artefacts and one with pottery from several periods. Excavations were limited to the Late Palaeolithic concentration, which is the theme of the author's masters thesis and which will be discussed in the following paper.

The character of the Late Palaeolithic assemblage of the open-air site of Jerxen-Orbke is very special. It contains typical Late Palaeolithic tools such as backed knives and truncated blades, as well as bipolar blade cores. These permitted a very preliminary and general dating to the end of the Late Palaeolithic. There were no 'fossils directeurs', however, such as tanged points (Ahrensburgian culture) or backed points (Azilian culture), which would have allowed a more specific cultural definition of the assemblage. Because of this, the main emphasis in the analysis was placed on technology, including the examination of scar patterns on the cores, as well as the analysis of the *chaînes opératoires* (Boëda *et al.* 1990), as a basis to allow comparisons with other well dated Late Palaeolithic sites. This paper will present brief descriptions of the geological and geographical aspects of the region of Detmold, Ostwestfalen-Lippe district, where the site is situated, a description of the open-air site itself and a fuller statement about the details of the technological analysis.

THE OPEN-AIR SITE OF JERXEN-ORBKE

Jerxen-Orbke is a district around Detmold, a small town situated in northwestern Germany, north of the Teutoburger Wald, in the European distribution area of the tanged points culture (Figure 1). This area spreads north-south from southern Scandinavia to the fringe of the German Mittelgebirge and east-west from the Russian-Polish border area to France. The site itself lies on a small rise, nearly 120 metres above sea-level. The rise is bordered on the south and on the west by two small rivers and consists of gravels of a ground moraine, covered by a layer of loess with a thickness of 2 metres. Both these layers overlie layers of sandstone and marl, dated to the Triassic.

THE GEOLOGICAL SITUATION

The stratigraphic development of the region where the site of Jerxen-Orbke is situated has been influenced by the first expansion of the ice shield during the Saale Glacial, the Drenthe stadial. During this glacial period, the North European ice shield reached the fringe of the Rheinisches Schiefergebirge. During the Drenthe stadial, the advancing ice shield reduced it's speed and stopped approximately 4 kilometres south of the area of excavation, due to a gradual rise in temperature. The formation of the moraine is a result of all these events (Farrenschon 1986: 89). The moraine consists of sandy, silty sediments with clay and boulder materials, consisting of different kinds of rock coming from Scandinavia, northern Germany, the region around the Baltic Sea (Baltic flint) and from nearby areas (Farrenschon 1986: 90). The concentration of large boulder material from the north in the area where the site of Jerxen-Orbke is situated marks the border of the northern ice shield. The end moraine has mostly been destroyed by erosion but, in spite of this, it reached a thickness up to 30 metres (Farrenschon 1986: 91).

The Saale Glacial was followed by a period of rising temperature, the Eem Interstadial, which was itself followed by a new cooling phase, the Weichselian Glaciation. The ice shield of this glacial stopped much further to the north, reaching almost as far as the Elbe. At this time, the region around Detmold was characterised by a periglacial climate with extreme fluctuations. At the beginning of the Weichselian Glacial the riverbeds and terraces in this region developed their present day structures, caused by a sinking of the erosion base in the North Sea. This was the time when loess and shifting sands were blown out of the riverbeds and spread over the older sediments (Farrenschon 1986: 87).

THE STRATIGRAPHY

The stratigraphic structure of the open-air site of Jerxen-Orbke was identified by means of a geological section 4 metres long and 1 metre deep, at the northern edge of the

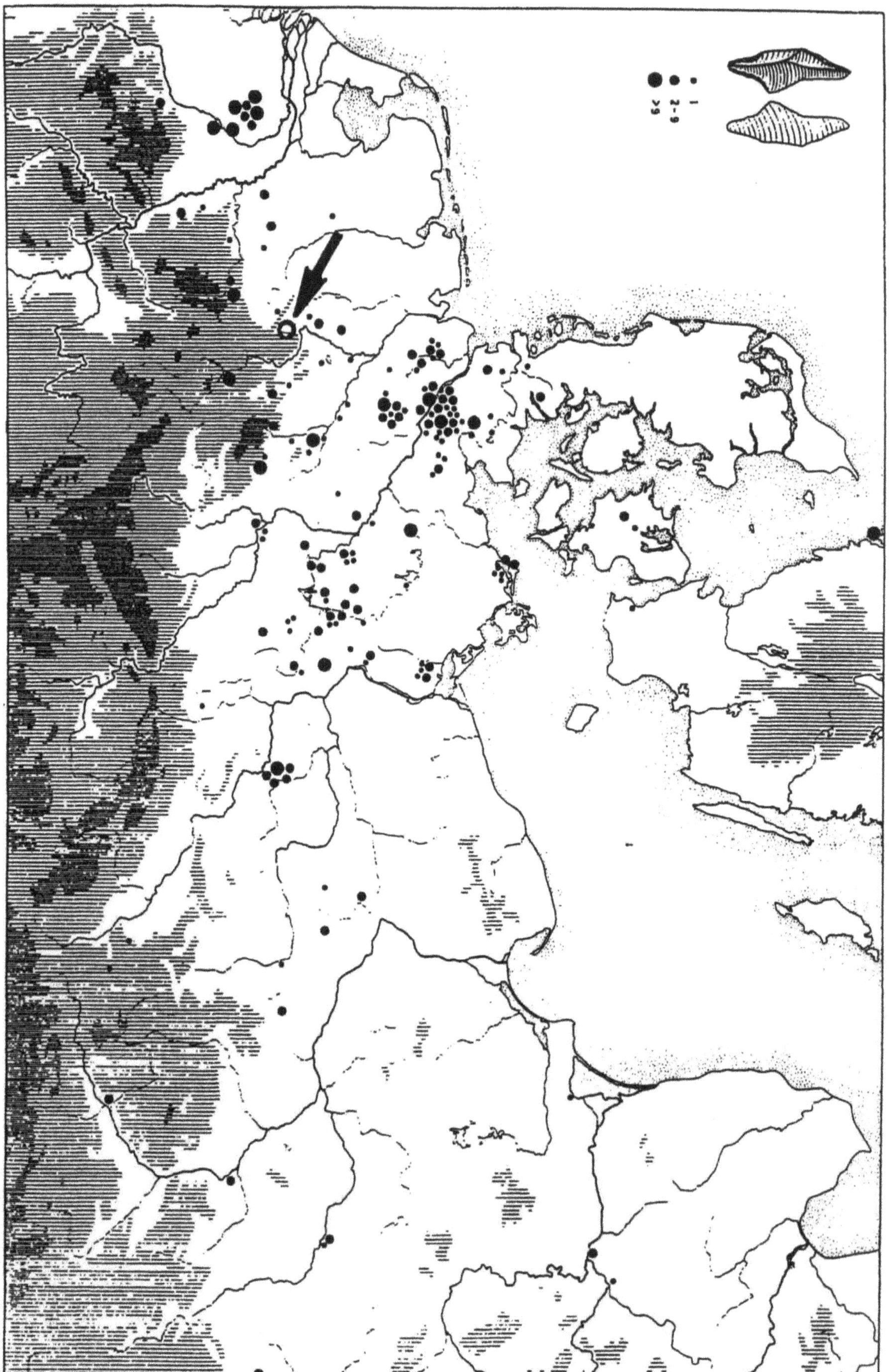

Figure 1: The open-air site of Jerxen-Orbke, Detmold. The map shows the distribution of the Ahrensburgian sites in the northern European plain (after Taute 1968)

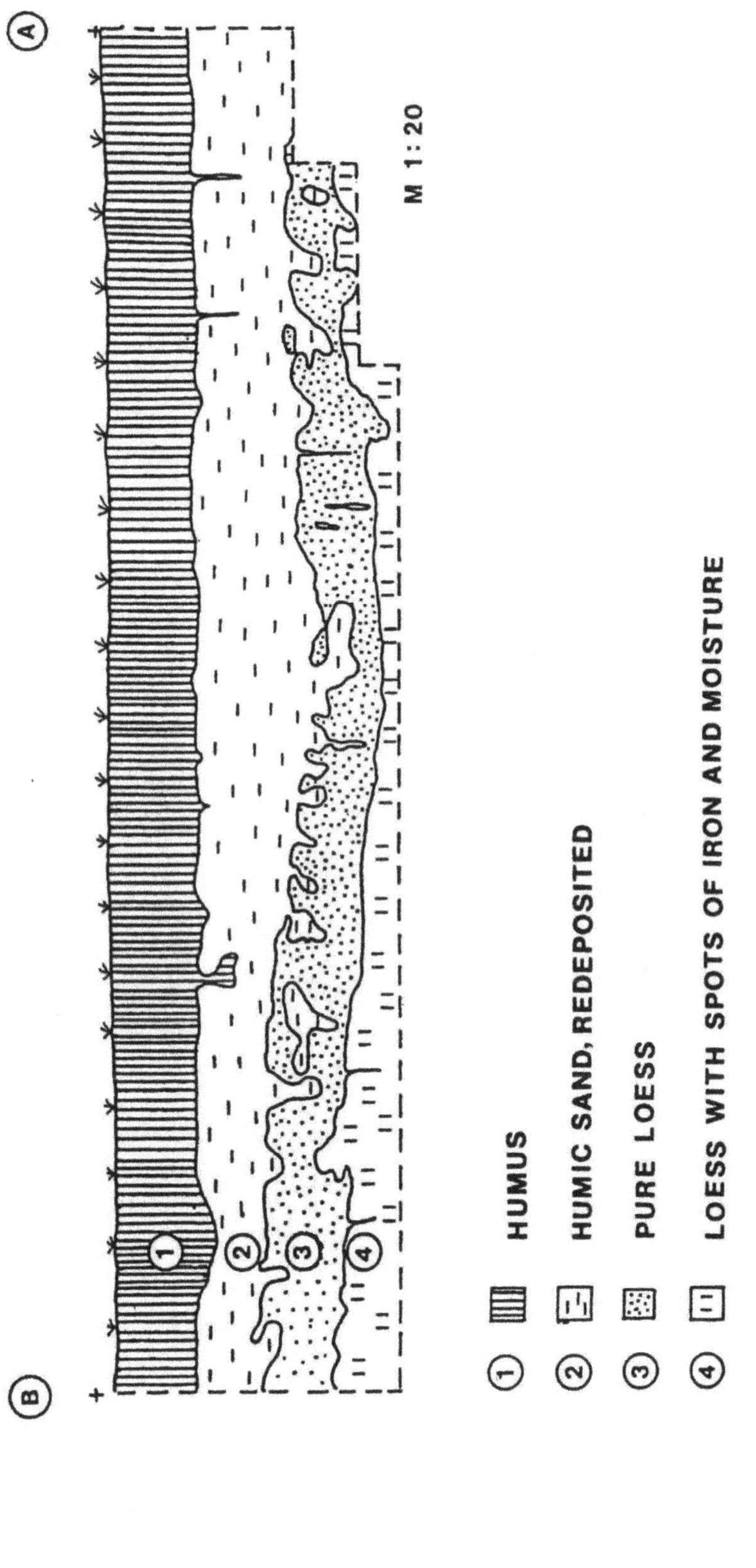

Figure 2: Geological section at Jerxen-Orbke, Detmold

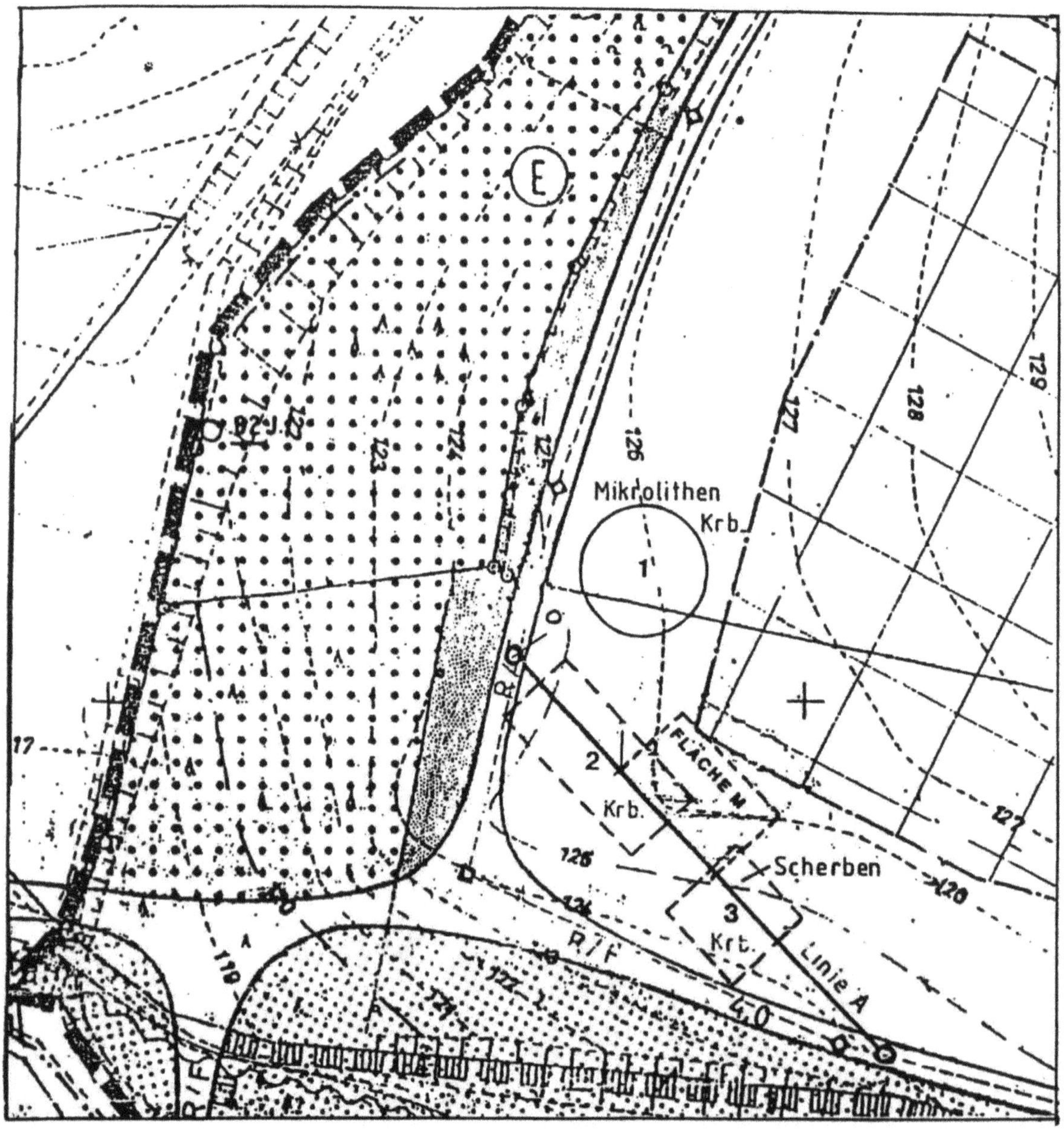

Figure 3: Concentration of finds from the surface

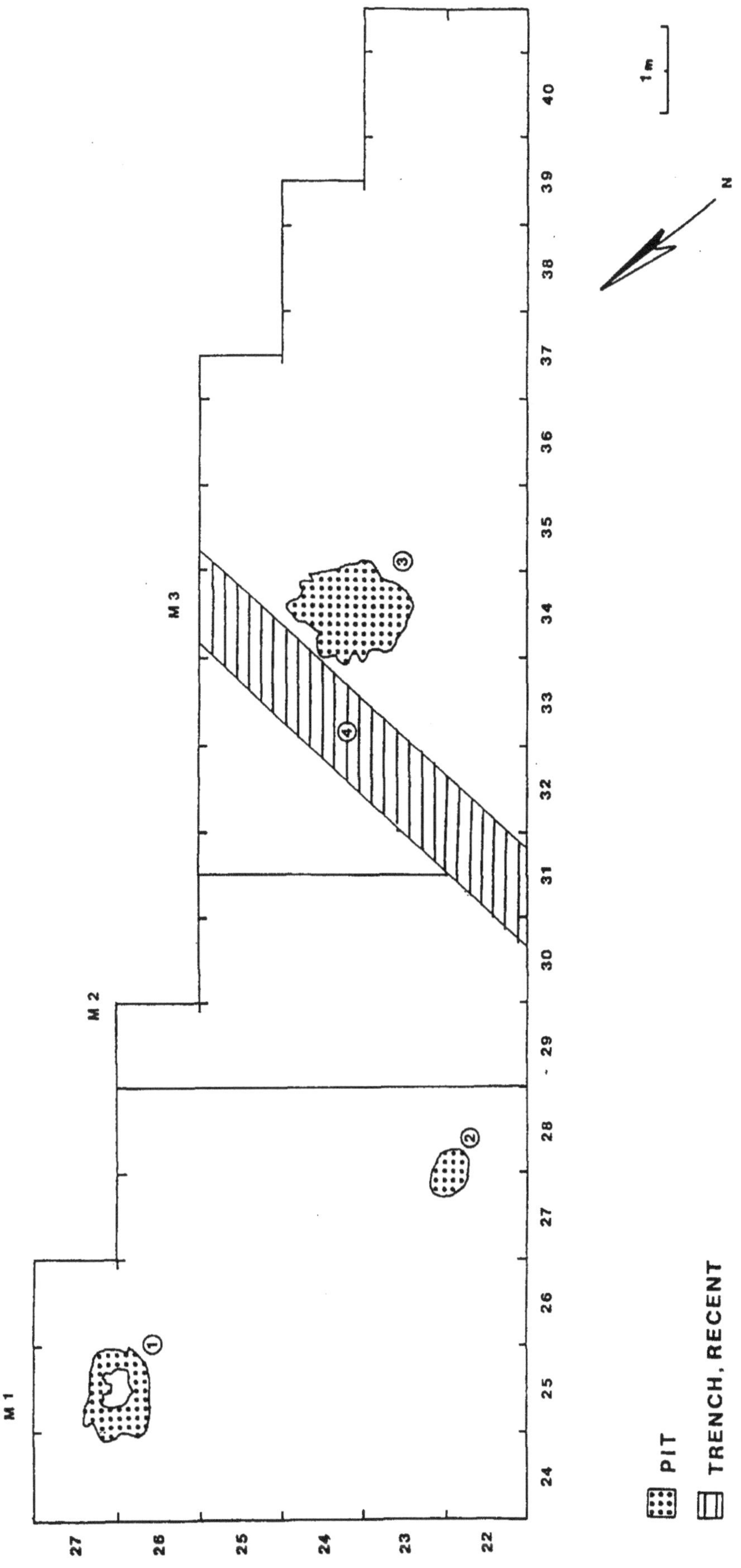

Figure 4: Structures in the excavation area M 1-3

excavation area. The stratigraphy consists of four layers which are described as follows (Figure 2):

— Layer 1: Humus (0.3 metres).

— Layer 2: Humic, silty sand of a light brown colour which has been redeposited from higher elevations (0.3 metres).

— Layer 3: Pure sandloess (0.4 metres), a glacial sediment in which the archaeological layer was imbedded.

— Layer 4: A horizon of sandloess with spots of iron and moisture (0.2 metres).

ARCHAEOLOGICAL ACTIVITIES AND THEIR RESULTS

There were three steps of archaeological activities leading to the result that is presented in this paper. The first was the discovery of the site by the collection of a lot of stone artefacts from the surface. This was carried out by an employee of the Lippisches Landesmuseum in Detmold. This collection contained patinated Mesolithic artefacts (isosceles triangles and microblade cores) and unpatinated Late Palaeolithic artefacts, such as scrapers, burins, truncations, backed knives (all with very sharp edges), flakes with facetted striking platforms and bipolar blade cores. There were also separate concentrations of pottery from three periods: Prehistoric, Middle Ages and Modern.

The second step was an intensive survey, resulting in the identification of three concentrations, one Mesolithic (1), one Late Palaeolithic (2) and one with pieces of pottery dating to different periods (3) (Figure 3). During the third step, the excavation, the main emphasis was placed on the concentration of Late Palaeolithic artefacts. The excavations began with an area of 10 square metres (area M1) and was enlarged during the following campaigns to 70 square metres (areas M2 and M3). A level was recognised for each sedimentary unit containing artefacts; altogether there were three levels, the third one of which contained the original archaeological horizon. The result was the discovery of an open-air site with many traces of flintknapping, including flakes, blades (partially retouched), truncations, chips and cores, as well as tools like endscrapers and backed knives. All the artefacts were unpatinated and had extremely sharp edges, like the surface finds. Thus, it seemed obvious that both assemblages, the surface one and the excavated one, belonged together.

Apart from the stone artefacts, three depressions and a modern, 1 metre wide trench were found, the latter caused by construction work for the drain (Figure 4). The trench (Figure 4.4) ran from the northeast to the southwest, across area M3. The first of the three depressions (Figure 4.1) was located in the northern section of area M1. It was formed by an oval-shaped ring and measured approximately 1.5 x 2 metres and was 15 centimetres deep. The second depression (Figure 4.2) was situated in the southern section of area M1, was round and half a meter in diameter and was only 0.5 centimetres deep. Both depressions contained some lithics, as well as small pieces of charcoal. The third depression (Figure 4.3), which lay in the middle section of area M3, measured 1.5 x 1.9 metres and was 26 centimetres deep. It contained only some small pieces of charcoal.

There are two possible interpretations of these depressions: the first, that they are traces of fallen trees, can be rejected in view of the fact that P.B. Kooi's description of traces of fallen trees (Kooi 1974: 57) does not apply to these depressions at Jerxen-Orbke. He considered two typical attributes for this sort of phenomenon: traces of fallen trees always have a ring-shaped structure, and in profile the layers from the top are turned on their side at a 90 degree angle. This second attribute also did not apply at Jerxen-Orbke. In this case, the depressions were filled with a homogeneous, light brown coloured sediment which stood out against the surrounding sediments. The second interpretation, that of traces of former huts, seems to be more reasonable for the first and third depression, in view of the fact that there was a significant spatial relationship between the lithic concentrations and these depressions. C.-J. Kind (1985: 134) has shown similar examples in his research on the relationships between huts and concentrations of artefacts; traces of huts are always located in the marginal zones of lithic concentrations. The smallest depression is obviously too restricted to be a hut and may represent a very minor pit. A detailed description of the distribution of the artefacts and their concentrations at Jerxen-Orbke will be given in the following section.

THE ASSEMBLAGE

The assemblage of Jerxen-Orbke contains a total of 5009 pieces, 1614 pieces from the surface, 212 from level 1, 217 from level 2 and 2966 from level 3, the original archaeological horizon.

Level 1 contained the following tools: endscrapers made on short, rounded flakes with cortex platforms (Plate 1: 1). The endscrapers have flat, irregular retouch, with traces of frequent use. Two distal truncations on blade fragments were found, the proximal pieces of which were missing (Plate 1: 2-3). One of the fragments is partially retouched inversely and shows some notching (Plate 1: 3).

Level 2 has one distal truncation on a blade (Plate 1: 4) and eight cores. One of the latter is a bipolar blade core in a state of advanced exploitation. It shows traces of preparation by the removal of some flakes from the flanks perpendicular to the direction of blade removal (Plate 1: 5).

Level 3: Eight burin spalls were found, all restricted to a small concentration near the depression in the southern

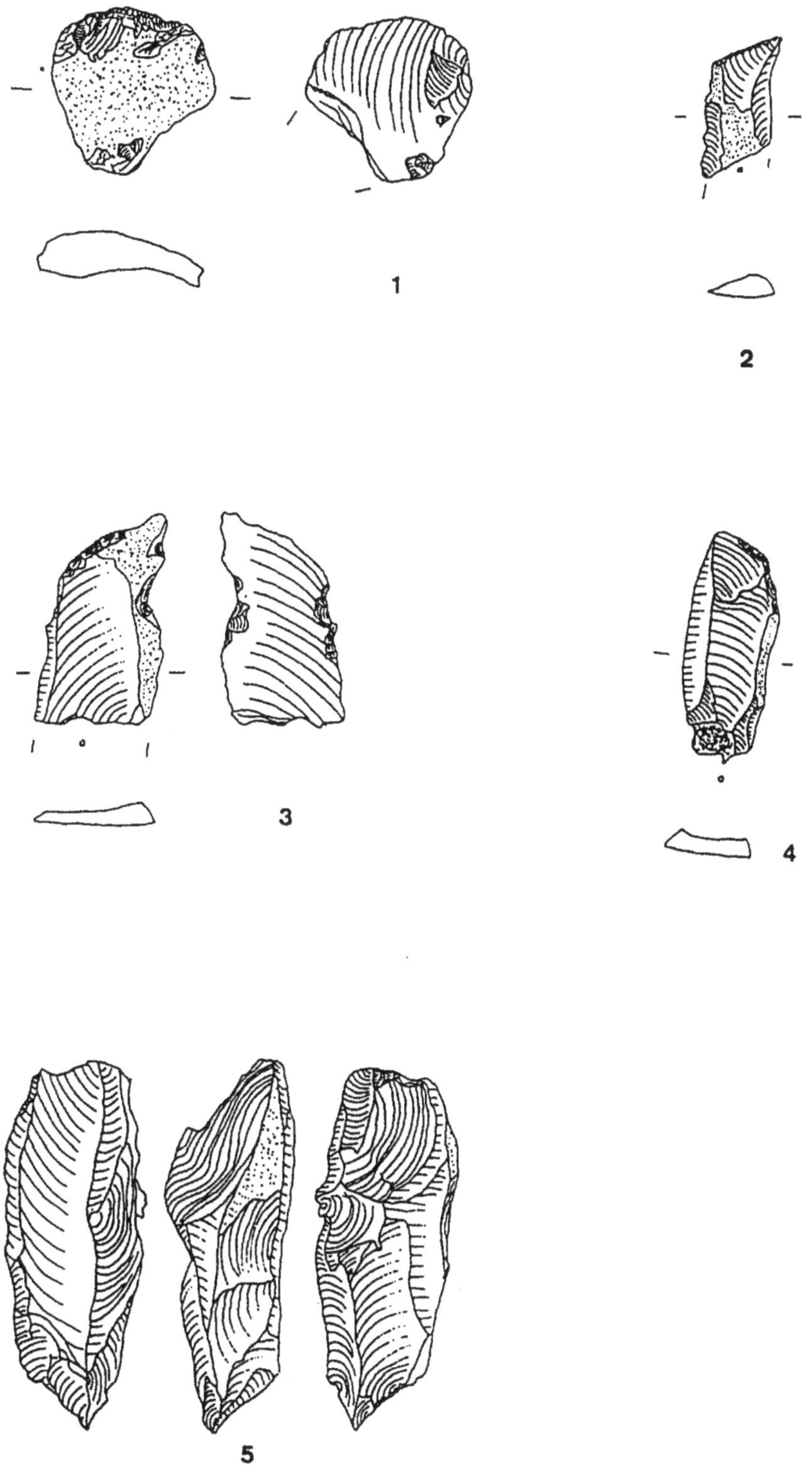

Plate 1: 1 cortical endscraper (level 1); 2-3 distal truncated blade fragments (level 1); 4 distal truncated blade (level 2); 5 core with prepared flanks (level 2) (full size)

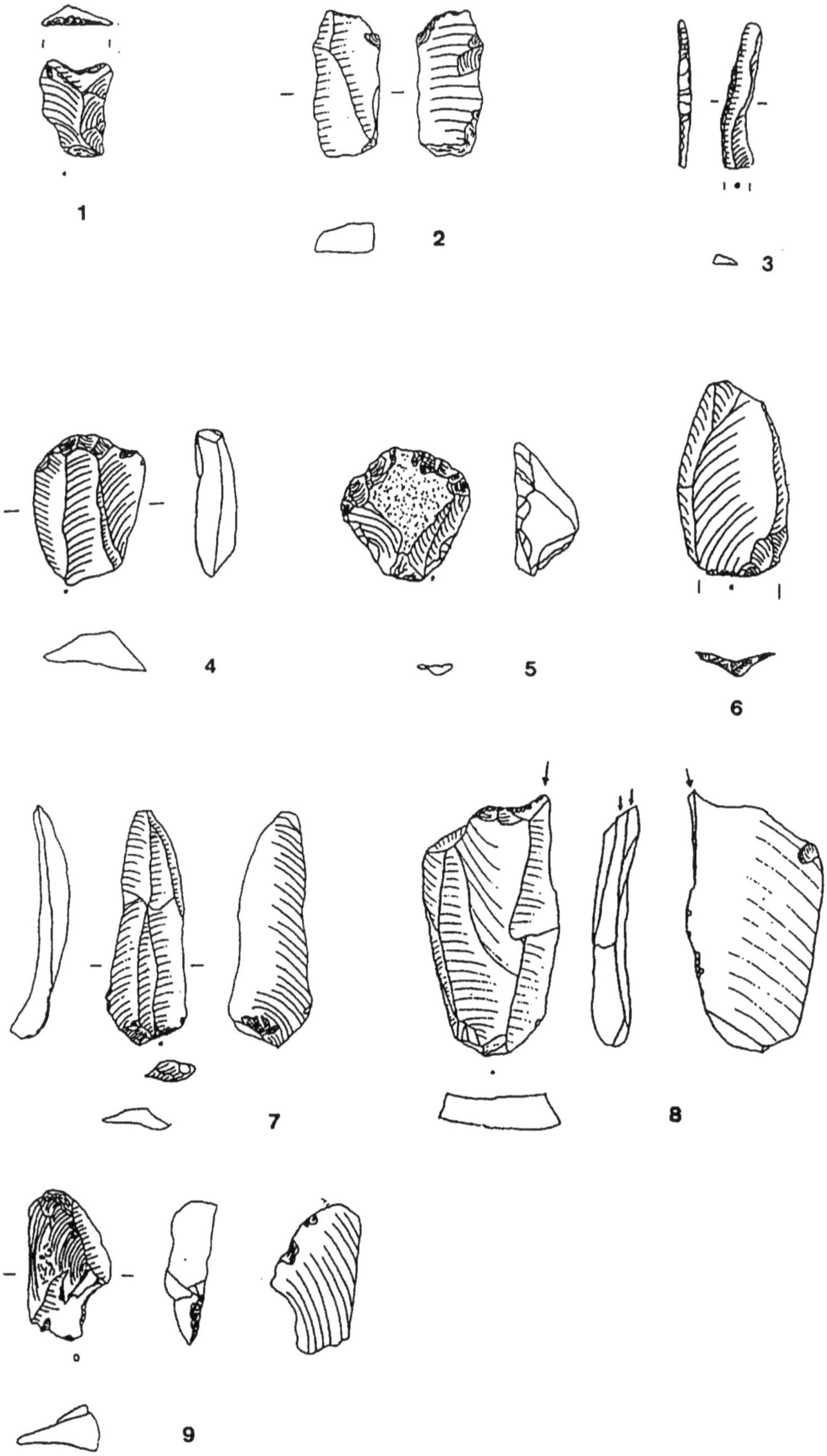

Plate 2: 1-2 distal truncated flakes; 3 backed bladelet; 4-5 endscrapers; 6 flake with a facetted striking platform; 7 blade; 8 burin; 9 lateral retouched flake (all artefacts from level 3) (full size)

section of area M1. The tool assemblage contains two endscrapers made on short flakes with a flat retouch, one with cortex on the dorsal side (Plate 2: 4-5); two distally truncated small flakes; one of the truncations is markedly concave (Plate 2: 1-2). In addition, there is an unretouched flake with a facetted striking platform (Plate 2: 6). Several blades were found, one of which has dorsal reduction and which has been struck off-axis (Plate 2: 7). This is an attribute which has been described by S. Hartz as typical for artefacts of the Hamburgian culture, as well as for the tanged points groups (Hartz 1987: 26). The assemblage also contains two flakes with partial lateral retouch (Plate 2: 9), one burin (Plate 2: 8) and a distal fragment of a small backed bladelet (Plate 2: 3).

THE DISTRIBUTION OF ARTEFACTS IN THE ARCHAEOLOGICAL HORIZON (LEVEL 3)

It seems quite striking that the highest density of artefacts occurs surrounding the depressions. It was possible to identify three clear artefact concentrations in the excavation area (Figure 5). The first concentration measured 2.5 x 2 metres and was located in the northern section of area M1, near the northern edge of the excavation. The highest number of artefacts in this concentration reached 35 pieces per quarter square metre unit.

The second and main concentration was situated in the southeastern section of area M1, becoming more and more invisible in area M2 toward the east. It measured 2 x 1.5 metres, with a maximum artifact density in quarter square metre unit 28/23d which contained 331 pieces, most of them chips. The second concentration is the largest found during the excavations.

The third concentration, measuring 2.5 x 2 metres was found at the northeastern edge of area M3. The majority of the flints are pebbles with rounded edges, mixed with a few chips, one tool and many unretouched artefacts. The edge of the third concentration nearly coincides with the western and southern edges of the quarter square metre units 35/24-35/25 and 35/24-37/24 and, presumably, it extends to the north in an area which has not been excavated. The southern edge of this concentration was destroyed by the construction work. Only surface artefacts from this area had been saved. There are also some quarter square metre units in the areas M1 and M3 without any artefacts at all. In light of this, it is remarkable that nearly all of the concentrations were in the immediate neighbourhood of the depressions.

A DETAILED TECHNOLOGICAL ANALYSIS OF TOOL PRODUCTION

In view of the fact that there were no culturally diagnostic artefacts for any specific Late Palaeolithic culture group, the main emphases were placed on a detailed analysis of the technology of the production sequences of flakes and blades and the comparison of these results with production sequences from other well dated Late Palaeolithic sites. Two steps were necessary to get the desired results: the first was the examination of the assemblages from the surface and from the excavation for concurrence of raw material units which had to precede the second step, the analysis of the technology of artefact production sequences, using the *chaînes opératoires* approach developed by E. Boëda, J.-M. Geneste and L. Meignen for Lower and Middle Palaeolithic assemblages (Boëda *et al.* 1990). The results of this work is presented below.

THE RAW MATERIAL UNITS

Without exception, the artefacts at Jerxen-Orbke are made on Baltic flint which comes from the nearby moraine. In preparation for the following technological analysis, the assemblages were divided into different raw material units, following the "workpiece analysis" method of K.H. Rieder (1992: 458). According to this method, each raw material unit is identified as an individual workpiece: that is, a single block or nodule of a raw material different from all other raw material in the assemblage.

The identification of individual raw material units was made with the help of analysing the colour, inclusions and the consistency of the cortex. It is remarkable that there is a very high number of different raw material units in this rather small artefact sample: a total of 58. The colour range of the individual units ranges from a dark brown, translucent, shiny, glass-like material to a less translucent, grey flint with a shiny surface. Other groups include a brittle, cream or light grey coloured, opaque flint with a lot of inclusions and different coloured zones. Some pieces are covered by cortex with a thickness of 1 millimetre, which has a white-yellow and, in some cases sulphur coloured, chalk-like consistency. In other cases, the cortex is dark grey and very compact. Some of the nodules show traces of abrasion, which may have been caused by fluvial activities or by the moving of the ground moraine. The individual raw material units show a lot of differences, indicating that the material came from a secondary deposit.

REFITTING THE STONE ARTEFACTS

The refittings of the stone artefacts from Jerxen-Orbke should provide answers to several questions: 1) are there any connections between the artefacts which had been collected from the surface and the artefacts from the excavation? 2) what does refitting tell us about the spatial distributions of artefacts within and between the concentrations? and 3) what individual production sequences can be identified by using this method?

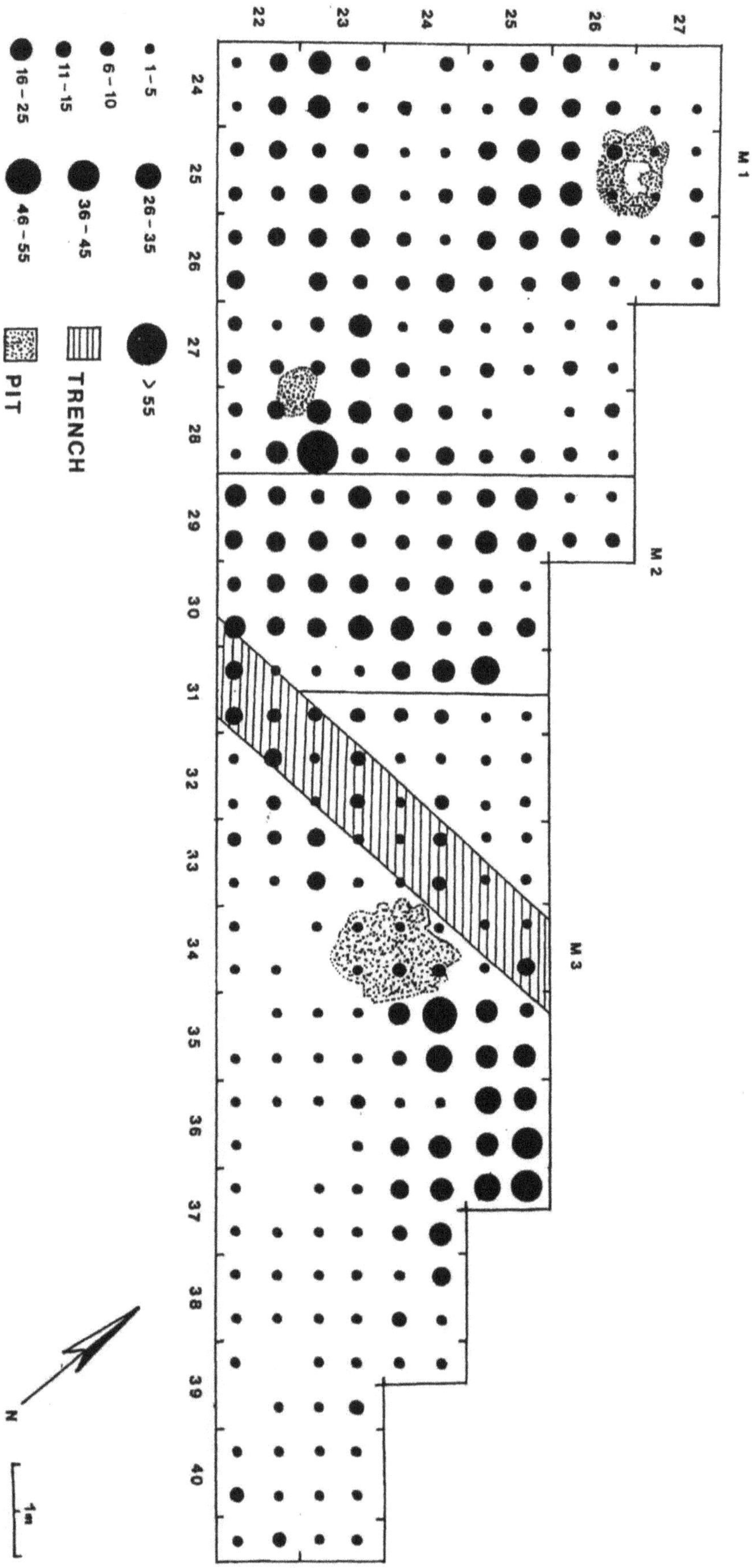

Figure 5: Distribution of artefacts in area M 1-3

At the beginning of this examination, all the artefacts from the surface as well as those from the excavations were grouped by their quarter square meter unit to restrict the initial attempts at refittings to areas as small as possible. This step produced 21 refits of the products of flintknapping and 14 refits of broken pieces, with 23 flakes, five blades, three plunged blades, three cores and 13 fragments. The second step involved rearranging the artefacts by the different raw material units and then examining them for further refits. H. Thieme (1990: 543) gives an example of this method for studying the Middle Palaeolithic assemblage of Rheindahlen. This step resulted in 11 refits of artefacts from the surface with artefacts from the excavations, as well as refits of artefacts only from the surface (5 cases). All together 51 refits were made from an assemblage of 5009 artefacts.

THE CONNECTION BETWEEN ARTEFACTS FROM THE SURFACE AND ARTEFACTS FROM THE EXCAVATIONS

In view of the fact that Jerxen-Orbke was discovered through a series of surface finds, it was necessary to prove the connection between these finds and the assemblage of the excavations by refittings. This aspect of the study was carried to answer two questions: 1) was there any relationship between the surface artefacts and the assemblage from the excavation in relation to different raw material units? and 2) would artefacts from the surface and the excavations fit together? As part of this effort, the materials were sorted by raw material units, keeping the surface (Sa) and the excavation (Ea) samples separate (Table 1).

This analysis led to the following results: in five cases, the quantity of artefacts from the excavations exceeds 50% of the raw material unit compared with the surface artefacts. In one case, the relationship is 50/50, while 16 raw material units contained only artefacts from the surface and 7 units contained exclusively artefacts from the excavations. In two cases, two nodules with a single scar each were identified. Altogether, 33 of the 58 raw material units (56%) contained artefacts from the surface as well as from the excavated assemblage. This provides a first step to demonstrate an original unity of both assemblages.

Each raw material unit was examined for possible refits to confirm the hypothesis that there would be refits between the surface and the excavation material. In ten raw material units it was possible to carry out refittings; in three raw material units, artefacts from the surface and from the excavations were refitted (Table 2). In all three cases flakes and blades as well as plunged blades were refitted to cores.

Table 1

Unit	Sa	Ea	total	% Sa/Ea
1	12	2	14	86/18
2	9	-	9	100/-
3	1	2	3	33/67
4	1	2	3	33/67
5	3	7	10	30/70
6	54	8	62	87/12
7	-	2	2	-/100
8	6	-	6	100/-
9	4	1	5	80/20
10	7	-	7	100/-
11	-	3	3	-/100
12	8	1	9	89/11
13	35	3	38	92/8
14	38	22	60	63/37
15	12	2	14	86/14
16	4	-	4	100/-
17	3	1	4	75/25
18	87	11	98	89/11
19	18	2	20	90/10
20	12	8	20	60/40
21	65	6	71	91/9
22	44	4	48	92/8
23	36	9	45	80/20
24	6	2	8	75/25
25	48	14	62	77/23
26	4	-	4	100/-
27	2	2	4	50/50
28	2	16	18	11/88
29	18	3	21	86/14
30	8	1	9	89/11
31	20	2	22	95/5
32	24	5	29	83/17
33	22	-	22	100/-
34	7	-	7	100/-
35	7	2	9	100/-
36	3	1	4	75/25
37	8	1	9	89/11
38	-	7	7	-/100
39	2	4	6	33/67
40	8	-	8	100/-
41	7	1	8	88/11
42	4	-	4	100/-
43	5	2	7	71/29
44	6	1	7	86/14
45	3	1	4	75/25
46	22	-	22	100/-
47	23	4	27	85/15
48	5	-	5	100/-
49	80	21	101	79/21
50	5	-	5	100/-
51	3	-	3	100/-
52	-	12	12	-/100
53	-	1	1	-/100
54	-	1	1	-/100
55	-	2	2	-/100
56	2	-	2	100/-
57	4	-	4	100/-
58	2	1	3	67/33

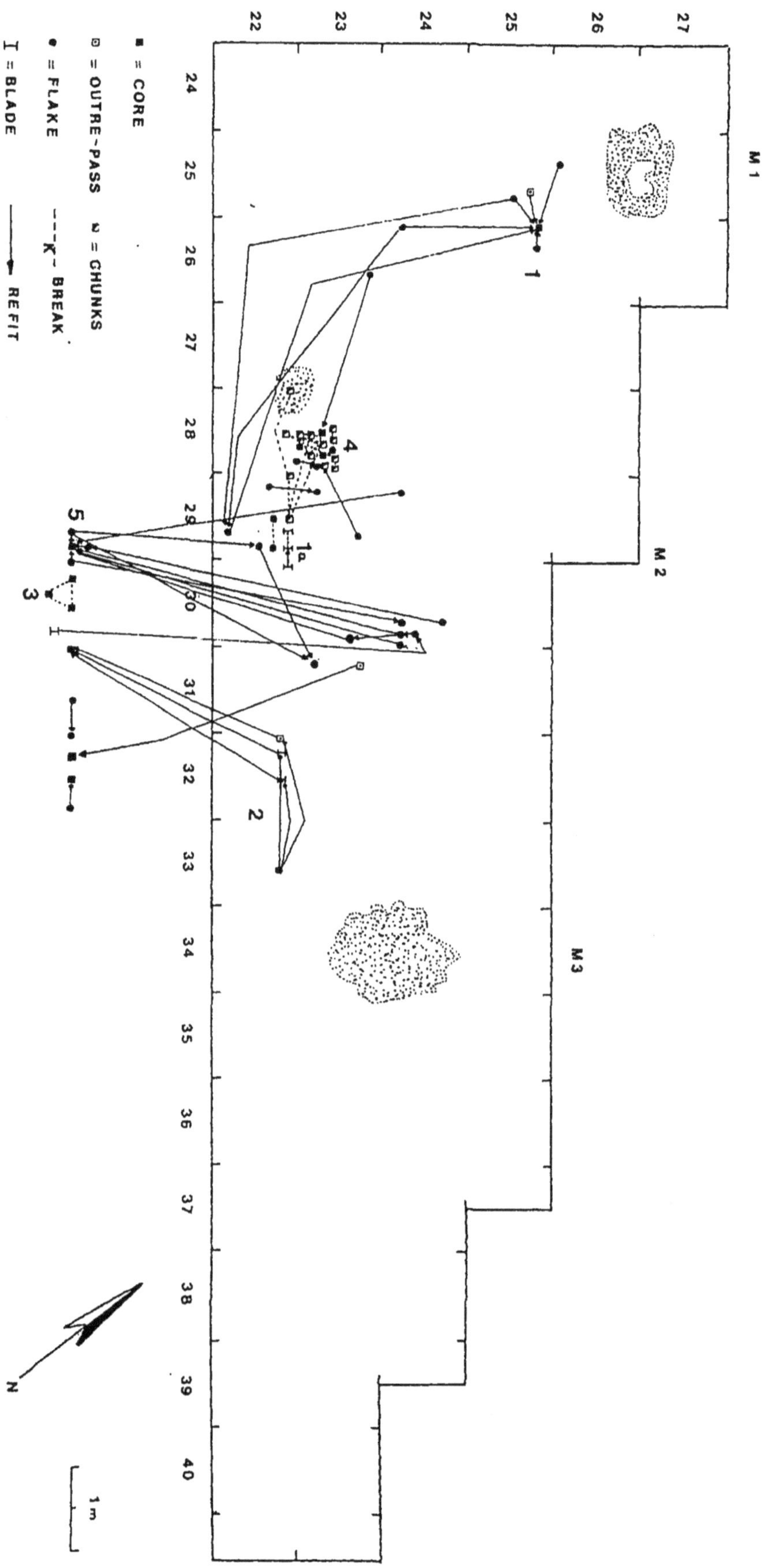

Figure 6: Distribution of refits in area M 1-3

Table 2. Combined Surface and *in situ* Refits by Raw Material Unit

Unit 14	Artefacts: 60	Refits: 13	(Sa 2/Ea 9)
Unit 19	Artefacts: 20	Refits: 2	(Sa 1/Ea 1)
Unit 38	Artefacts: 7	Refits: 5	(Sa 1/Ea 4)

Sa = Artefacts from the surface
Ea = Artefacts of the excavation

THE SPATIAL DISTRIBUTION OF THE REFITS FROM THE EXCAVATIONS

The distribution of the refits inside the excavation area is shown in Figure 6, following the method of E. Cziesla (1986: 251 and 1990: 9). With the help of this method, the author was able to reconstruct the following situation: most of the refits are concentrated in the northern section of area M1, as well as in the middle and southern sections of areas M2 and M3. No refits were located in the eastern section of area M3. In area M1, most refits are situated very close to the depressions at a distance of not more than 2 metres. A single exception is Refit 1, where two pieces involved in the refitting were found at distances of 1.5 metres and 5 metres from the main cluster of refitted artifacts and the nearest depression. Refit 1a, which was situated at a distance of 2 metres from the small depression at the south-western edge of area M2, could be identified as part of Refit 1 because of the consistency of the raw material, although no blades could be refitted to the core. In the south-eastern corner of area M1 the concentration of Refit 2 is clustered within 0.5 metres; only one artefact was located some 2 metres away (Figure 6). The greatest number of pieces involved in refitting was concentrated in the quarter square meter 28/23 and are almost all part of Refit 4 (Figure 6).

A third concentration of refitted artefacts, with a scattering radius of 2 metres, was located at the south-western edge of area M3. Although most of these pieces belong with Refit 5, there are some artefacts belonging to four different refitting pieces which are situated at a distance of not more than 1 metre from each other. Since some of the artefacts involved in these refits came from the surface, it is impossible to reconstruct their original position. In the plan (Figure 6) they were placed at the edge of the excavation area where they had been collected, according to the information supplied by the museum employee.

AN EXAMPLE OF A PRODUCTION SEQUENCE AT JERXEN-ORBKE

This section will present a detailed description of one refitting and the technological production sequence it represents. This is the most complex and complicated refitting. It is also the most important one, since it made it possible to identify a special procedure of artefact production which is characteristic of this site. The production sequence will be explained as a series of discrete stages, although this forms a generalisation of the actual, specific processes. The starting situation: the flint-knapper removed the cortex from the nodule. This is indicated by small flake scars covering the nodule's surface. Some remnants of cortex are still preserved; nevertheless, it was impossible to reconstruct the original shape of the nodule because of the very advanced state of exploitation (Figure 7). The process of core exploitation is still reconstructable and it can be divided into twelve individual steps:

— Step 1 (Figure 8.I): first, a striking platform (S1) is prepared (shown as dashed lines). In this case, it could not be ascertained if the striking platform was prepared by removing a flake or if it is a frostcleft surface.

— Step 2 (Figure 8.I): the exploitation surface used in the following production sequence is at a 90 degree angle to the striking platform. From this striking platform, a first blade (1) is removed along the first exploitation surface and terminates at a frostcleft plane. This step was reconstructed with the help of the blade scar; the blade itself is not contained in the assemblage. In the figure (Figure 8.I), it is marked by hatched lines.

— Step 3 (Figure 8.I): a big flake (2) is removed along the same exploitation surface as the blade (1) and reaches the end of the core, where it is also stopped by the frostcleft plane. In this case, the flake was found in the assemblage. After this sequence of production, the exploitation surface (A1) is not used any longer.

— Step 4 (Figure 8.II): following this, the core is rotated 90 degrees around the vertical axis and approximately 45 degrees around the transversal axis before the removal of a third big flake (3). This flake runs diagonally and is nearly bipolar to the first exploitation platform (A1). It was presumably removed to regulate the curvature of the exploitation surface A1' which is a lateral enlargement of the exploitation platform A1. The dorsal side of flake 3 is covered by scars of former removals which could not be reconstructed.

— Step 5 (Figure 8.II): before starting the production of a series of blades removals, a flake (4) was removed in the same way as flake 3, presumably for the same purpose, that is the regulation of the curvature of exploitation surface A1'. This flake (4) could also be identified only by its scar and it is marked by hatchings in the figure. Following this, a small series of blades and flakes (5) was produced, but the shape of the core does not allow any conclusions to be made concerning the exact number and shapes of the artefacts involved in this removal sequence. This is the reason why, while this sequence has been considered as a unit, the scars have been marked by hatchings. Only one arrow, symbolising the direction of exploitation, has been assigned to this sequence in the figure. The scar, which was caused by

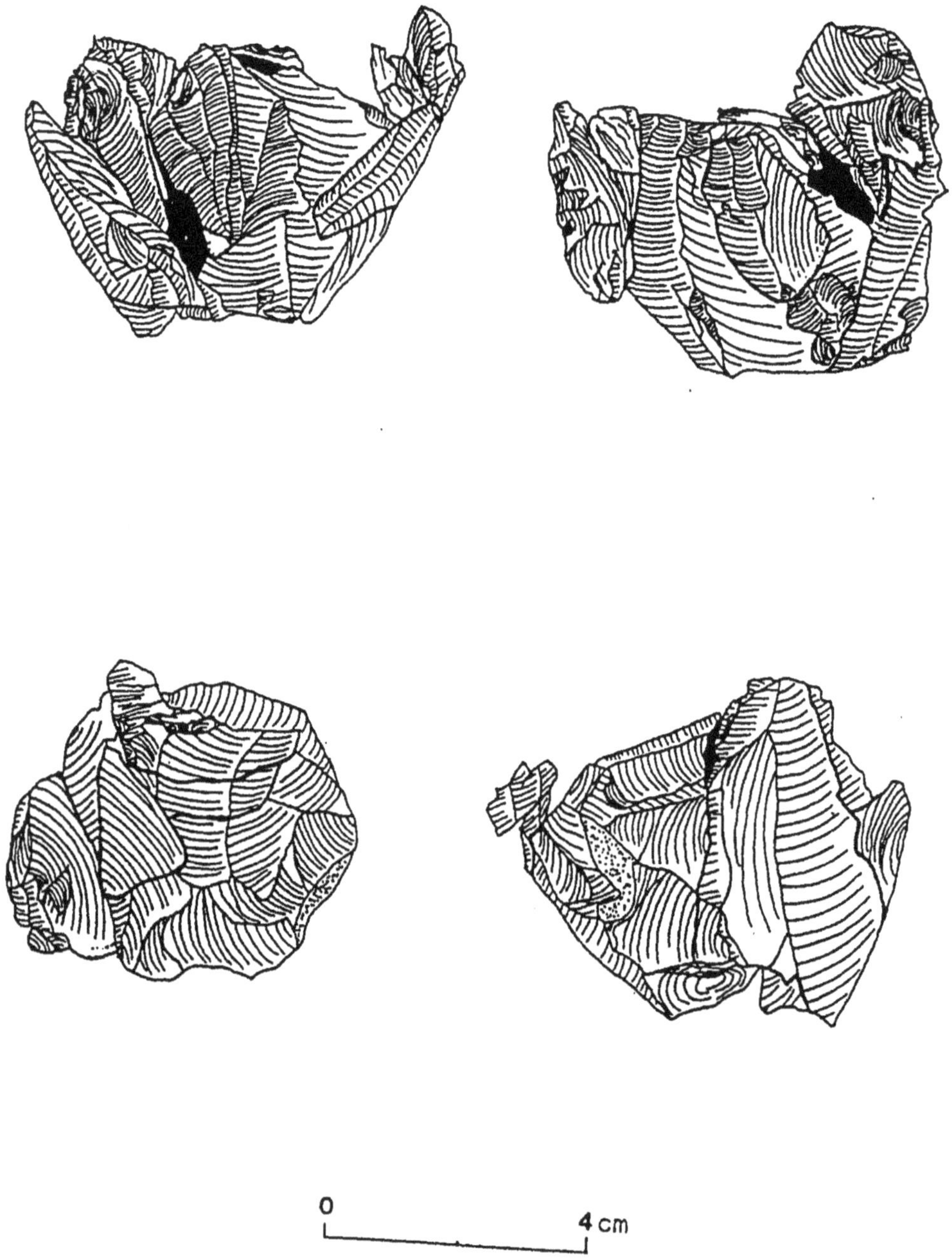

Figure 7: Refitted core

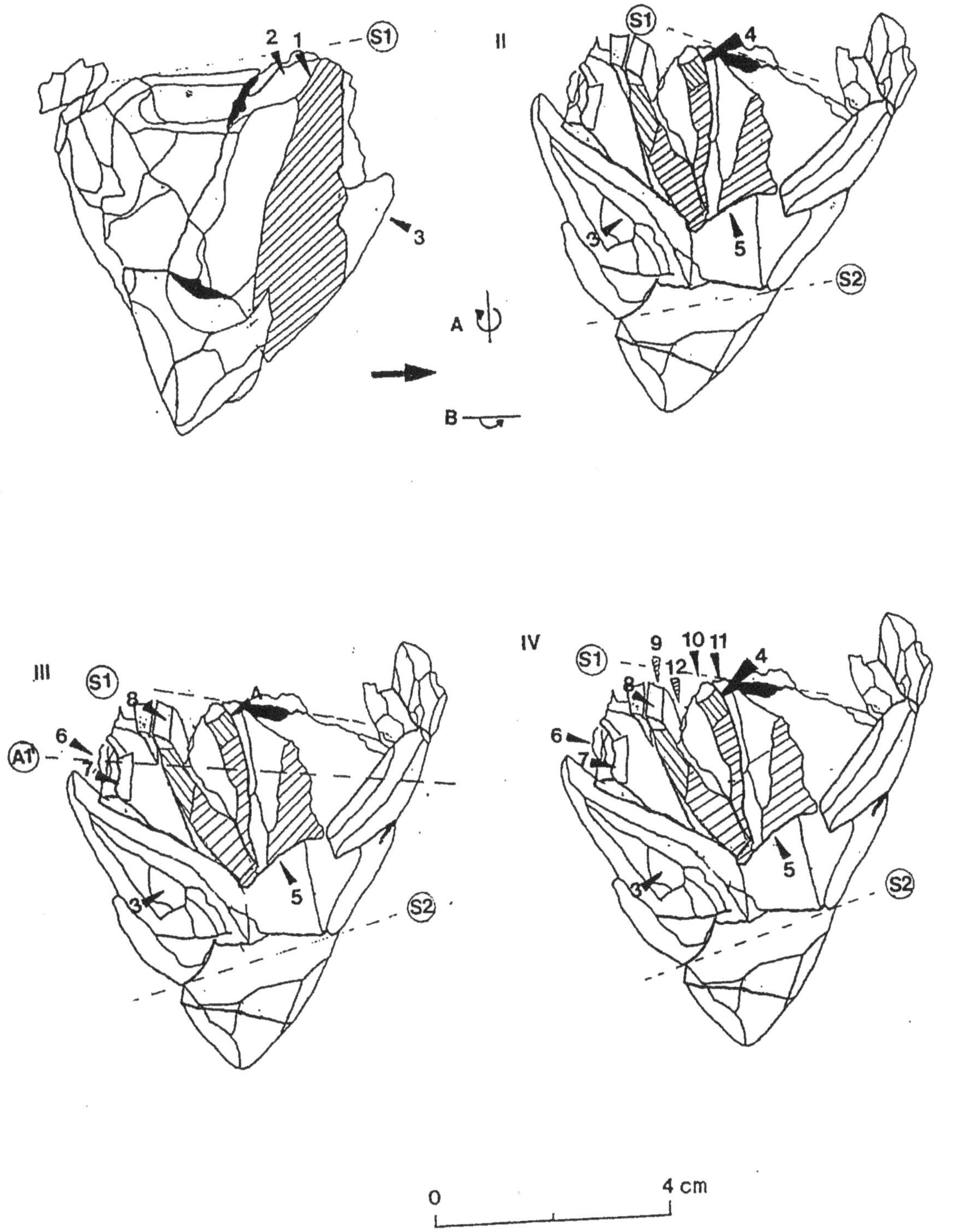

Figure 8: Production sequence steps 1-7

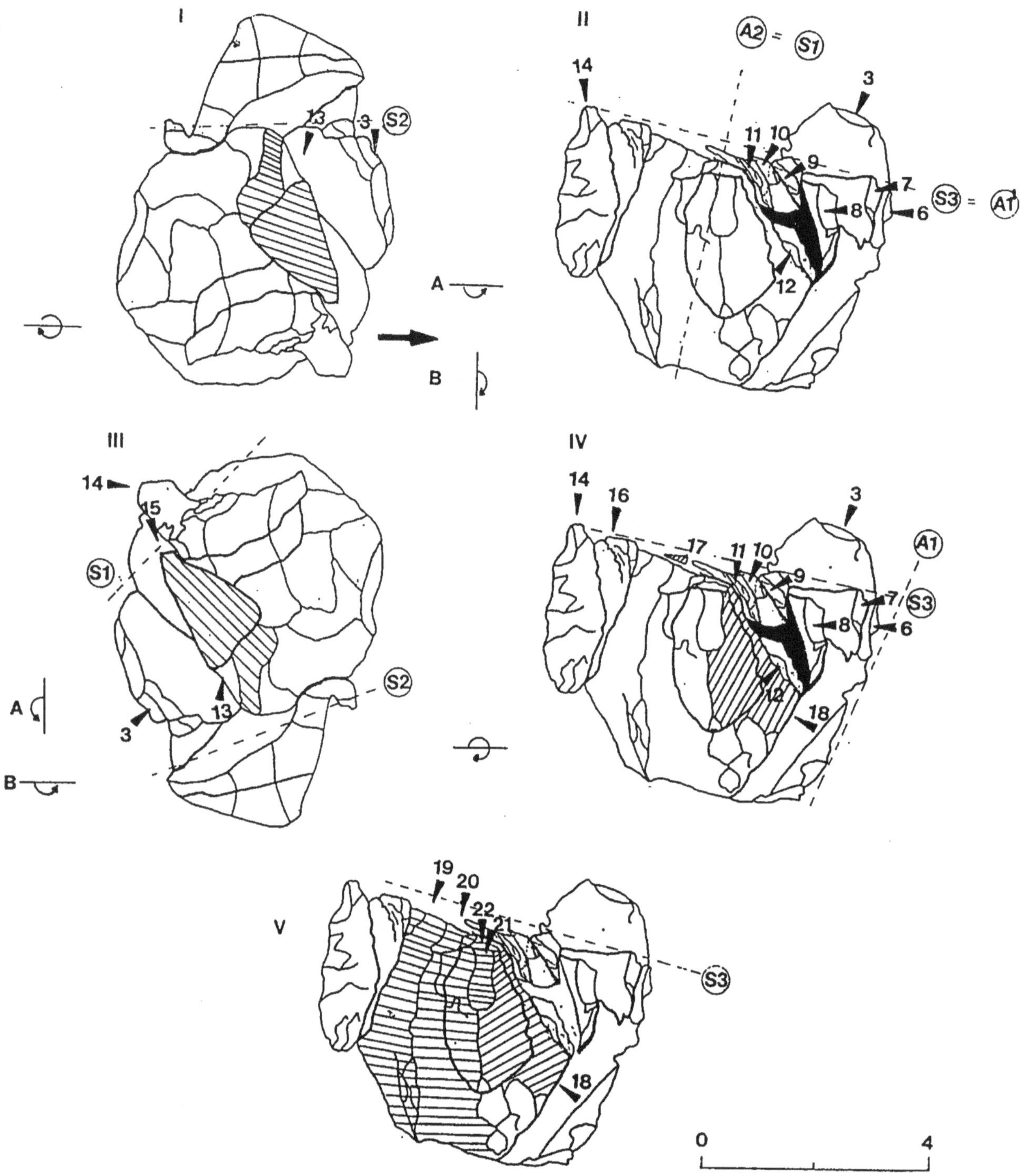

Figure 9: Production sequence steps 8-12

the removal of flake 2, was now used as a second striking platform (S2), opposite to the first striking platform (S1). The scars of series 5 run bipolar to the scars of flakes 1 and 2, which are situated at the back of the core in this figure.

— Step 6 (Figure 8.III): in view of the fact that the exploitation surface A1' was flattened by the production of a series of artefacts (5), a new guiding ridge was made to continue with the blank production by the removal of three flakes (6, 7, 8) in the same direction as the previous flake (3) producing two intersecting planes. For this purpose, three flakes (6, 7, 8) were removed along the same axis and direction as flake 3. The exploitation surface A1' was then used here as a striking platform.

— Step 7 (Figure 8.IV): following this, a new series of flakes (9-12) was removed alternately to each other. These are all distally terminated by the same frostcleft as described in steps 2 and 3. The series was removed from the striking platform S1, the direction of exploitation corresponding to that of flake 1. The arrows which symbolise flakes 9 and 12 are marked by hatchings, because they are not visible from this perspective.

— Step 8 (Figure 9.I): the core was rotated 180 degrees around the transverse axis and a big flake (13) was removed, using the striking platform S2 for a second time. The scars of flakes 3 and 13 converge.

— Step 9 (Figure 9.II): the core was now rotated 45 degrees around the transverse axis and 45 degrees around the vertical axis. The exploitation surface A1' was turned into a third striking platform S3. The striking platform S1 was then used as an exploitation surface (A2). From S3 a crested blade (14) was removed after the removal of several cortical flakes.

— Step 10 (Figure 9.III): a flake (15) was removed crosswise to flake 14. The former was stopped by a hinge.

— Step 11 (Figure 9.IV): a second crested blade (16) was removed from the striking platform S3 to regulate the curvature of the exploitation surface A2. The striking platform S3 was prepared by removing a flake (17), which is marked by a hatching in the figure. Following this, a second rejuvenation of the exploitation surface A2 was carried out by the removal of a flake (18), having been identified by its scar which runs crosswise to the direction of exploitation of S3. In this case, the exploitation surface A1 was used as a striking platform.

— Step 12 (Figure 9.V): the following production sequence, of a last series of blades (19-20), was identified by their scars. The last two removals of flakes (21-22) were both stopped by hinging. After this, all exploitation stopped.

This example of a production sequence was chosen among several others which had been reconstructed in the masters thesis, because it documents the pattern of core exploitation which is typical at this open-air site. The flint-knapper carried out a method of core exploitation, utilising changing orientations of the core, as is seen in steps 5, 6, 9 and 11. This special method could also be documented by the examination of the scars on some abandoned cores, as will be described in the following section.

INDIVIDUAL PATTERNS OF BLANK PRODUCTION DEVELOPED BY STUDIES OF SCARS ON CORES

In addition to the analysis of the refittings, the abandoned cores were examined for their flake scars to identify individual patterns of blank production at Jerxen-Orbke. The assemblage of Jerxen-Orbke contained 101 cores, 42 of which were suitable for analysis. The rest were either fragments or in a state of very advanced reduction, such that any information about production sequences was no longer available. To confirm the usefulness of such a method, they have been compared to the results obtained from the refittings. Actually, it was possible to identify five patterns of blank production which will be presented in this section because they are typical for the Late Palaeolithic, with one exception: the method of blank production utilising a change of core orientation. This method has been documented using both scars on cores, as well as by an analysis of the refittings, as discussed above.

— **Pattern A** (Figure 10): in this pattern, the core were not prepared at all (Figure 10.1). It shows a parallel, unipolar blade production which is carried out on a core with one striking platform (Figure 10.2). This pattern has been documented from five cores and one refitting. The abandoned cores (Figure 10.3) are medium-sized with an average length of 3.5 centimetres and are cylindrical-shaped or flat. Some are slab-shaped, with an average length of 7 centimetres In some cases, the blades were removed alternately to each other.

— **Pattern B** (Figure 11): this shows a bipolar blade production on a core with two opposed striking platforms. There were two different ways of preparing the striking platforms (Figure 11.1). The platform was prepared by removing either one or several parallel flakes at an angle of 60 or 90 degrees to the intended exploitation surface. In some cases, the direction of flake removals for the platform preparation varied (Figure 11.2). The first striking platform may have been prepared by removals from the exploitation surface, while the second striking platform may have been prepared by removing the preparation flakes from the sides of the core. In one case, the direction of removals was changed, while preparing the same striking platform.

Before starting blade production, in some cases the core sides were prepared either by removing long crested blades from them along the intended direction of exploitation (Figure 11.3), or by removing several parallel flakes from

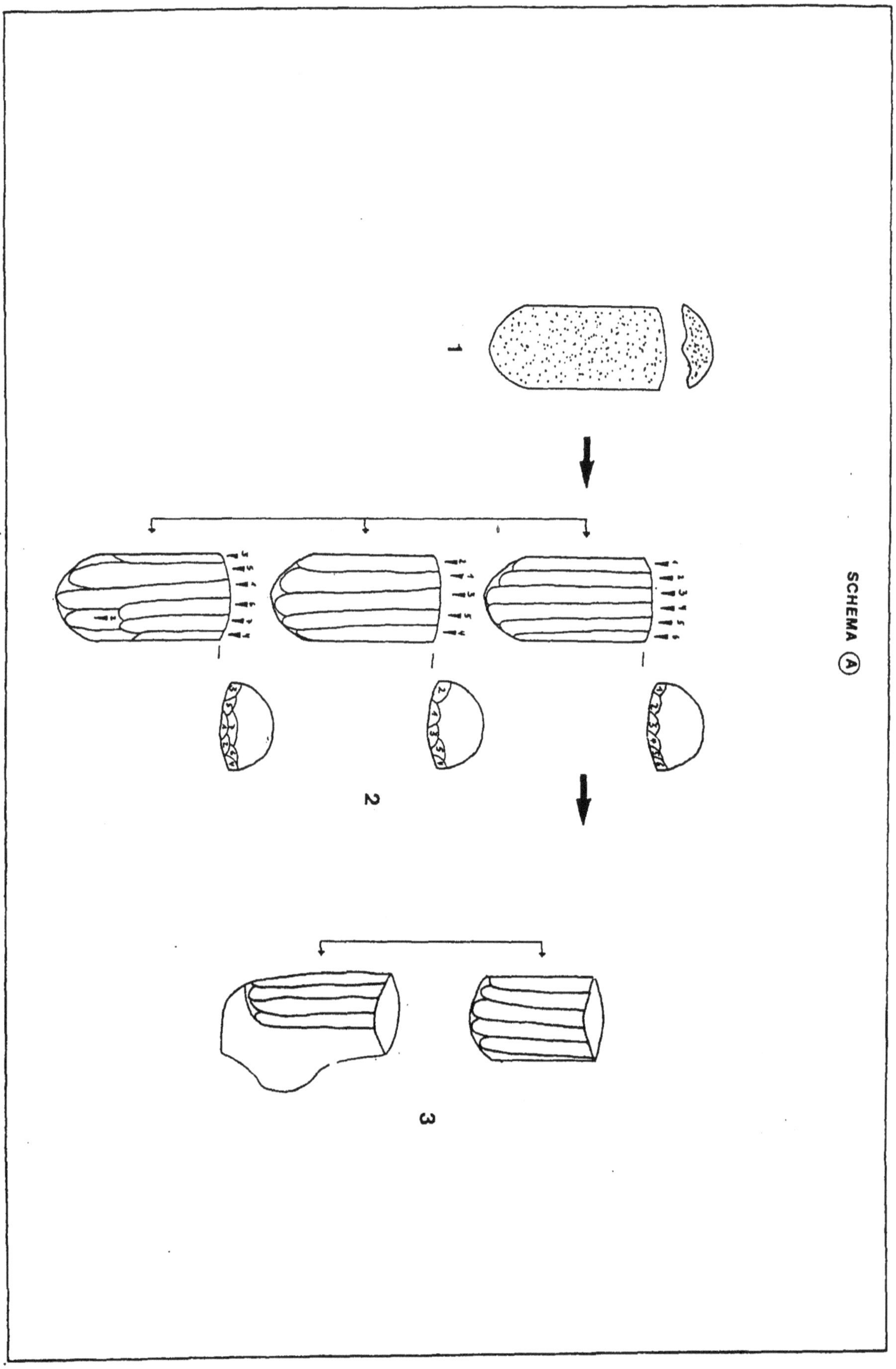

Figure 10: Pattern A of blank production at Jerxen-Orbke

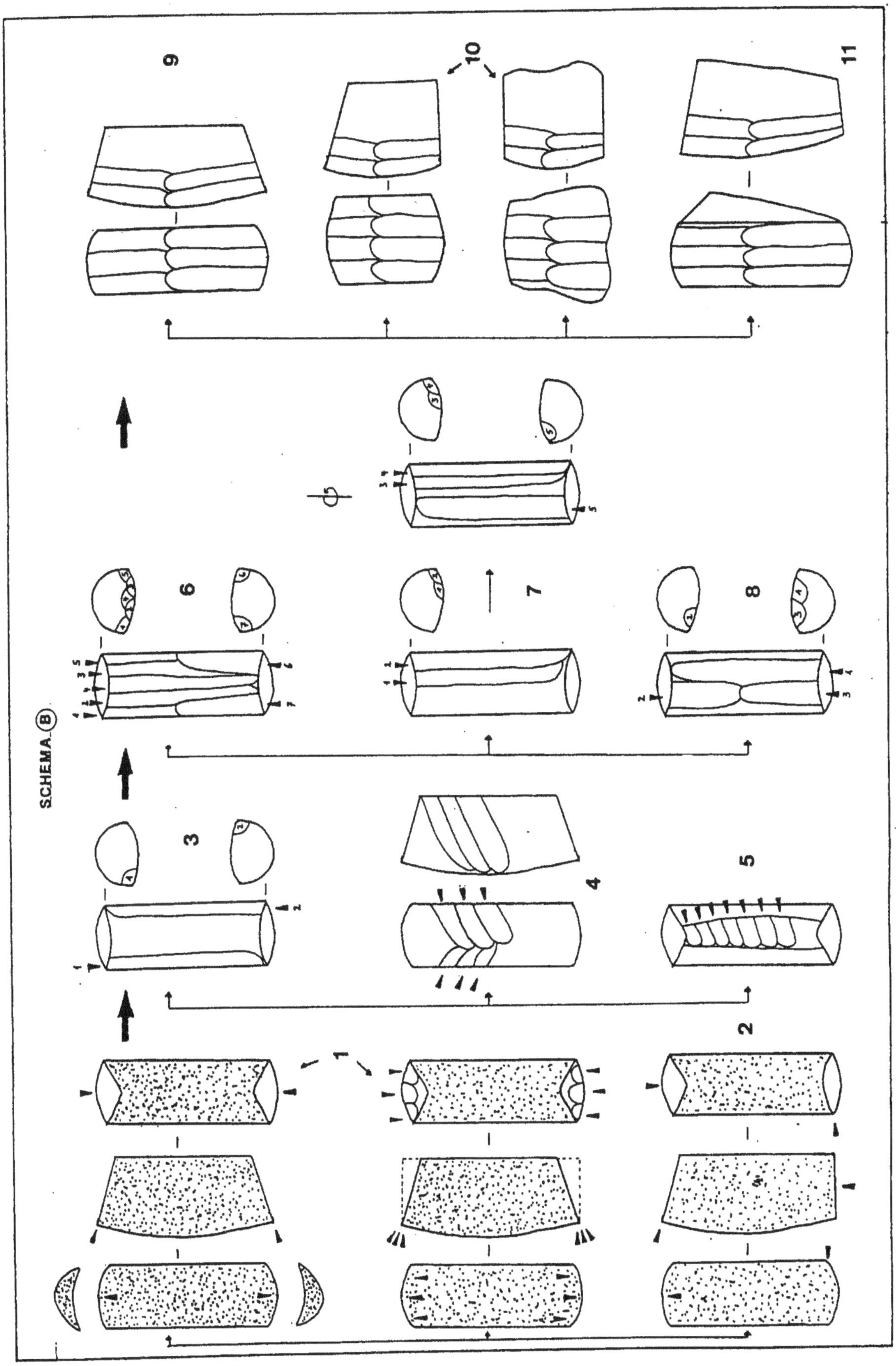

Figure 11: Pattern B of blank production at Jerxen-Orbke

the sides of the core, crosswise to the intended direction of exploitation (Figure 11.4), such that they intersect at their distal ends in the middle of the future exploitation surface. With these intersections, a guiding crest was created before starting the blade production. A very specific way of preparing the back of the core was also identified in this pattern of blade production, as well as in other patterns (Figure 11.5). A long, cylindrical-shaped core was prepared at its back by several parallel flakes, running crosswise to the direction of exploitation on the front. This method has also been described by N. Pigeot in her book about the Magdalenian site of Etiolles (Pigeot 1987: 35). Cores which have been prepared in this way are also found in assemblages of Late Palaeolithic sites from Poland, such as Swidry Wilkie I (Taute 1968: Pl.138 8; Sawicki 1935: Pl.V 7, 7a).

Blade production on cores was carried out in different ways:

1. The blades were removed parallel to each other until the exploitation surface had been exhausted (Figure 11.6). Following this, the core was rotated 180 degrees around the vertical axis and a second series of blades was removed from an opposite striking platform (Figure 11.6).

2. The blades were removed parallel or bipolar to each other, while utilising the back of the core to produce additional blades (Figure 11.7).

3. In this case, the blades were removed bipolar to each other. After the removal of every blade, the core was rotated 180 degrees around the vertical axis before the next blade was produced (11.8).

The abandoned cores are long, cylindrical-shaped with two opposite striking platforms which are both at 60 degree angles to the exploitation surface which have an average length of 6 centimetres (Figure 11.9). Other abandoned cores are sturdy, cylindrical-shaped, and also have an angle of 60 degrees between their striking platforms and the exploitation surfaces. Their average length is 3.8 centimetres (Figure 11.10). A third group of cores still has the form of the original nodule and approximately 60 % of their surfaces are still cortex covered. The angles between the striking platforms and the exploitation surfaces range between 60 degrees and 90 degrees. Their length ranges from 3.5 up to 7 centimetres (11.11).

— **Pattern C** (Figure 12): this pattern shows unipolar and bipolar blade production with a preparation of the exploitation surface. This pattern has been documented on 20 cores and from three refittings. First, the striking platform was prepared in one of four different ways (Figure 12.1). The first is the preparation of the striking platform by removing a flake at an angle of 60 degrees to the intended exploitation surface. In the second, the flake to prepare the striking platform is removed from the core side crosswise to the intended direction of exploitation (Figure 12.2). The third way shows the centripetal removal of several flakes from the intended striking platform (Figure 12.3). In this case, the striking platform lies at a 90 degree angle to the exploitation surface. In the fourth way, the striking platform is prepared by the removal of several parallel flakes at an angle of 60 degrees or 90 degrees to the exploitation surface (Figure 12.4).

Before starting blade production, the flanks of the core were prepared in one of two different ways: the first was to prepare the flanks by removing several parallel flakes, starting from the exploitation surface at an angle of 90 degrees (Figure 12.5). In the second way, a crested piece is removed from each flank of the core, along the direction of planned exploitation (Figure 12.6).

Actual blade production was carried out in one of two different ways:

1. The blades were removed from a single platform, parallel to each other (Figure 12.7).

2. The blades were removed unipolar and parallel to each other, utilising the back of the core as another exploitation surface to remove two blades in the opposite direction from that used on the main exploitation surface (Figure 12.8).

The main attribute of Pattern C is the way that the curvature of the exploitation surface was regulated. This was achieved by flakes being removed from the proximal end of the core, opposite to the direction of exploitation which is marked by small arrows in figures 12.7 and 12.8. The flakes which are produced in this process are either short and sturdy or long, small and flat. The abandoned cores are either cylindrical-shaped, with an average length of 4.5 centimetres (Figure 12.9) or wide and sturdy, with an irregular shape and an average length of 4.4 centimetres (Figure 12.10). One core is conical-shaped (Figure 12.11); other cores still show the original shape of the nodule with 60% of their surface covered with cortex (Figure 12.12). Their average length is 6.2 centimetres. Two cores had back preparation as defined above for Pattern B.

— **Pattern D** (Figure 13): this pattern exhibits unipolar blade production on the front side of the core, while the back of the core shows utilization for the production of a series of blades struck crosswise to the direction of exploitation on the front. This pattern was recognized on two cores, but it was not possible to carry out any refittings related to them. In one case a natural frostcleft was used as a striking platform without any preparations; in the second case, the striking platform had been prepared by the removal of a flake at a 60 degree angle to the exploitation surface (Figure 13.1). The blades were struck from a single platform (unipolar) along the exploitation surface (Figure 13.2). Following the exhaustive exploitation of the front flaking

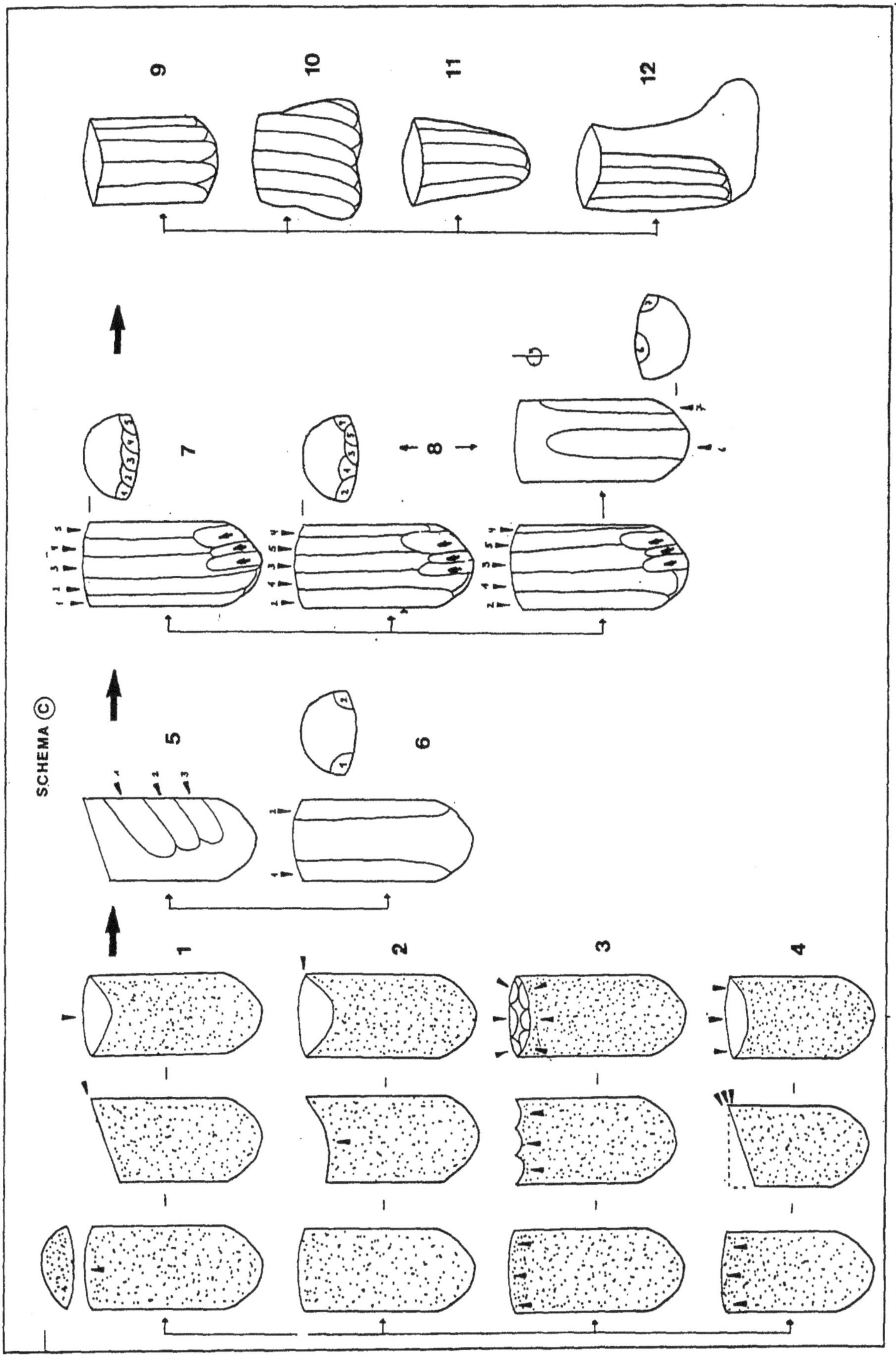

Figure 12: Pattern C of blank production at Jerxen-Orbke

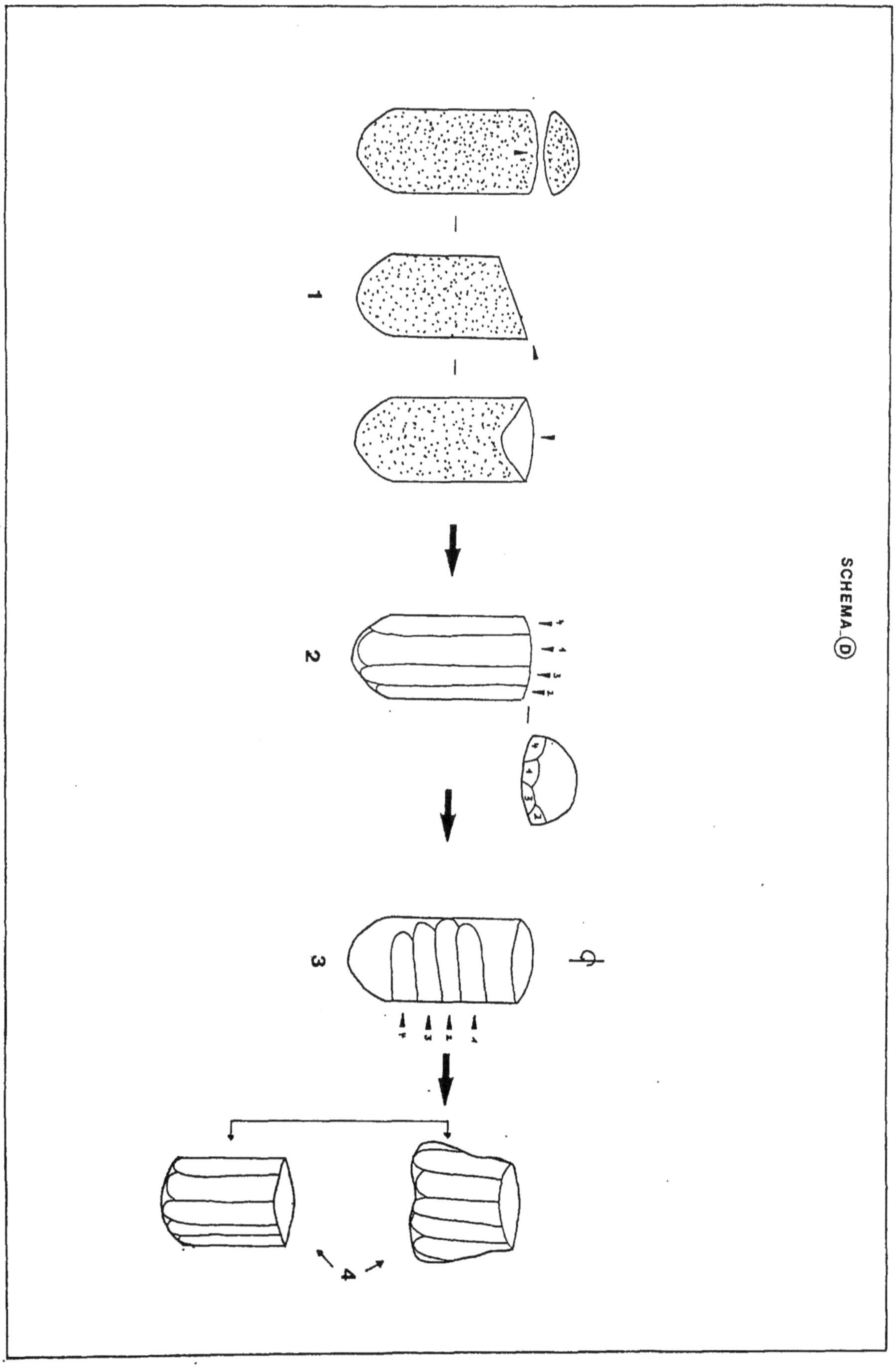

Figure 13: Pattern D of blank production at Jerxen-Orbke

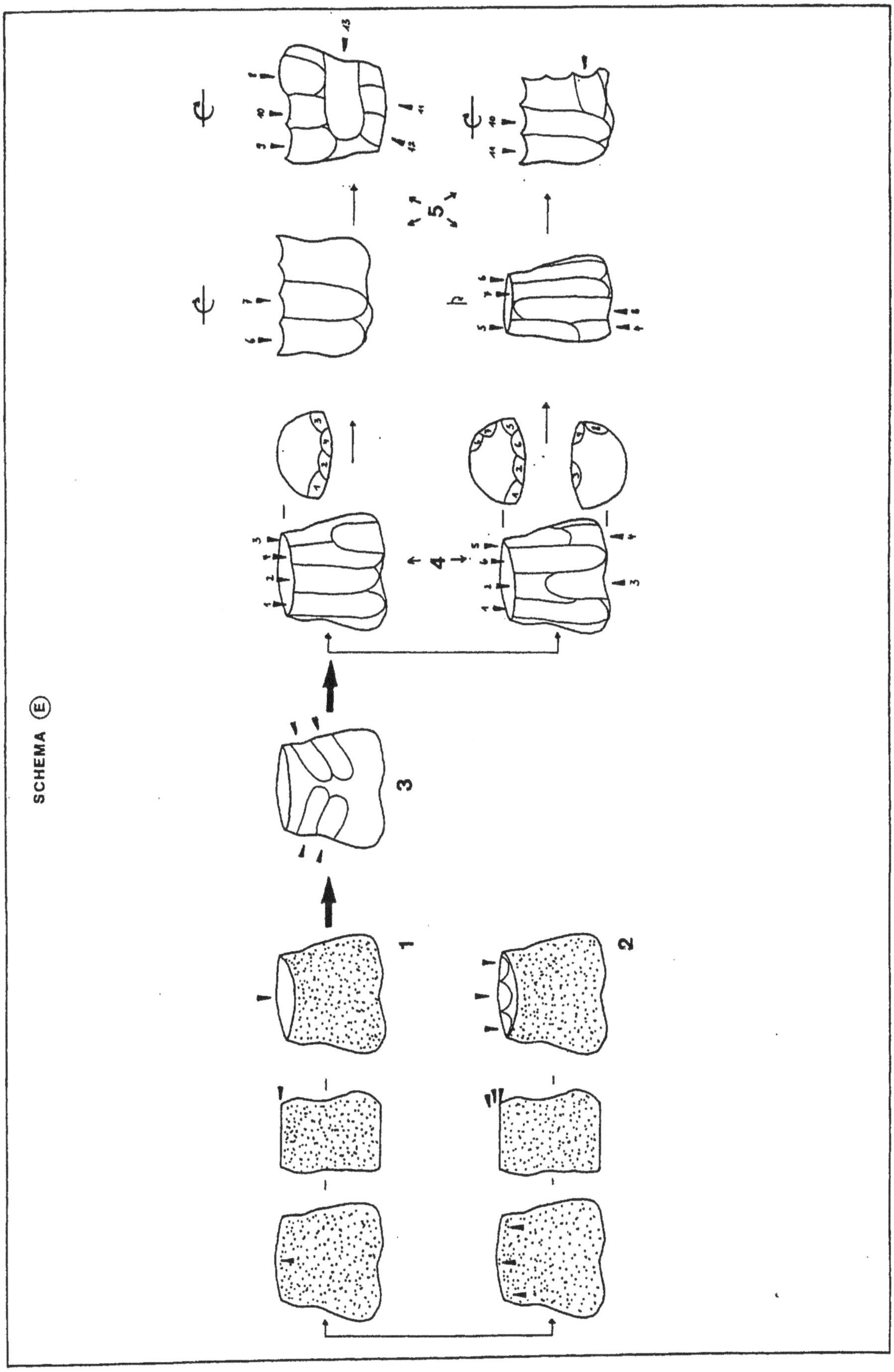

Figure 14: Pattern E of blank production at Jerxen-Orbke

surface, the core was rotated 180 degrees around the vertical axis and 90 degrees around the transversal axis to produce a second series of blades struck crosswise to the direction of the first exploitation surface on the front of the core (Figure 13.3). To do this, one flank of the core had been used as the second striking platform. The abandoned cores in this category are either irregular or cylindrical-shaped and sturdy (Figure 13.4).

— **Pattern E** (Figure 14): this is the most important pattern because it represents a very specific way of treating the cores during blade production. It involves an alternate unipolar and bipolar blade production, as well as a change in the core orientation, in order to use the original exploitation surface as a striking platform. Pattern E could be observed on nine abandoned cores and was present in one refitting. The reduction sequence was as follows:

1. First, a striking platform was prepared in one of two different ways: a) one cortical flake was removed from the nodule at 90 degrees to the flaking surface (Figure 14.1) and its scar was utilised as the striking platform for the production of the first series of blades; b) alternatively, the striking platform was prepared by removing several flakes parallel to each other at the same 90 degree angle to the intended exploitation surface (Figure 14.2).

2. In some cases, the core flanks were prepared by removing several flakes parallel to each other from the flanks, running crosswise to the direction of the blade removal (Figure 14.3).

3. At this point, blade production started: a) a first series of blades was removed from the striking platform (Figure 14.4), either uni- or bipolar; it is likely that although there was only a single striking platform, the blades were removed from the opposite end of the core in order to remove the cortex, although this could not be proved by the artefacts in the assemblage; b) following this, the core was rotated 90 degrees around the transversal axis and the first exploitation surface was turned into a second striking platform (Figure 14.5) from which a new series of blades was removed. The process of rotation was repeated several times during the exploitation of the core, using the scars of previous blade production as new striking platforms for the following production sequences.

The cores, having undergone this type of exploitation, always appear small and irregular. In all cases, the original nodule was no longer recognisable, indicating extensive reduction and exploitation.

A PECULIARITY: CORE EXPLOITATION WITH CHANGED ORIENTATION

There is a total of four patterns of core exploitation (Patterns A through D) which are well known for the Late Palaeolithic. The fifth pattern (Pattern E), core exploitation with changed orientation, has to be regarded as a very specific way to achieve exhaustive exploitation of the raw materials at Jerxen-Orbke. It seems that the flint-knappers at Jerxen Orbke aimed at a most effective production of high quality blades and flakes, in connection with intensive exploitation of the available raw material. To achieve this, they chose a method of core reduction utilising changing orientations of the exploitation surfaces on the same core.

In view of the fact that the assemblage at Jerxen-Orbke was initially dated only very generally to the Late Palaeolithic, because there were no typical tools that permitted dating to any specific Late Palaeolithic culture group, the reconstruction of the production sequences provides another possible avenue of linking the site with other, well dated Late Palaeolithic sites with the same reduction strategies. In fact, there are sites showing parallels with the exploitation strategy seen at Jerxen-Orbke. One example is a site of the tanged point groups at Reingsen, near Iserlohn (Western Germany) which was studied and described by R. Blank. In his work, he mentions the presence of several cores with two or more striking platforms (Blank 1985: 289). Unfortunately he does not give a detailed description of these cores; furthermore they are not illustrated. Another example is given by B. Ginter in his work on the extraction, production and distribution of raw materials and flint products in the Late Palaeolithic in northern Europe. He uses the expression "cores with changed orientation" (Ginter 1974: 81) for some cores in the assemblages of several sites which he examined but, like R. Blank, he does not go into further detail. Nevertheless, there are some well dated Late Palaeolithic assemblages containing cores with several striking platforms, a point in need of further attention.

ACTIVITY AREAS AT JERXEN-ORBKE

Aside from the results of the technological analysis, there was a second very important aspect of Jerxen-Orbke that had to be addressed: the possibility of defining different activity areas within the excavated area. With the help of distribution maps of individual groups of artefacts, it was possible to define a pattern of several activity areas and their connections to each other. Four kinds of activity areas have been identified (Figure 15):

1) Two areas of habitation (huts);
2) Three areas of supply;
3) Three zones of consumption (production);
4) One depression (of unknown function)

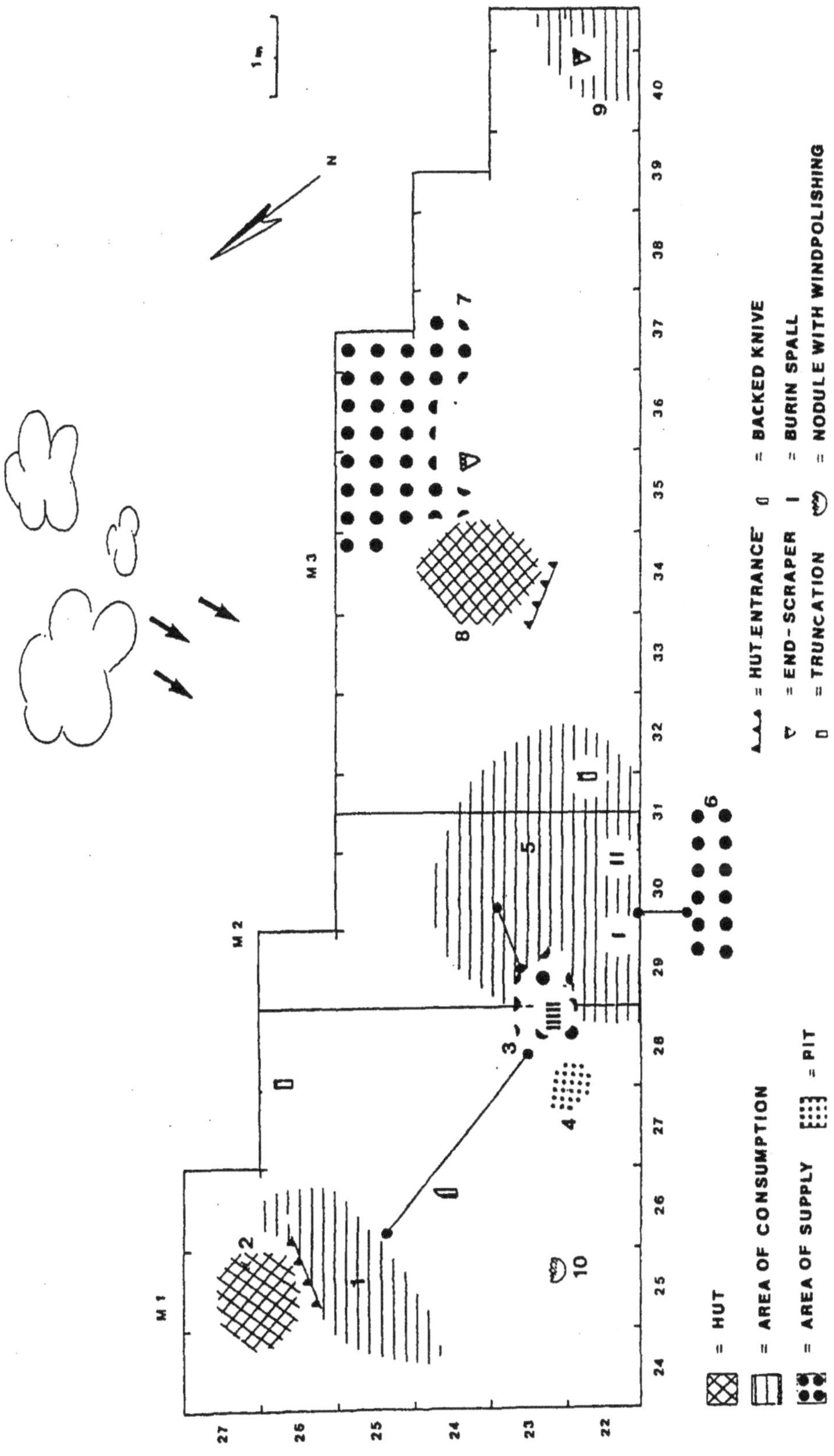

Figure 15: Activity areas at the site of Jerxen-Orbke

The function of the activity areas were defined as follows:

1) The definition of a habitation area depends upon the size, structure and the relationship of the depression to the surrounding concentrations of artefacts.

2) An area of supply is a concentration of nodules, chunks and cores. Except for testing the quality of the raw material by smashing the nodules coming from the moraine before they were transported partly to the areas of consumption, no other reduction of them took place here.

3) An area of consumption (production) is a concentration of artefacts, indicating intensive core reduction through the production of blanks. It is a predominant concentration of cores, debris, blades, flakes and tools.

4) All the areas which are too small to be interpreted as a trace of a former hut are defined as simple depressions.

Following this, the relationship between the different areas was examined. The three areas of consumption are equally distributed all over the whole of area M, as shown in the distribution map (Figure 15). The first area (1) is situated in the northern part of area M1, directly adjacent to one of the two huts (2), with it's entrance turned towards the consumption area. This pattern was defined with the help of the distribution maps of several artefact groups, retouched as well as unretouched, and the distribution map of the chips. All these show a scatter of artefacts between areas 1 and 2, which was presumably caused by walking.

The consumption area (1) was supplied from the zone of supply (3), which is situated near the small depression (4) in the south-eastern part of area M1 and which spreads over the edge of area M2. This area of supply contained a total of eight raw material units, four of which could also be identified in the consumption area (1).

The second consumption area spreads mainly over the middle and southern part of area M2 and approximately 1.5 metres into area M3. It contained a number of cores, debris, basic artefacts, three burin spalls and one truncation. It had been supplied by supply area 3 and 6, which were identified with the help of the refitting map. A third supply area (7) was located in the north-eastern part of area M3, directly next to the second hut (8). In this area (7), some basic artefacts and a large number of chunks, presumably belonging to a single nodule, were found. In this case, a relationship to the other zones in area M could not be demonstrated.

The third consumption area (9) was situated at the eastern edge of area M3, isolated in the same way as supply area 7. Both areas contained a high number of burned pieces, indicating a relationship to fireplaces that were probably located outside the excavated area. A fragment of a nodule (10), that had been found in the southern part of area M1, is remarkable because it has wind polish on one side which indicates that during the Late Palaeolithic phase of settlement the wind mainly came from the east. In view of the fact that there are no demonstrable relationships among areas 7, 8 and 9 and the remaining zones, this can be interpreted as an indication of different, short phases of settlement with individual huts and concentrations of artefacts.

The main activity at the site was presumably the mining and primary exploitation of raw materials coming from the nearby ground moraine. But it cannot be excluded that other activities like hunting or working on skins were carried out at this place because a lot of endscrapers were identified in the assemblage. Unfortunately no traces of hunting activities could be identified because the occupation layer was embedded in a layer of loess without any lime, so that no organic materials like bones were preserved.

TYPOLOGICAL PARALLELS WITH WELL DATED LATE PALAEOLITHIC SITES

Apart from the technological analysis, an attempt was made to find some typological parallels with other well dated Late Palaeolithic sites to permit assigning Jerxen-Orbke to some specific Late Palaeolithic culture group. The typological examination concentrated on the tools, most of which are endscrapers (64%). Normally they are made on short, flat, and in some cases cortical, flakes. Except for six examples, the endscrapers have a flat, broad, fan-shaped edge. Of the former, two small endscrapers are thumbnail, one is made on a long and slender blade, and two are made on elongated flakes with extremely steep retouched edges. The latter are very unusual for the endscrapers of Jerxen-Orbke. The sixth endscraper is made on a relatively large and broad flake. Since most of the endscrapers at Jerxen-Orbke are made on short and flat flakes, a general dating to the Late Palaeolithic seems obvious (Hahn 1991: 179).

The two thumbnail endscrapers indicate a younger phase of the Late Palaeolithic. Normally, they are found in Mesolithic assemblages, but in some cases they also known from Late Palaeolithic assemblages (Hahn 1991: 179). The two examples from Jerxen-Orbke are not patinated, so they certainly belong with the Late Palaeolithic artefacts.

A comparison between the Jerxen-Orbke tool assemblage and several well dated Late Palaeolithic sites showed a very close correspondence to a site of the tanged point groups in Northern Germany, Deimern 45, which was excavated by W. Taute in 1959 and published in 1968 (Taute 1968: 19-38). Most of the endscrapers from Deimern 45 were made on flat, short fan-shaped flakes (Taute 1968: Plate 13), like the pieces from Jerxen-Orbke. This is an attribute considered by N. Arts as typical of the Ahrensburgian, as seen in the endscrapers from the Ahrensburgian site of Vessem-Rouvwen in the Netherlands (Arts 1985: 305 and

Figure 18.4-5). Another unusual endscraper from Jerxen-Orbke shows ventral retouch, similar to that on the dorsal face of the endscraper. There are two comparable examples: one from Deimern 45 (Taute 1968: Plate 14.3) and the other from Remouchamps (Dewez 1987: Plate 138.32). Furthermore, the Jerxen-Orbke assemblage contained two truncations, corresponding very closely to one example from Deimern 45 (Taute 1968: Plate 5.7).

Another unusual artefact from Jerxen-Orbke ends in a distal point with a distinct ridge running right through the middle of the dorsal side, dividing it into two sections. The right part is covered by broad retouch at the distal end. The proximal end is missing. A comparable example had been described by Z. Sulgostowska in the tanged point assemblage of Cekanowo, a site in Poland (Sulgostowska 1989: 111 and Plate I.7). She places this piece among the points, a position that cannot be given for the example from Jerxen-Orbke because it is broken. The Jerxen-Orbke assemblage contains, furthermore, some flakes with facetted striking platforms that are nearly identical to one piece from Deimern 45 (Taute 1968: Plate 14.4).

The typological parallels seem to be very striking arguments to assign Jerxen-Orbke to the sites of the tanged point groups, although no tanged points were contained in the assemblage. Jerxen-Orbke, however, is not the only site exhibiting this phenomenon. Another example is known at Höfer, a site in Northern Germany, which was excavated and published by S. Veil (Veil 1987: 311). This suggests that sites of the tanged point groups, but without any tanged points, are a facies that should be given more attention in the future.

SUMMARY

The open-air site of Jerxen-Orbke (Stadt Detmold) was discovered by an employee of the Lippisches Landesmuseum in Detmold, Northern Germany, in 1988. She had carried out a series of surveys after flint artefacts and pieces of pottery came to the surface during construction work in 1987. Three surface concentrations were found during her surveys, one of which contained fresh Late Palaeolithic materials. In 1990 a series of excavations started, carried out by a team of the Prehistoric Institute of the University of Cologne. These led to the discovery of a Late Palaeolithic open-air site with several artefact concentrations, one flint-knapping place and two traces of former huts. The occupation layer was identified at a depth of between 50 and 70 centimetres below the surface, embedded in a layer of limeless loess, so that no organic remains were preserved.

The assemblage of Jerxen-Orbke contained some endscrapers, truncations and backed knives which are typical of the Late Palaeolithic, but there were no tools which would have allowed this site to be assigned to any specific Late Palaeolithic culture group. Thus, the main emphasis of the analysis was placed on a detailed examination of the technological aspects of artefact production. With the help of this analysis, it was possible to identify five patterns of core exploitation, four of them being characteristic of general Late Palaeolithic methods. These patterns include various unipolar as well as bipolar core reduction strategies with different methods of core preparation.

The fifth pattern, core exploitation with changing flaking orientations by transforming former exploitation surfaces into striking platforms, is a main attribute of this flint-knapping site. This shows some parallels with a few well dated Late Palaeolithic sites of the tanged point groups. A further result of the technological analysis was the identification of different activity areas. They indicate several different activities, such as intensive production of blanks and the testing of flint nodules collected from the nearby moraine. There are some indications of behavioural connections between individual activity areas which indicate several short phases of occupation.

The dating of Jerxen-Orbke was made possible with the help of technological as well as typological comparisons with other well dated sites of the groups of tanged points like Deimern 45 (Northern Germany), Remouchamps (Belgium) and some sites of the Ahrensburgian culture in Poland, like Swiedry-Wilkie and Cekanowo. All these parallels indicate a date to the transition between Dryas III and the early Preboreal.

Acknowledgements

Work at Jerxen-Orbke was initiated and encouraged by the late Wolfgang Taute who supervised my M.A. thesis about this site. My special thanks are due to Anthony E. Marks for adapting the English text and for constructive comments. I should also like to thank all those who took part in the field campaigns (1990-1992) and all those who contributed advice and critique, especially Thorsten Uthmeier and my husband, Jürgen Richter.

REFERENCES

Arts, N.

1985 A survey of Final Paleolithic Archeology in the Southern Netherlands. In M. Otte (ed.) *De la Loire a l'Oder. Les civilisations du Paléolithique final dans le nord-ouest Européen.* Koll. Liège 1985, BAR 444, Oxford, 287-356.

Blank, R.

1985 Ein Fundplatz der endpaläolithischen Stiel-spitzen am nördlichen Mittelgebirgsrand. *Archäologisches Korrespondenzblatt* 15, 287-292

Boeda, E., J.-M. Geneste & L. Meignen

1990 Identification de chaînes opératoires lithiques du Paléolithique ancien et moyen. *Paleo* 2, 43-80.

Cziesla, E.
1986 Über das Zusammenpassen geschlagener Steinartefakte. *Archäologisches Korrespondenzblatt* 16, 251-261.
1990 On refitting of stone artefacts. In E. Cziesla, S. Eickhoff, N. Arts and D. Winter (eds.) *The Big Puzzle. International Symposium on Refitting Stone Artefacts*. Bonn, 9-44.

Dewez, H.
1987 *Le Paléolithique superieur recent dans les grottes de Belgique*. Publications d'Histoire de l'Art et de l'Archeologie d'Université catholique de Louvain 57.

Farrenschon, J.
1986 *Erläuterungen zur geologischen Karte von Detmold, Blatt 4091*. Krefeld.

Ginter, B.
1974 The extraction, production and distribution of raw materials and flint products at the Late Paleolithic in the northern part of Central Europe. *Nabitka* 22, 5-122.

Hahn, J.
1991 Erkennen und Bestimmen von Stein- und Knochenartefakten. Einführung in die Artefaktmorphologie. *Archeologica Venatoria* 10, Tübingen.

Hartz, S.
1987 Neue spätpaläolithische Fundplätze bei Ahrenshöft, Kr. Nordfriesland. *Offa* 44, 5-52.

Kind, C.-J.
1985 *Die Verteilung von Steinartefakten in Grabungsflächen. Ein Modell zur Organisation alt- und mittelsteinzeitlicher Siedlungen*. Tübingen.

Kooi, P.B.
1974 De Orkaan van 13. November 1972 en het Ontstaan van "Hoefijzervormige Grond-sporen". *Helinium* XIV, 57-65.

Pigeot, N.
1987 Magdalénien d'Etiolles. Economie de debitage et organisation sociale. *XXV'e supplément à Gallia Préhistoire*. Paris.

Rieder, K.H.
1992 *Kritische Analyse alter Grabungsergebnisse aus dem Hohlen Stein bei Schambach aus der Sicht der Profiluntersuchungen von 1977-1982*. Dissertation Tübingen.

Sawicki, L.
1935 L'industrie swidrienne de la station Swidry Wielkie I. *Obitka z "Przegladu Archeo-lodicznego"* V, zesz 1.

Sulgostowska, Z.
1989 *Prahistoria miedzyrzecza Wisly, Niema i Dniestru u schylku plejstocenu*. Warzawa.

Taute, W.
1968 Die Stielspitzengruppen im nördlichen Mitteleuropa. Ein Beitrag zur Kenntnis der späten Altsteinzeit. *Fundamenta A* 5, Köln-Graz, 19-38.

Thieme, H.
1990 Wohnplatzstrukturen und Fund-platzanalysen durch das Zusammensetzen von Steinartefakten; Ergebnisse vom mittelpaläolithischen Fundplatz Rhein-dahlen B1 (Westwandkomplex). In E. Cziesla, S. Eickhoff, N. Arts and D. Winter (eds.) *The Big Puzzle. International Symposium on Refitting Stone Artefacts*. Bonn, 543-568.

Veil, S.
1987 Ein Fundplatz der Stielspitzengruppen ohne Stielspitzen bei Höfer, Ldkr. Celle. *Archäologisches Korrespondenzblatt* 17, 311-322.

RAW MATERIAL ECONOMY AT THE MAGDALENIAN SITES OF HAUTERIVE-CHAMPRÉVEYRES AND NEUCHÂTEL-MONRUZ (SWITZERLAND)

Marie-Isabelle Cattin

INTRODUCTION

The originality of the Magdalenian sites at Champréveyres and Monruz (on the shore of Lake Neuchâtel) lies not only in the diversity of the allochthonous raw materials used but also in the large quantities in which they are present in the assemblages. The technological study and the refitting made it possible to reconstruct the mode of introduction of the different types of raw material to the sites and their respective economic management. Among the raw materials exploited, only one of them has a strictly local provenance (with outcrops situated less than 5 kilometres from the site). The other raw material sources are located along the Jura, from the Bellegarde/Seyssel region (department of Ain, France; southern flint) to Olten (canton of Soleure, Switzerland; northern flint). These two regions represent the extreme limits of the raw material procurement area and are situated about 100 kilometres to the south-west and 60-80 kilometres to the north-east of the sites respectively (Affolter 1989).

CONTEXT

Located one kilometre away from each other, the sites of Champréveyres and Monruz are located at the north-western extremity of the lake of Neuchâtel (Figure 1). They were discovered along the route of the A5 highway and were the object of rescue excavations carried out by the *Service Cantonal d'Archéologie de Neuchâtel* (Affolter *et al.* 1994; Benkert *et al.* 1984; Egloff 1989: 29-35; Leesch 1997; Moulin 1991). Close to the shore, they lie only a few hundred metres from the lower slopes of the Jura. The site at Champréveyres has three distinct zones (sectors 1, 2 and 3) where Palaeolithic settlements dated to the Magdalenian and Azilian have been found, although the Azilian level has only been identified in sector 1. At the site of Monruz there is a Magdalenian and a small Epipalaeolithic settlement (Affolter *et al.* 1994), although at present the cultural attribution of the latter horizon is uncertain since the toolkit mainly consists of short endscrapers on flakes but with no associated microliths. At both sites the archaeological levels are preserved only in natural depressions which characterised the landscape after the retreat of the Rhone Glacier; thus the total extension of the sites is unknown and it is likely that the occupied area along the shore of the lake was much larger. This article will focus on the levels attributed to the Magdalenian in sector 1 at Champréveyres and at the site of Monruz.

The Magdalenian settlements date to the end of the last glaciation, and more precisely to around 13,000 BP according to the radiocarbon dates obtained on wood charcoal and pollen analysis (Affolter *et al.* 1994; Gaillard & Moulin 1989; Leesch 1993: 201). According to the palaeobotanical data, in this period the vegetation was dominated by herbaceous plants together with dwarf birch and willows, in a steppe landscape without trees (Gaillard & Moulin 1989). Among the hunted fauna horses prevail over reindeer, marmots, hares, polar foxes and ibex (Morel & Müller 1997).

The lithic and bone industries are attributed to the recent Magdalenian, and the sites are characterised by an abundance of backed bladelets (> 50%) and the absence of curved backed points, angular backed points and shouldered points (Leesch 1993). Although the two settlements are comparable from the point of view of the kinds of animals hunted and the lithic industries, Monruz differs markedly from Champréveyres because of the presence of sunken hearths (*foyers en cuvette*) and ornamental objects (Affolter *et al.* 1994; Egloff 1995). The two sites are connected by the refit of two blades, one from each site, which demonstrates their strict contemporaneity (Cattin 1992).

THE SITE AT CHAMPRÉVEYRES

The flint artifacts collected at the site amount to 5562 pieces larger than 1 centimetre (total weight 16.4 kg). These objects fall into three broad categories: unretouched blanks (75.7%), tools (10.6% defined according to the type list and 4.4% whose use has been demonstrated by use-wear analyses) and waste derived from tool production, rejuvenation and use (9.3%) (Figure 2). Although the assemblage is dominated by unretouched blanks, the abundance of tools and their waste suggests that not all of the production process took place on-site. In contrast, there is on-site evidence for all the activities carried out in relation to the exploitation of hunted game. Tool production was mainly oriented towards the production of bladelets, transformed into backed bladelets, and blades, used unretouched for butchering activities.

The petrographic analysis of the flint (J. Affolter) led to the identification of 18 types, in addition to a group of unidentified raw material (MP 000) of allochthonous origin, but not attributed to one of the known types due to surface alterations (firecracking, thick patina etc.). The quantities of

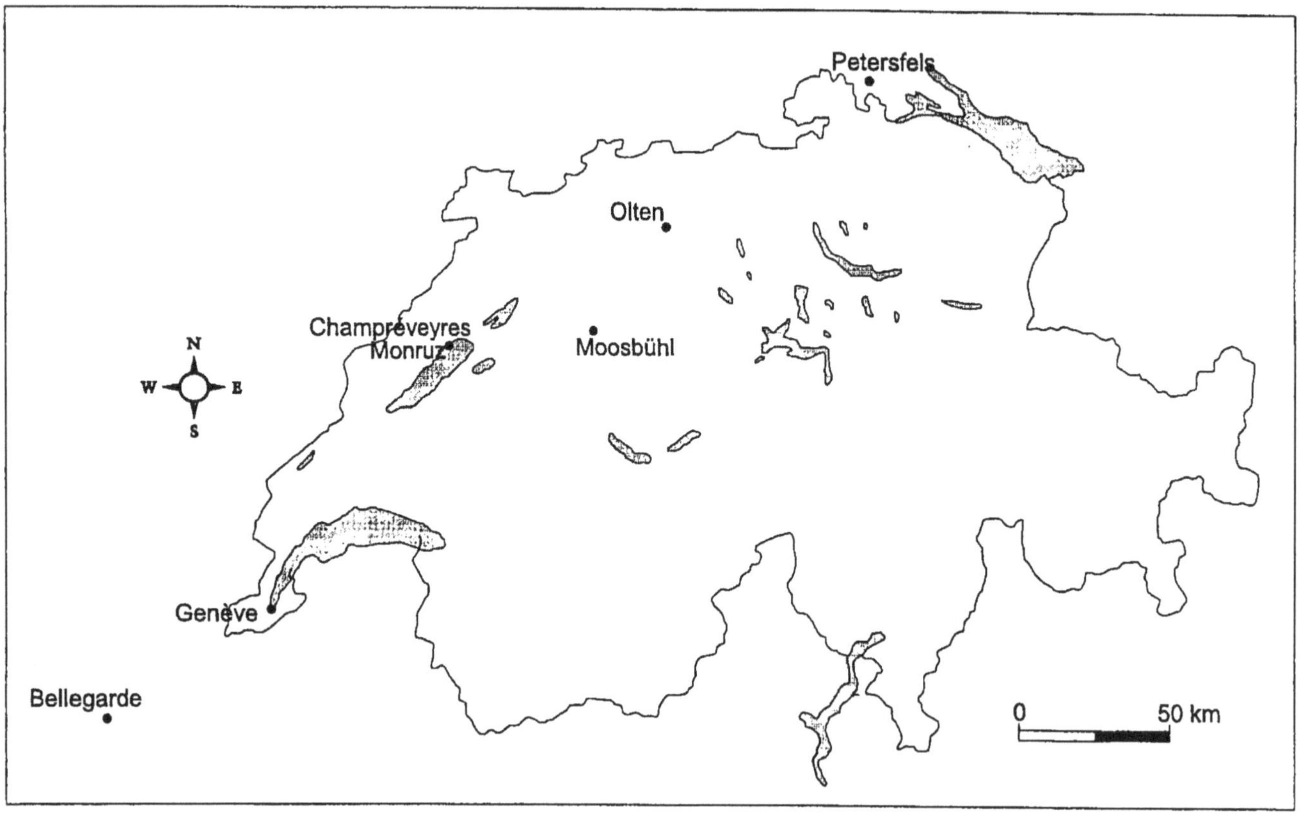

Figure 1: Map showing the places mentioned in the text

the different types of raw material are very variable and range between 1 and 2220 pieces in number weighing between 2.7 and 12,300 grams in weight (Figure 3). The local raw material (Hauterivian, MP 401) is dominant in this assemblage with a value of 40%; it is immediately followed by the allochthonous flint from the south (Turonian-Lower Senonian, MP 201) and from the north (Kimmeridgian-Portlandian, MP 101), which are present with values of 19.8% and 17.4% respectively. Finally the other raw materials, which were found in much lower quantities, are each represented by less than 4%.

If one compares the quantity of allochthonous flint with the local raw material, there is a predominance of the former over the latter (Figure 4a). On the other hand, local flint outweighs the allochthonous flint (Figure 4b). This reversal in the situation can be explained by the mode of introduction of the raw materials onto the site (size of transported nodules, preformed cores and/or partially prepared cores, transport of blanks). In fact the blocks of local flint are of a larger size than those of allochthonous raw material. This is due to the relative proximity of the outcrops, which also allowed for a less rigorous selection, as is indicated by the presence of cores which were rejected after the removal of only a few flakes. In contrast, the Magdalenians seem to have paid much greater attention to their choice of raw material blocks from more distant sources.

RAW MATERIALS AND SITE OCCUPATION

The bone remains indicate that the horses and reindeer were killed in the spring and in the autumn (Morel & Müller 1997); therefore it is possible to suggest that Champréveyres was occupied during a series of short term visits which are not, however, revealed by the stratigraphy (Leesch 1997). A single long term occupation (from the spring until the autumn) can be excluded because the distribution of the lithic artifacts and faunal remains do not show any marked accumulations (Conkey 1992). The equal proportions of flint coming from opposite directions, one from the south (MP 201) and the other type from the north (MP 101) (Figure 3), may be an indicator that the site was occupied in two distinct seasons. However, the spatial distribution of these raw materials does not suggest that certain structures were related to a particular season as both materials had been exploited around the hearths where the knapping waste of at least three cores was found. It is unlikely that during a single occupation a group of hunter-gatherers would have brought equal quantities of the two types of flint, unless they obtained it by means of exchange or specific expeditions done by special work groups (Binford 1980). If one envisages a route which would have passed through one of the two source areas, which lie 200 kilometres apart, before arriving at Champréveyres, we would expect one of the raw material types to predominate over the other. In order to justify the

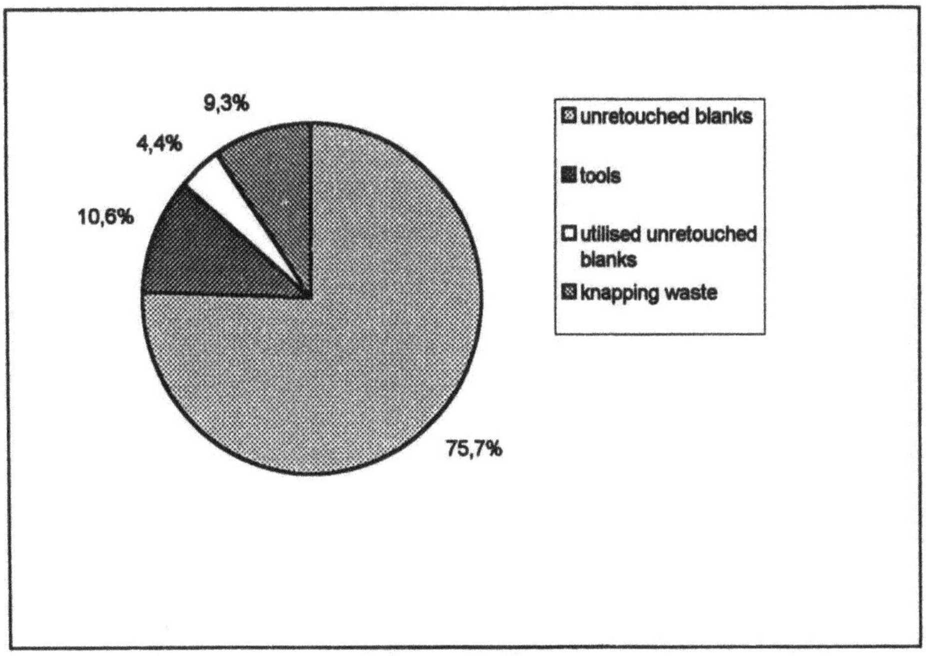

MP	unretouched blanks	tools	utilised unretouched blanks	knapping waste	total
401	2026	75	62	57	2200
201	799	136	54	115	1104
101	609	146	73	157	985
301	151	25	14	10	200
501	111	33	7	16	167
105	78	46	8	25	157
304	122	25	1		148
109	25	15	6	42	88
305	25	8	1	7	41
503	15	6	6	5	32
226	19		2		21
108	2	11	3	5	21
306	5			10	15
504	4	4	2		10
203	4				4
407	3				3
204	1				1
303		1			1
000	210	60	5	69	344
Total	4209	591	244	518	5562

Figure 2: Composition of the different assemblages for each raw material type

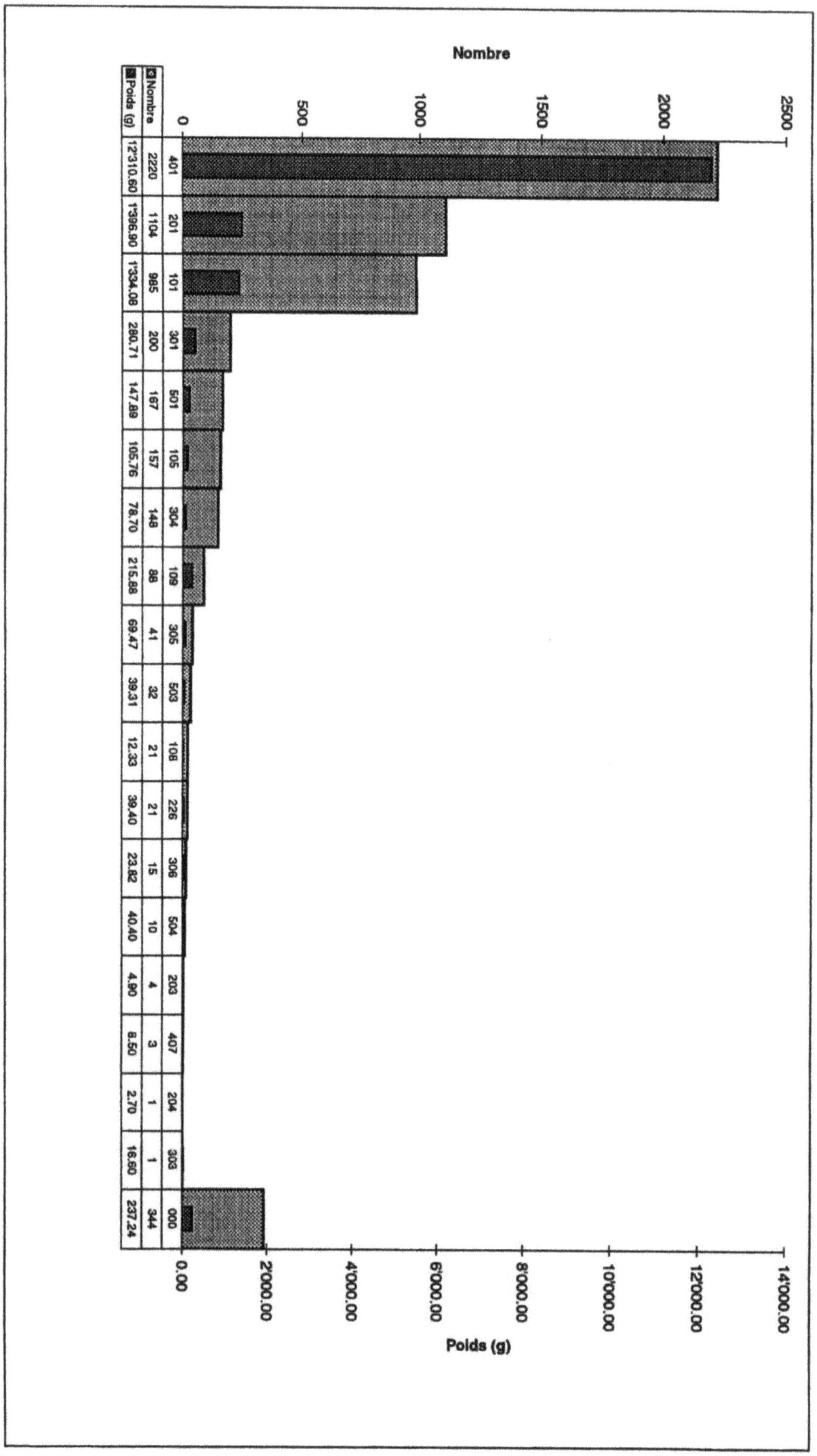

	401	201	101	301	501	105	304	109	305	503	108	226	306	504	203	407	204	303	000
Nombre	2220	1104	985	200	167	157	148	88	41	32	21	21	15	10	4	3	1	1	344
Poids (g)	12'310.60	1'396.90	1'334.08	280.71	147.89	105.76	78.70	215.68	69.47	39.31	12.33	39.40	23.82	40.40	4.90	8.50	2.70	16.60	237.24

Figure 3: Proportion of the different types of raw material by number and weight

Origin	Number	Number %	Weight (g)	Weight %
Local	2220	39.91	12310.60	75.22
Allochthonous	2998	53.90	3817.35	23.33
Undetermined	344	6.18	237.24	1.45
Total	5562	100.00	16365.19	100.00

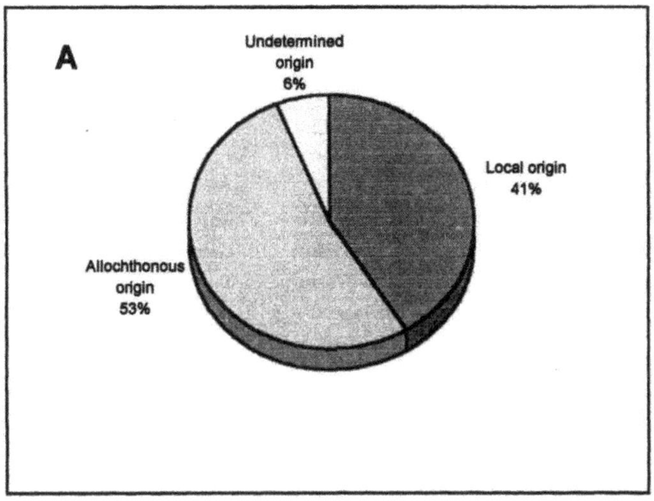

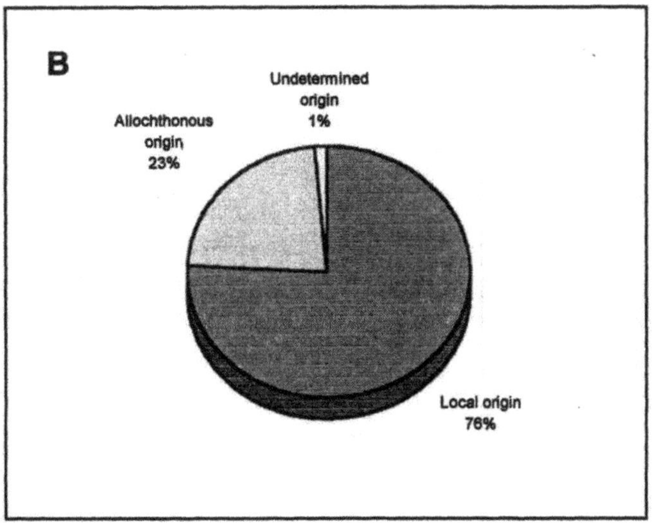

Figure 4: Quantitative (A) and ponderal (B) proportions of the raw materials according to their origin

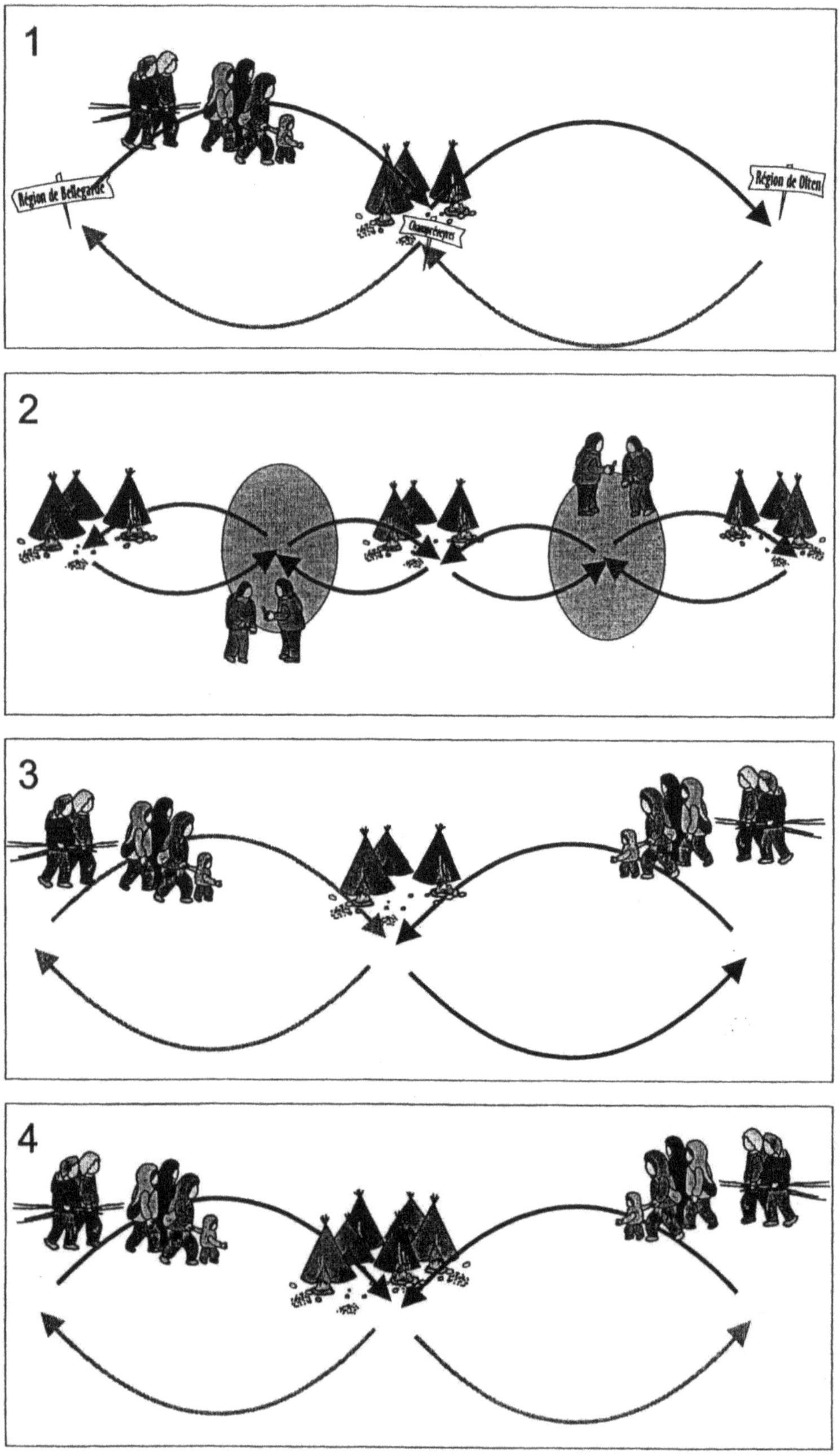

Figure 5: Hypotheses for the occupation of the site

presence of equal proportions of the two raw material types during the same occupation, we must conceive either the concomitant presence of two groups (one coming from the north and the other from the south) and the practice of exchange.

In contrast, if one attributes the flint from the north to one occupation and the flint from the south to another, the following systems of raw material procurement can be hypothesised (Figure 5):

- the equal quantities of raw material could be explained by the movements of a single group which came from the north and stopped at Champréveyres before moving south; later, returning to the north, they would have stopped again at this site; the territory covered by the group would therefore have extended over a distance of 200 kilometres;
- if the territory frequented by a single group was smaller than in the previous case, one could envisage the practice of exchanges in common territories towards the north and south, where the Magdalenians from Champréveyres met other groups;
- if the site was occupied by two or more groups, one could conceive that Champréveyres corresponded to one of the halts on a circuit chosen by mobile groups of hunter-gatherers, one towards the south and the other towards the north; the movements would not have been more than a hundred kilometres in length;
- one could also suggest the meeting of groups coming from the north and from the south at different times of the year.

Although the sequence of occupations finds convincing arguments through archaeological data, the number of groups that occupied the site, their trajectories and their social relationships all remain hypothetical. In addition, to explain the presence of both raw material types near some of the hearths, one could envisage the re-use of these structures. This conjecture can be corroborated by the superimposition of undisturbed knapping areas on others where the artifacts were well disturbed.

As regards the modes of raw material procurement, one could still question the significance of the abundance of raw material from very distant sources (northern and southern flint) while the allochthonous flint from closer sources is less represented. The latter flint sources, which were located along the routes of the groups, were not very important and perhaps seemed less interesting to the hunter-gatherers. The assemblage of raw materials utilised might indicate that they filled their baggage on sites where flint was abundant, and collected the rest along the route towards their destination in an opportunistic fashion. The presence of other allochthonous raw material may therefore represent the result of procurement at these halts. These hypotheses all suggest direct procurement of the raw material, although they may also indicate exchanges of one or two cores only, which would explain their scant representation.

In Switzerland there are many sources of flint though most are of limited extension. They are mainly found in the Jura Mountains (especially in the Malm levels), and are quite rare in the Alpine massif and on the molasse plateau. Furthermore, in these last two areas, the tectonic-modified raw material means that it is not of good flaking quality (Affolter 1995). This relative rarity of sources and the irregular quality of the raw materials influenced the procurement strategies. At Champréveyres the abundance of allochthonous raw materials seems to demonstrate that the Magdalenians kept stores of good quality raw material for future uses (especially in cases when they knew that they would find a more mediocre quality, as at Champréveyres). A similar strategy has been found at the site of Moosbühl (canton of Berne), which is located in a region where raw material sources are rare or even absent (Barr 1973). The following examples illustrate the different practices with regard to raw material.

MODES OF INTRODUCTION OF THE RAW MATERIALS

The separation of the artifacts into three groups, unretouched blanks, tools and tool production waste, according to raw material types gives a first impression of the condition in which they arrived at the site (Figure 6), acknowledged by the very diverse proportions of these groups. Refits were then used to specify more precisely the modes of introduction. It goes without saying that the local raw material was brought onto the site in an unmodified or only slightly modified form, which is not the case with the allochthonous materials.

The local Hauterivian flint: immediate consumption and storage

This coarse grained flint outcrops on the slopes of the Jura Mountains in the surroundings of the site. Despite its mediocre quality, it was intensively exploited; this is indicated by the high quantity of knapping waste products (91.3 %) and refits (32 %) (Figure 7). In contrast, there are very few tools (6.1 %); one can conclude that there was either little transformation of the blanks produced or a selection of blanks to be used at other sites.

Introduced in the form of unmodified blocks or perhaps just tested for their quality at the source, the blocks demonstrate that the reduction process was oriented towards the production of both blades and bladelets; however, there is a slight preference towards the former for the production of tools. The dominant tool class is that of the unretouched blanks (45.26%) whose use was established by the use-wear analysis, and it is likely that this raw material was above all used to make blanks, and in particular blades, which were destined for immediate use (Cattin 1994). This type of flint was mainly used for butchering animals, but it was also used

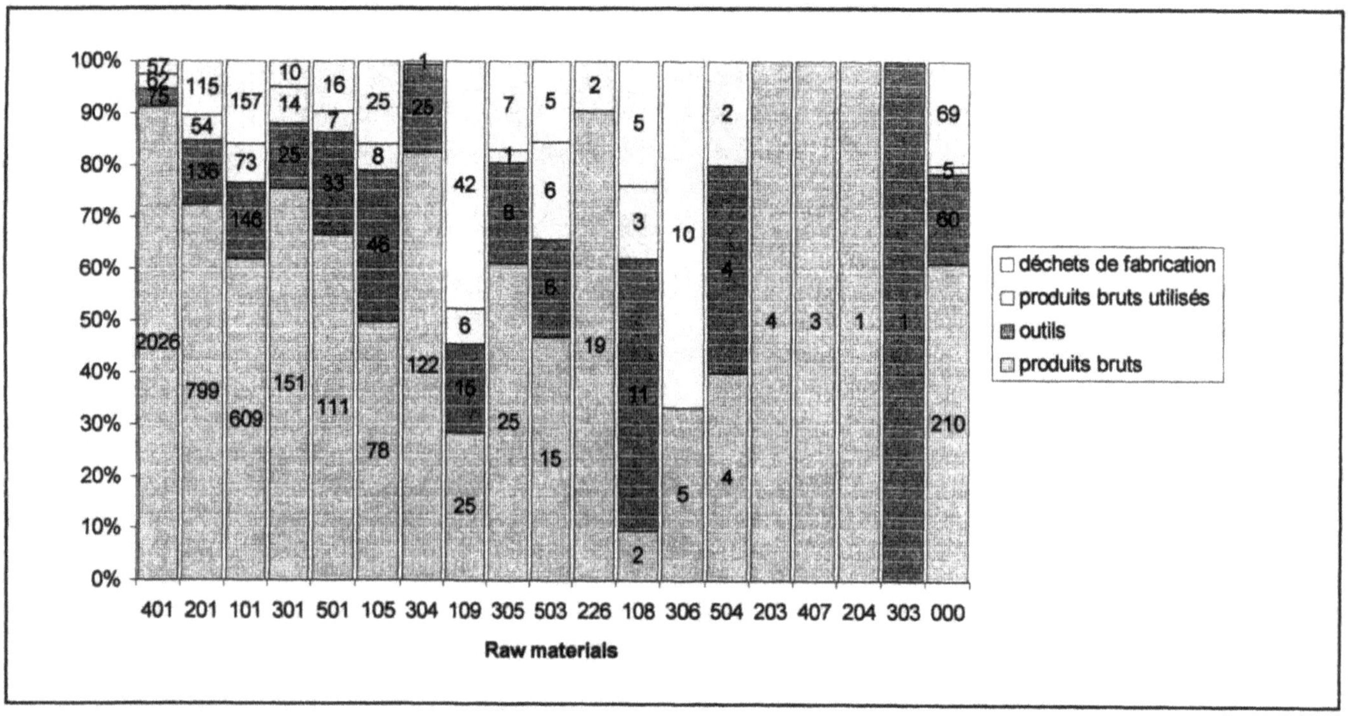

Figure 6: Proportion of the different categories of artifact for each raw material type

to constitute stores for use later in the movements of the group of hunter-gatherers. This is illustrated by a refit which shows the preparation of a raw material block designed for the production of blades. The absence of waste from core reduction and the absence of the core at the site indicates that the prepared block was taken off-site, to be used elsewhere (Cattin 1990, 1994).

The southern flint (MP 201): stores of blades and bladelets

This good quality flint of Cretaceous age from the region of Bellegarde-Seyssel (department of Ain, France) was used for the production of blades and bladelets. Average sized nodules were introduced onto the site (about 10 cm in length) which permitted a mixed reduction strategy. In fact it is one of the rare raw materials where the cores show the transition from blade to bladelet production.

Largely exploited at the site, this raw material was introduced in the form of nodules, as the refits show, which had probably been previously tested; all the phases of the reduction sequence are documented at the site (Figure 8). However, one cannot exclude the possibility that some of the artifacts were introduced ready made, in the form of blanks, since there is a high percentage of tools and tool production waste. The majority of the tools were made on bladelets, while blades and flakes were used in more modest quantities.

The northern flint (MP 101): a store of bladelets

The source of this fine grained flint is situated in the Olten region (canton of Soleure, Switzerland), where the Kimmeridgian strata outcrop. At the source nodules of all sizes are found (between 5 and 15 cm in length), but the refits show that at Champréveyres it was mainly small nodules that were brought in. They were imported in the form of blocks which had probably been tested, and the large quantity of knapping waste of this raw material leaves no doubt that the reduction process was carried out on-site; the equally high proportion of tools and tool production waste also suggests an import or a great transformation of the tools. The composition of this assemblage shows a representation of elements belonging to each stage of the *chaîne opératoire* (Figure 9).

The reduction process was above all oriented towards the production of bladelets and, to a lesser extent, of blades. The latter were largely used for the manufacture of tools (53 %) and the single blade core found is not sufficient to explain their abundance. It therefore seems that the blade blanks had been imported to the site. All the other cores are of very small dimensions and can only have been used to produce bladelets. These bladelet cores, and the correspondence between the knapping areas and concentrations of backed bladelets, suggest that these small nodules constituted stores

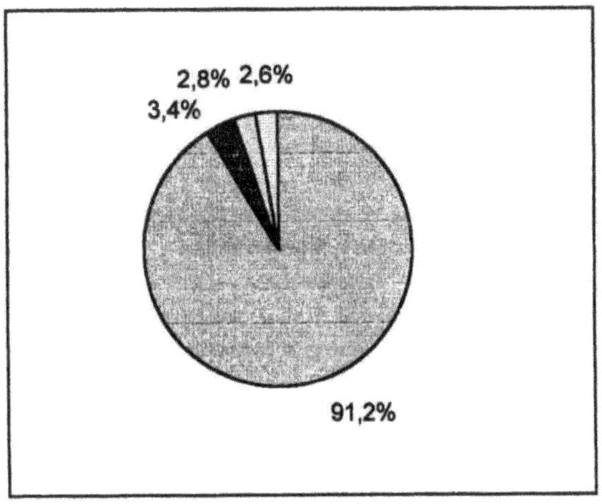

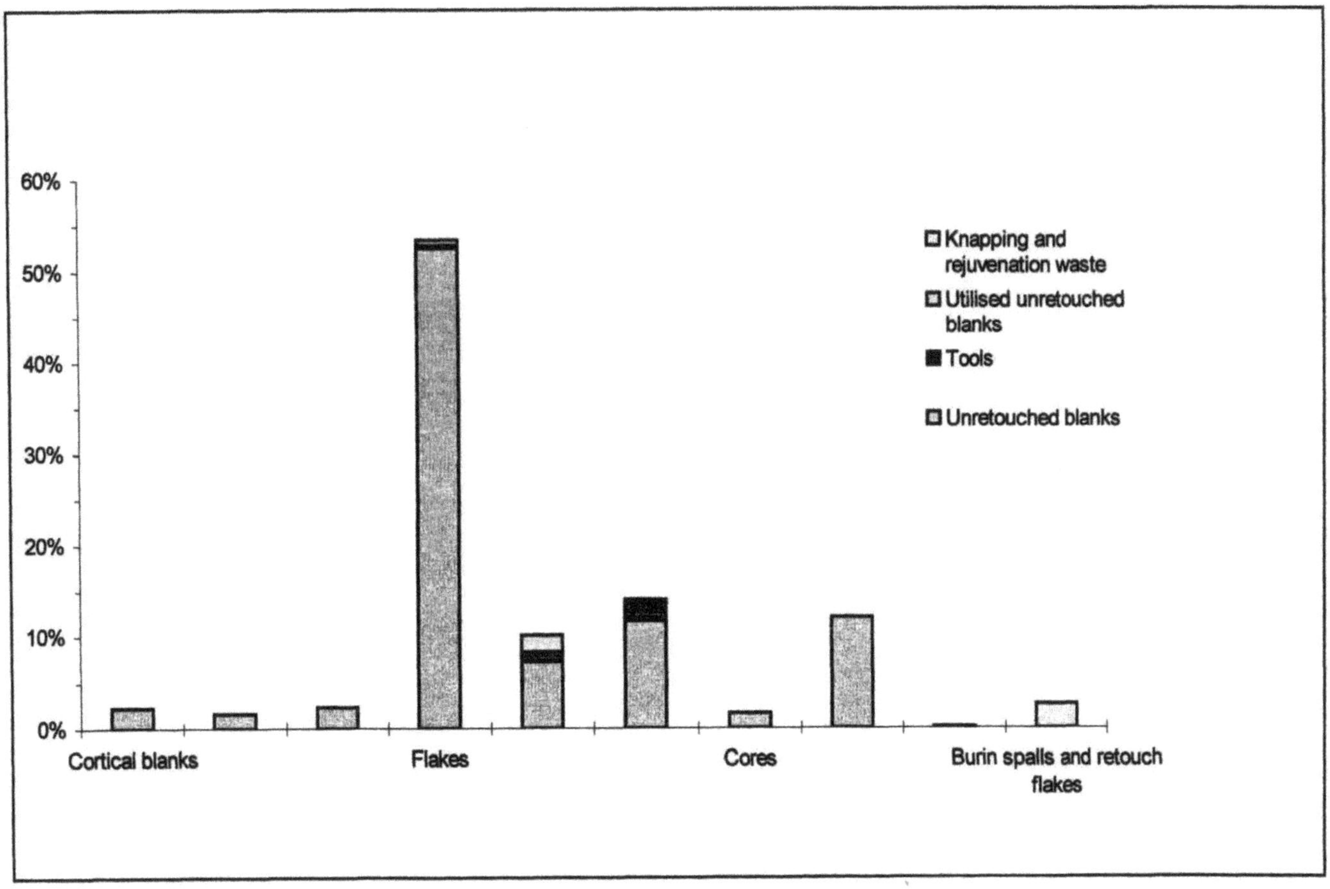

Figure 7: Distribution of the artifacts made of local Hauterivian flint (MP 401) according to their typo-technological group

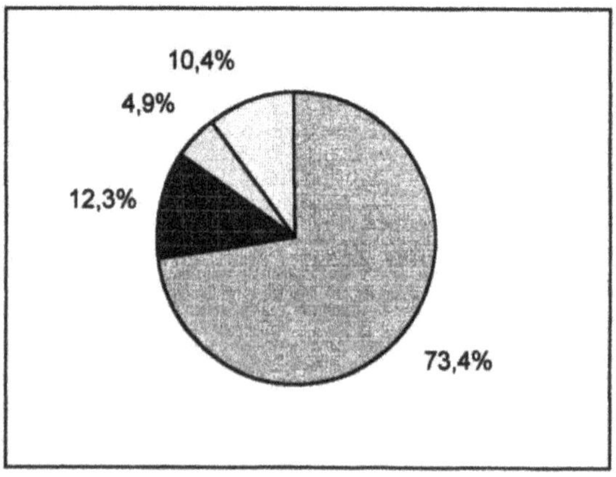

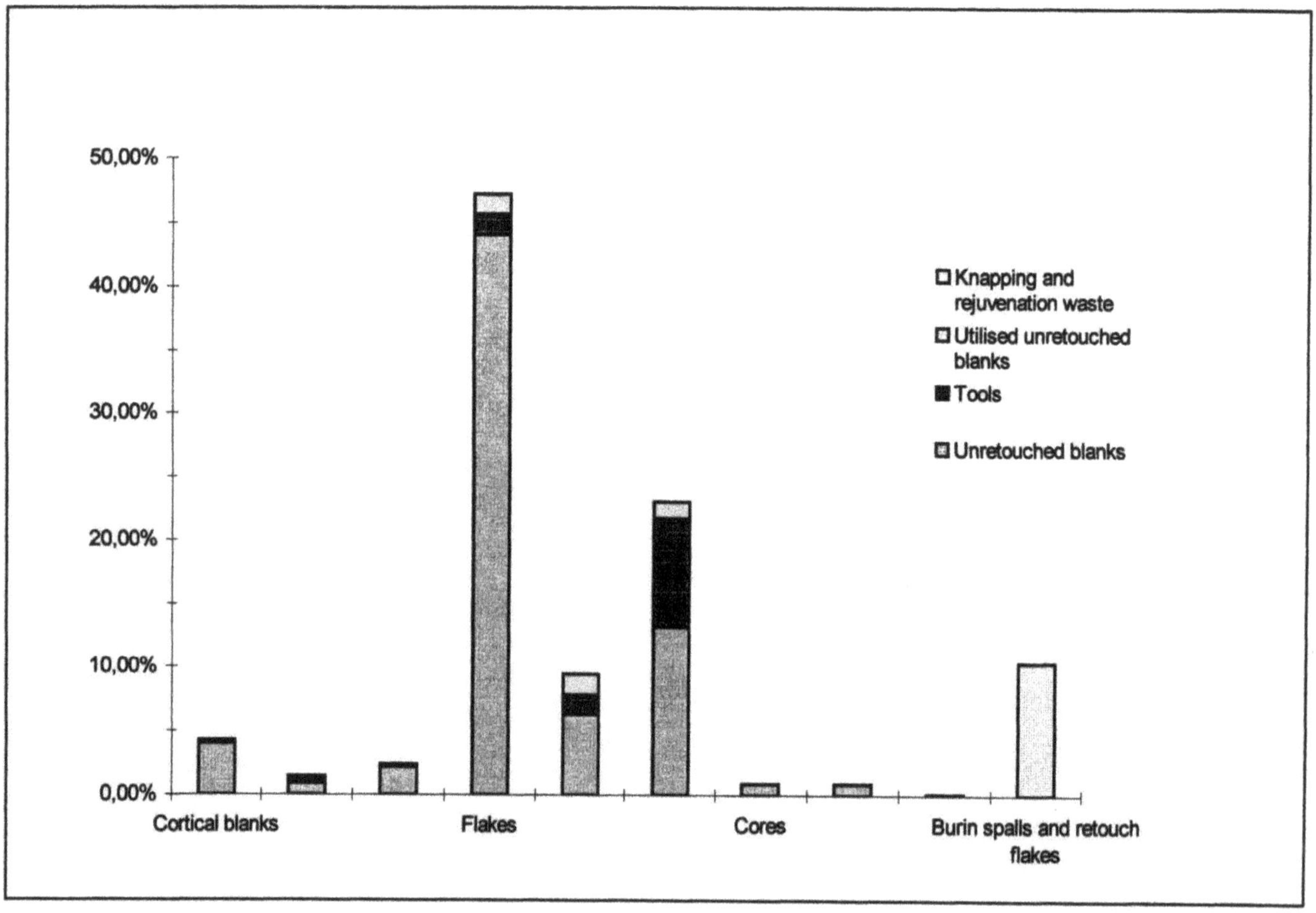

Figure 8: Distribution of the artifacts made of allochthonous southern flint (MP 201) according to their typo-technological group

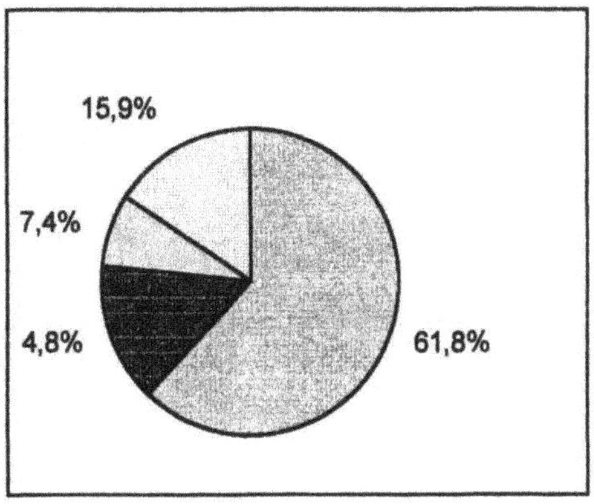

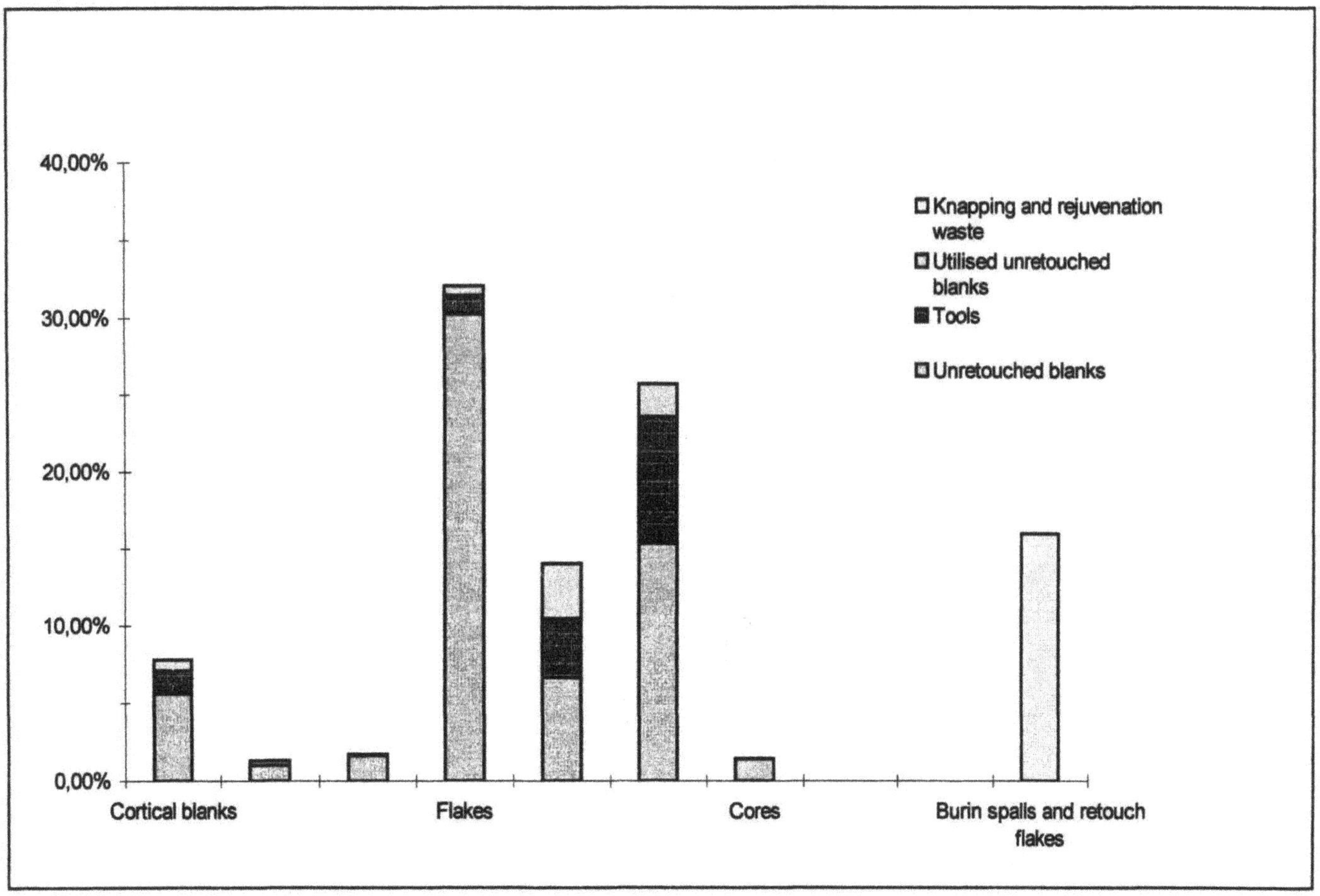

Figure 9: Distribution of the artifacts made of allochthonous northern flint (MP 101) according to their typo-technological group

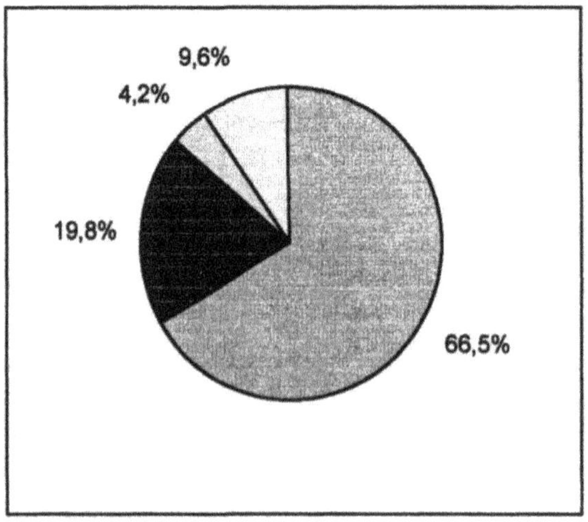

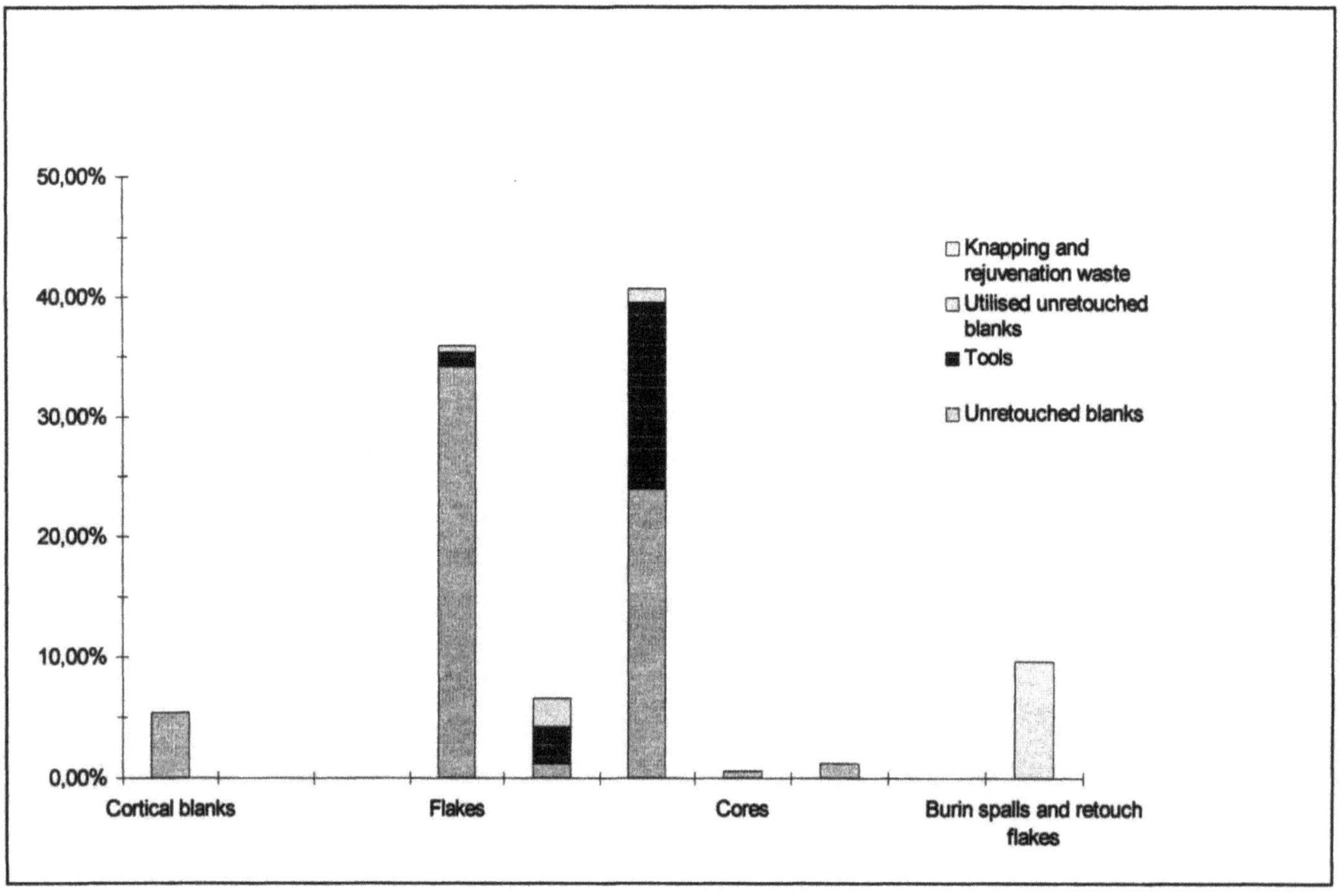

Figure 10: Distribution of the artifacts made of allochthonous flint (MP 501) according to their typo-technological group

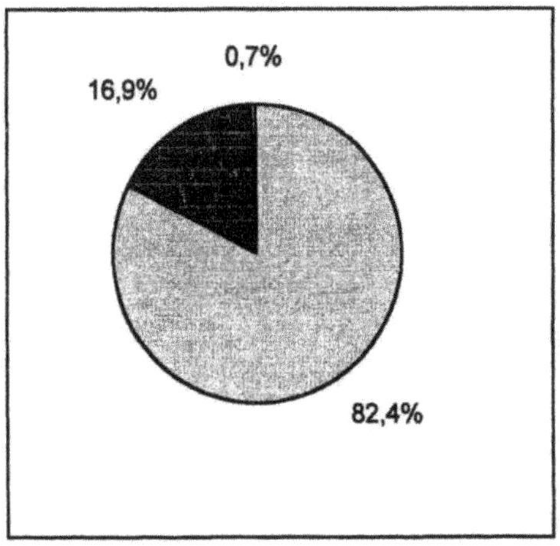

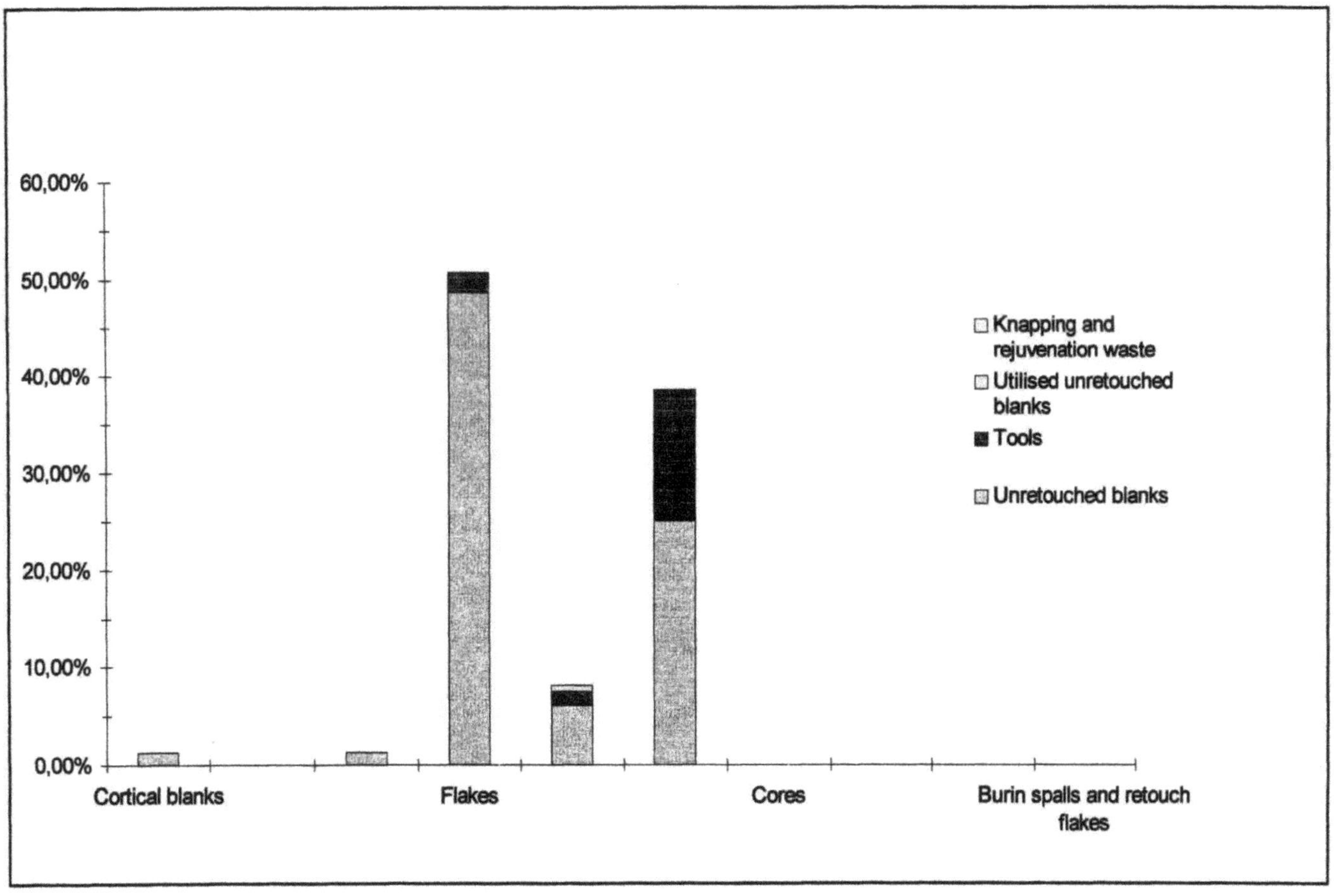

Figure 11: Distribution of the artifacts made of allochthonous flint (MP 304) according to their typo-technological group

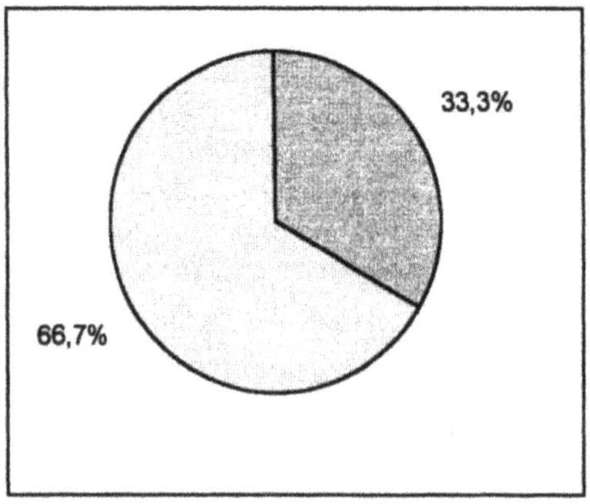

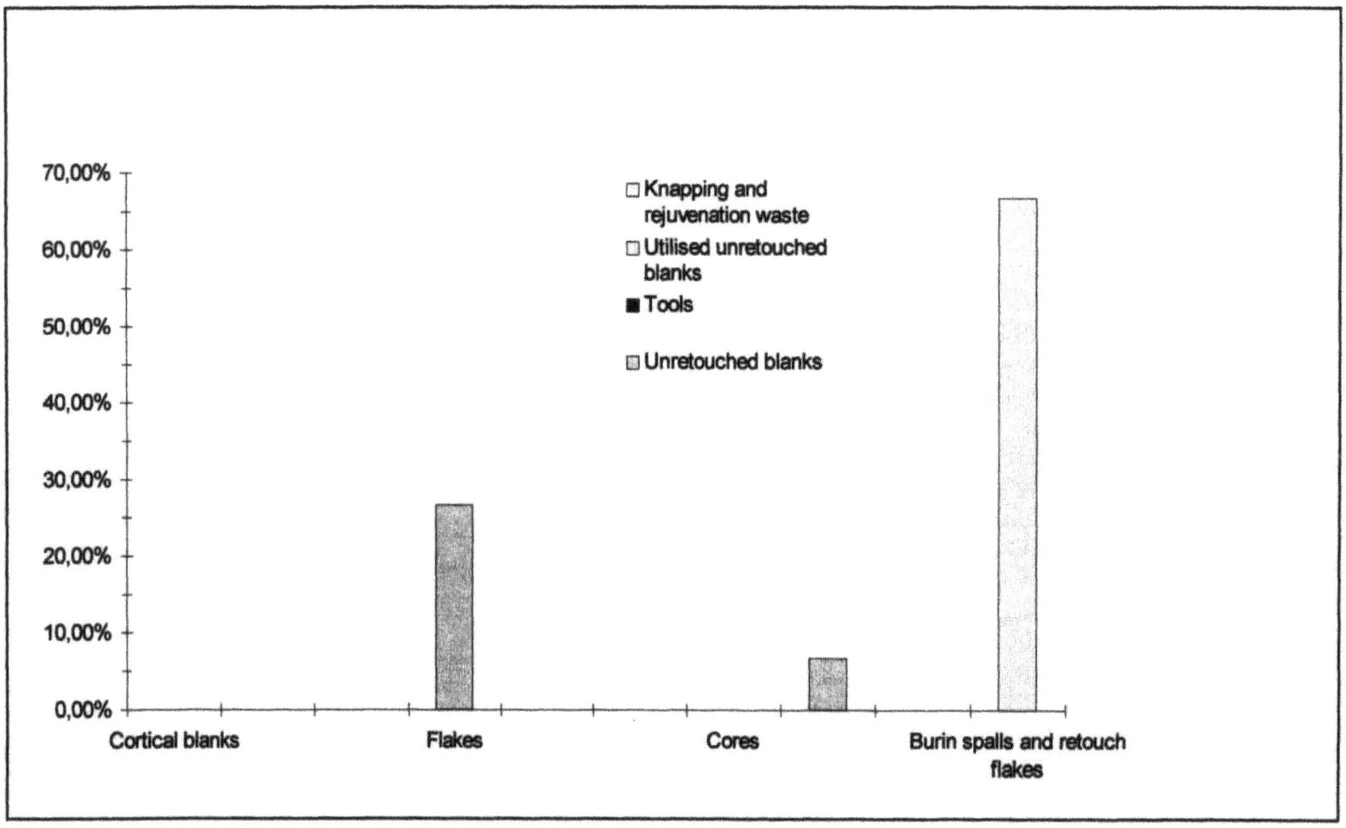

Figure 12: Distribution of the artifacts made of allochthonous flint (MP 306) according to their typo-technological group

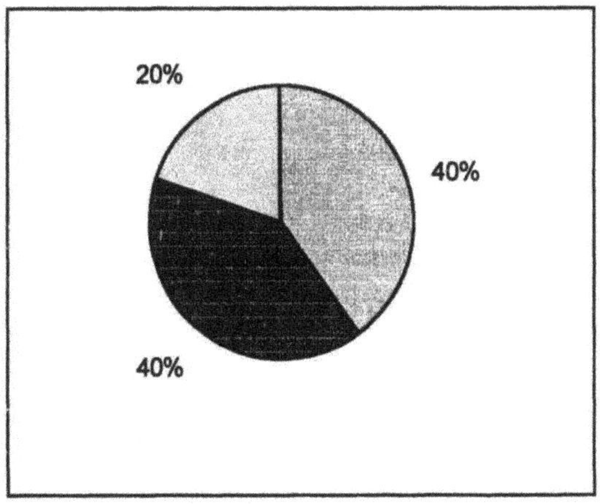

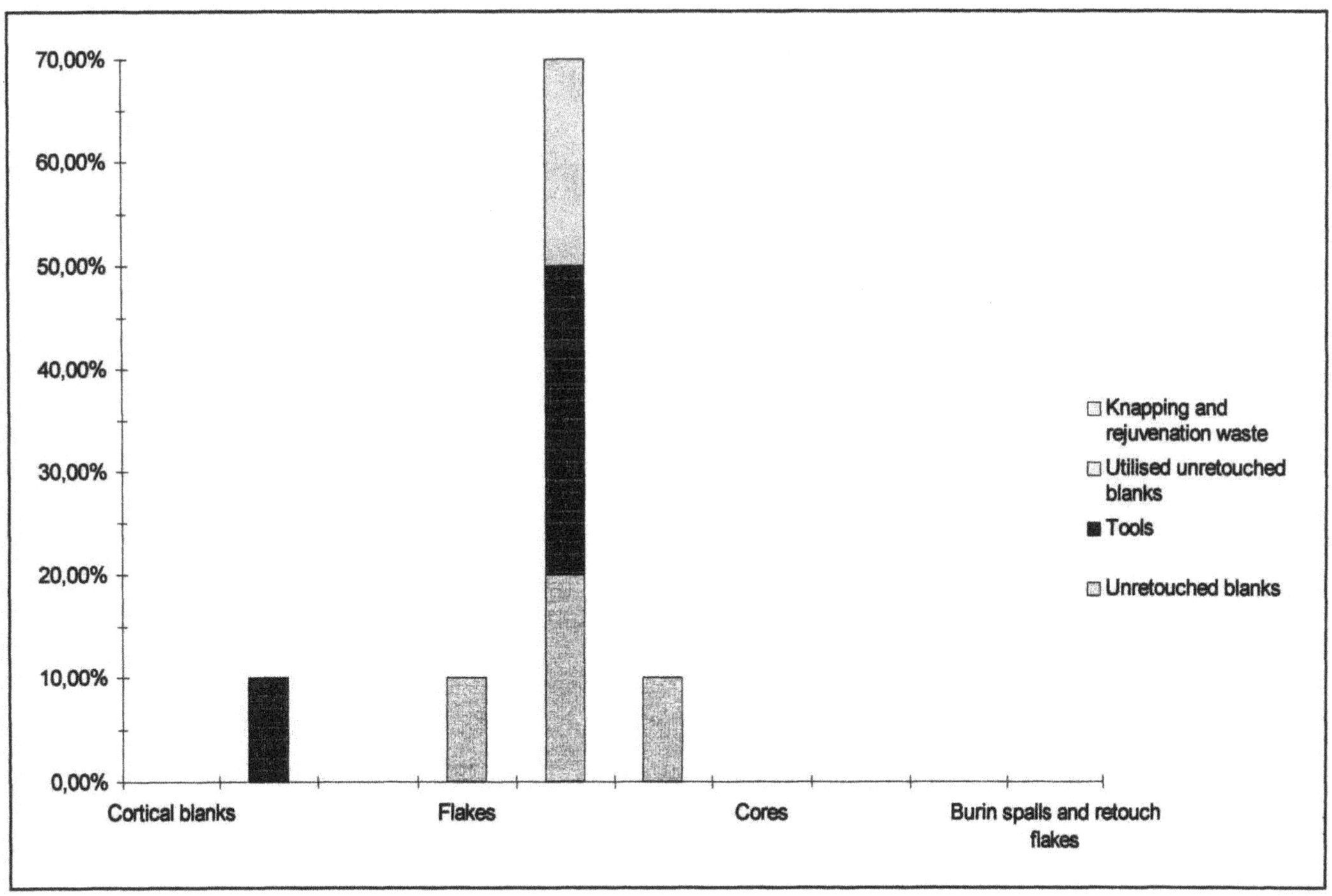

Figure 13: Distribution of the artifacts made of allochthonous flint (MP 504) according to their typo-technological group

of raw material for the manufacture of armatures (Cattin 1990, 1994).

Other allochthonous raw materials

The other allochthonous raw materials are much less abundant. They arrived at the site both in the form of nodules as well as unretouched blanks and tools.

To illustrate the first case, we have chosen raw material types MP 501 and MP 304. The former is a fine grained flint which comes from a source dated to the Hauterivian and which has yet to be located, though it definitely lies to the south of Neuchâtel where the lower Cretaceous strata outcrop. A single nodule which had been partially preformed was imported onto the site and knapped there, as is documented by the large concentration of knapping waste found (76%). The core was subsequently abandoned at the site. The reduction process was oriented towards the production of short blades and bladelets, a large quantity of which were then transformed into tools (Figure 10).

The allochthonous raw material MP 304 probably comes from a source of Jurassic age which has yet to be located, but which is no doubt situated more than 30 kilometres to the north of the Champréveyres. At the site, it is characterised by a high concentration of knapping waste related to the exploitation of a single core only (Figure 11). The low number of products with cortex indicates that the core was prepared away from the camp, and was introduced after this phase in the reduction process. Subsequently knapped on the site, the core has not been found. One could therefore hypothesise that it was not exhausted and that it was taken away from the site when the group departed. The flaking waste is dominated by flakes and bladelets, and there is only a small quantity of blades. The tools were preferentially made on bladelets, which indicates that the reduction process was oriented towards their production. The presence of isolated cores can be explained in various ways:

- opportunistic behaviour involving the collection of nodules during hunting trips or during migrations for example;
- the practice of exchange;
- the cores might represent the last elements of an initially larger store;
- the sources of these raw materials only furnished a limited quantity of blocks which could be exploited.

The introduction of tools and unretouched blanks is illustrated by the raw materials MP 306 and MP 504. The first, of middle Oligocene age, comes from a source which has yet to be identified. It is only represented by a very small number of pieces which are mainly burin spalls and which are not associated with any burin (Figure 12). The small quantity of knapping waste means that it was not exploited in the preserved area. This flint above all testifies to the import of one or more burins, which must have then been exported after their use. In contrast, the raw material type MP 504 documents the import of blanks which were used and then abandoned at the site. The source of this raw material is located some distance from the site and corresponds with the Séquanian level at Lampenberg (canton of Bâle Campagne) about 100 kilometres to the north-west of Champréveyres. The few artifacts made of this raw material all appertain to the phase of core reduction and are characterised by a predominance of blades, nearly all of which were transformed into tools (Figure 13). These artifacts, which do not refit, suggest that they had not been knapped at the site; they were therefore probably imported as unretouched blanks or in the form of finished tools. These two examples confirm the observations made on the rest of the assemblage, and show that the majority of the imported blanks and/or tools were laminar products. The Magdalenians seem to have transported large nodules destined for this production only occasionally, and preferred to import blanks. These objects suggest equally different behavioural hypotheses:

- the distant sources of the raw materials could be an indication of the practice of exchange;
- the transport of blanks manufactured elsewhere;
- the relatively long duration of use of certain artifacts.

Therefore there were two distinct modes of managing the allochthonous raw materials: the import of cores which were mainly destined to the production of bladelets; and the import of unretouched blades and finished tools. Some artifacts were also exported, such as cores and tools.

In summary, it seems that the raw materials procured from distant sources were introduced onto the site in different forms, ranging from nodules to blanks, and that the distance from the source does not seem to have favoured a particular mode of introduction. On the other hand, it is undeniable that the stores of bladelets arrived in the form of nodules, and that this occurred more rarely in the case of the blades. The weight of the transported objects therefore seems to have been of some importance. But the mediocre quality of the local raw material available at Champréveyres also explains the important role played by allochthonous raw materials, since its coarse grain was not suitable for the production of backed bladelets. The importance of hunting and butchery activities at the site therefore seems to have conditioned the procurment and exploitation of the raw materials.

COMPARISON WITH THE SITE OF MONRUZ

The site of Monruz is characterised by a great abundance of lithic artifacts (more than 46,000 pieces larger than 1 centimetre) and the site, which is about twice the size of that at Champréveyres, has produced 9 times as many lithic artifacts. Monruz is still being studied, and consequently the quantities of artifacts and the hypotheses are only provisional. Waste products are very abundant (93% of unretouched blanks), while tools and tool manufacturing waste represent 3% and 4% of the assemblage respectively.

	Raw material	Champréveyres	Monruz
Local flint	401	+	+
Plateau flint	108	+	+
	305	+	+
	601		+
Allochthonous flint	101	+	+
	105	+	+
	109	+	+
	201	+	+
	203	+	
	204	+	+
	224		+
	226	+	
	301	+	
	303	+	+
	304	+	
	306	+	+
	317		+
	404		+
	407	+	+
	501	+	+
	503	+	
	504	+	+
	615		+
Flint of undetermined origin	000	+	+

Figure 14: Comparison between the sites of Champréveyres and Monruz of the presence (+) of the different types of raw material

Although the intensity of the knapping activities and the density of the artifacts may be attributed to a series of successive occupations, they may equally suggest longer term occupations. In addition, and in contrast to the site at Champréveyres, a large part of the tools was manufactured at the site itself.

The raw materials used are as varied as they are at Champréveyres and they come from the same sources (Affolter *et al.* 1994). However, some of them only occur at either one site or the other, which suggests different relationships or trajectories (Figure 14). Contacts with the south-west of Germany are demonstrated by the presence of small female statuettes made of lignite which are identical to those found at Petersfels, and by Tertiary molluscs imported from south German sources (Affolter *et al.* 1994; Egloff 1995).

At the current stage of research, the knapping activities which took place at the site have mainly been reconstructed from the study of the cores, which number 183 (compared with 68 at Champréveyres). They reveal a more intensive exploitation of the local raw material than at Champréveyres, but with an analogous aim: the production of blades (Figure 15). The cores of allochthonous raw materials were also used for making blades, although they were preferentially used for the manufacture of bladelets. This preference accorded to the allochthonous raw materials seems to have been identical at the two sites and can be explained by the good quality of this flint. Various cores indicate a mixed reduction strategy involving the production of both blades and bladelets, and in larger proportions than at Champréveyres. This aspect is even more interesting since it concerns, among others, the Kimmeridgian flint from the north (MP 101); the Magdalenians at Monruz therefore chose nodules of all dimensions of this raw material. It is also known that some of the blades made of this raw material found at Champréveyres were actually manufactured at Monruz as a result of the two pieces which refit, one of which was found at Champréveyres and the other at Monruz (Cattin 1992).

The analysis of the cores at Monruz shows that the nodules were brought to the site as stores for the production of backed bladelets, although this practice is not as evident as it is at Champréveyres. The occupants at Monruz imported larger nodules which enabled them to produce both blades and bladelets. The importance of the production of laminar blanks was undoubtedly related to the activities which were carried out at the site, and in particular the exploitation of the carcasses of the hunted horses. It also certainly reflects the probably different functions and duration of occupation at the two sites.

As at Champréveyres, certain raw materials are represented by only very few nodules or even by only one nodule; among

CHAMPRÉVEYRES

Product	Local flint		Non-local flint		Total	
Bladelets	6	8.96%	20	29.85%	26	38.31%
Blades	16	23.88%	7	10.45%	23	34.33%
Mixed blades/bladelets	4	5.97%	2	2.99%	6	8.96%
Other	10	14.93%	2	2.99%	12	17.91%
Total	36	53.73%	31	46.27%	67	100.00%

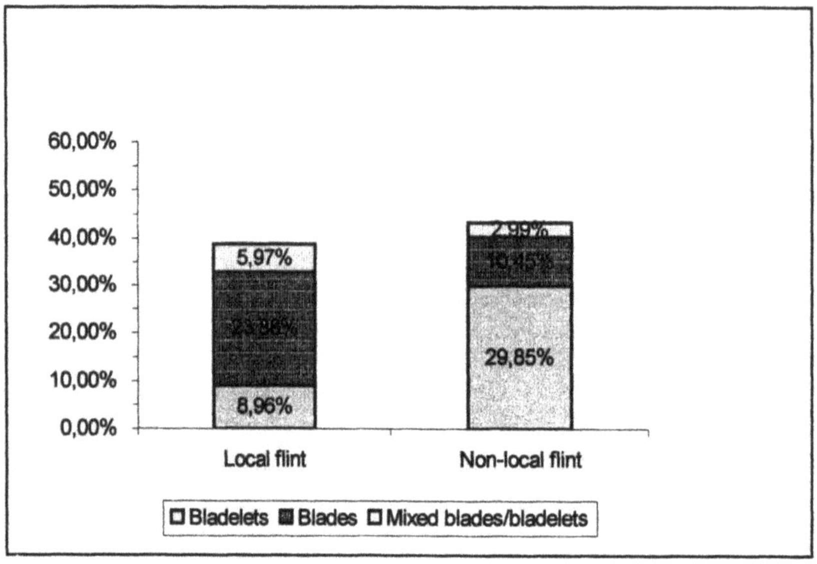

MONRUZ

Product	Local flint		Non-local flint		Total	
Bladelets	14	7.65%	39	21.31%	53	28.96%
Blades	76	41.53%	13	7.10%	89	48.63%
Mixed blades/bladelets	9	4.92%	17	9.29%	26	14.21%
Other	11	9.80%	2	2.80%	13	7.10%
Total	112	61.20%	71	38.80%	183	100.00%

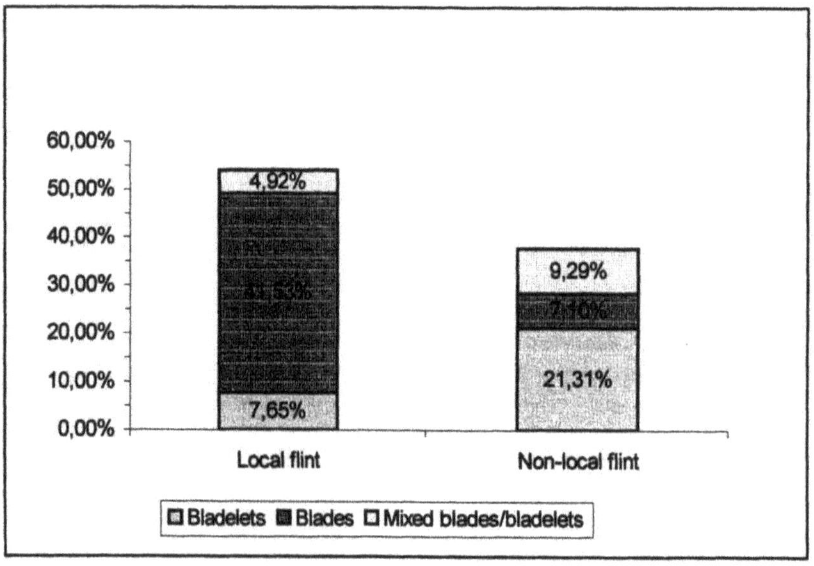

Figure 15: Knapping products and origin of the cores from the sites of Champréveyres and Monruz (provisional count for the latter)

these is the allochthonous raw material MP 501. As regards the MP 504 raw material, it can be concluded that at Champréveyres this flint arrived in the form of prefabricated products. In contrast, the preliminary examination of the material from Monruz revealed the knapping of this raw material in different locations at the site. A deposit of knapping waste was found in a small ditch with a diameter of about 25 centimetres and a depth of 12 centimetres; the dimensions of these waste flakes indicate that the core must have been quite large (estimated length about 15 centimetres). Here this raw material was without doubt exploited with the aim of producing blades. Since all the cores found (4 in total) show that they had been used for the production of bladelets, one can conclude that this raw material was used first for the production of blades and subsequently for the production of bladelets. These data suggest that the artifacts found at Champréveyres may have been manufactured at Monruz, since the majority of the tools made from this type of flint are burins and two such burins were foud in the assemblage at Champréveyres.

CONCLUSIONS

The modes of introduction and exploitation of the raw materials reveal some points in common between the two sites, although there are notable differences. Champréveyres seems to have been a hunting camp where the reduction process was oriented towards rearming the projectile points and to the butchery of the hunted animals. In fact other types of tools were produced in much lower quantities, and a large part of them must have been imported to the site. At Monruz, the exploitation of the lithic raw material seems to have been centred on the production of all types of blanks, and the import of prefabricated tools was less significant compared with at Champréveyres. The importance of the debitage, as well as the presence of sunken hearths, indicates a long term occupation. Monruz seems to have been a base camp where, among other things, the tools necessary for hunting expeditions were manufactured. At present the relationships between the two sites are only vague and have been hypothesised by means of certain modes of behaviour with regard to the raw material. Their comprehension constitutes one of the aims of the study of the site at Monruz, and in particular by means of the success, or not, with the inter-site refitting.

Acknowledgements

I would like to thank Sarah Milliken for having invited me to contribute a paper to this volume, and also for her translation of my text. I would also like to thank my colleagues, and in particular Denise Leesch, Dominique Robert-Bliss and Nigel Thew, for their constructive comments.

BIBLIOGRAPHY

Affolter, J.

1989 Première approche des gîtes de silex et leur exploitation préhistorique. *Minaria Helvetica* 9: 55-60.

1995 Matière première. In Schweizerische Gesellschaft für Ur- Frühgeschichte, *La Suisse du Paléolithique à l'aube du Moyen-Age. II. Néolithique*. Bâle: 122-124.

Affolter, J., M.-I. Cattin, D. Leesch, P. Morel, N. Plumettaz, N. Thew & G. Wendling

1994 Monruz, une nouvelle station magdalénienne sur les rives du lac de Neuchâtel. *Archéologie Suisse* 17(3): 94-104.

Barr, J. H.

1973 *The Late Upper Paleolithic site of Moosbühl: An attempt to analyze some of its problems.* Unpublished doctoral thesis, Université de Berne.

Benkert, A., J. Reinhard & F. Schifferdecker

1984 Chasseurs de rennes et paysans des temps lacustres dans la baie de Champréveyres. *Archéologie Suisse* 7(2): 42-53.

Binford, L.R.

1980 Willow smoke and dogs' tails: hunter-gatherer settlement systems and archaeological site formation. *American Antiquity* 45 (1): 4-20.

Cattin, M.-I.

1990 Silex-economy on the magdalenian site Hauterive-Champréveyres (Switzerland): spatial organization and technology by refitting. In E. Cziesla, S. Eickhoff, N. Arts & D. Winter (eds.) *The Big Puzzle. International Symposium on Refitting Stone Artefacts, Monrepos 1987*, 363-369. Studies in Modern Archaeology. Holos, Bonn.

1992 Un raccord entre deux sites magdaléniens. *Revue de Préhistoire Européenne* 1: 35-42.

1994 Approche techno-économique des silex magdaléniens de Hauterive-Champréveyres. *Annuaire de la Société Suisse de Préhistoire et d'Archéologie* 77: 127-129.

Conkey, M.

1992 Les sites d'agrégation et la répartition de l'art mobilier, ou: y a-t-il des sites d'agrégations magdaléniens? In *Le peuplement magdalénien. Paléogéographie physique et humaine. Colloque de Chancelade, 10-15 octobre 1988*, edited by A. d. c. d. Chancelade, pp. 19-25. Documents Préhistoriques, 2. Editions du C.T.H.S., Paris.

Egloff, M.

1989 Des premiers chasseurs au début du christianisme. In *Histoire du Pays de Neuchâtel. De la Préhistoire au Moyen Age*, pp. 11-160. vol. 1. Attinger, Hauterive (Suisse).

1995 Les figurines féminines magdaléniennes de Neuchâtel (Suisse). In *La Dame de Brassempouy. Actes du colloque de Brassempouy (juillet 1994)*, edited by H. Delporte, pp. 71-87. E.R.A.U.L. 74. Université de Liège, Liège.

Gaillard, M.-J. & B. Moulin

1989 New results on the Late-Glacial history and environment of the Lake of Neuchâtel (Switzerland). Sedimentological and palynological investigations at the Palaeolithic site of Hauterive-Champréveyres. *Eclogae Geologicae Helvetiae* 82(1): 203-218.

Leesch, D.

1993 Cadre chronologique et faciès industriels. In: Höneisen,M.,Leesch, D., Le Tensorer J.-M. Le Paléolithique supérieur récent. In *La Suisse du Paléolithique à l'aube du Moyen-Age. I. Paléolithique et Mésolithique*, edited by S. G. f. U.-u. Frühgeschichte, pp. 153-164, Bâle.

1997 *Hauterive-Champréveyres, 10. Un campement magdalénien au bord du lac de Neuchâtel.* Archéologie neuchâteloise 19. Musée cantonal d'archéologie, Neuchâtel.

Morel, P. & W. Müller

1997 *Hauterive-Champréveyres 11. Un campement magdalénien au bord du lac de Neuchâtel. Archéozoologie.* Archéologie neuchâteloise 23. Musée cantonal d'archéologie, Neuchâtel.

Moulin, B.

1991 *Hauterive-Champréveyres 3. La dynamique sédimentaire et lacustre durant le Tardiglaciaire et le Postglaciaire.* Archéologie neuchâteloise 9. Editions du Ruau, Saint-Blaise.

MAGDALENIANS-EARLY AZILIANS IN THE CENTRE OF THE PARIS BASIN: A FILIATION? THE EXAMPLE OF LE CLOSEAU (RUEIL-MALMAISON, FRANCE)

Pierre Bodu

INTRODUCTION

This article deals with the evolution of the Late Glacial industries in the Ile-de-France by means of the study of a site where the Azilian is represented by three occupation levels and where a slightly younger assemblage has recently been found. Here the accent will be particularly placed on the early phase of the Federmesser groups which seems to have close relationships with the Magdalenian assemblages. The aim of this article is therefore to present the data which highlight the close relationship which exists between the two industries.

In France, the Paris Basin is famous for the excellent state of preservation of its Magdalenian sites. Since the beginning of the 1960s, Professor André Leroi-Gourhan and his team were working to define better the Magdalenian tradition in the north of France and in particular the palaeoethnological aspects of this period. Based on the exemplary excavations of now famous sites (Etiolles, Marsangy, Pincevent, Verberie), the research has been concerned with the technical, economic, spatial and social aspects of this cultural entity.

The recent discovery of sites in this vast region where the lithic industries have different technological and typological characteristics incited us to question the presence of other groups apart from the Magdalenians. This series of discoveries has consequently enriched our knowledge of these assemblages which appertain to the final phase of the Late Glacial and are therefore later than the Magdalenian.

Given the partial nature of the discoveries, consisting only of lithic artifacts, and the bad state of preservation of the sites, it was difficult to define more that the technological characteristics of the lithic industries. Before the beginning of the 1990s only surface collections were available, or assemblages from small scale excavations (Fosse *et al.* 1990). Therefore we only had limited stratigraphic data and often no means for obtaining absolute dates (since neither fauna nor charcoal were preserved). This fact is clearly in contrast with the majority of the research carried out in the north of France, Belgium or Germany, where Late Glacial sites are particularly well documented (De Bie 1997; Baales and Street 1996).

This situation has recently improved with the discovery in the Ile-de-France of a site attributed to the Federmesser groups tradition, Le Closeau at Rueil-Malmaison. The integration of the results of this rescue excavation with the research programme on "Late Glacial habitats and populations in the south of the Paris Basin" has proved to be particularly fruitful from the point of view of enriching our knowledge of the last Final Palaeolithic populations in the Ile-de-France. Before the construction works for the A86 at Rueil-Malmaison (Hauts-de-Seine), the excavations uncovered a vast camp site of the Federmesser groups tradition which is contemporary with the Allerød oscillation. Various Azilian occupation levels were found fossilised in the middle of an ancient channel of the river Seine. The study of these enabled us to hypothesise a local evolution of the populations. Other Late Glacial assemblages with different technological and typological characteristics pose the question of a continuity of occupation at the site at the very end of the Allerød and during the younger Dryas.

At Le Closeau the early phase is the best preserved as far as the faunal remains are concerned as well as from the point of view of the spatial organisation of the habitation units discovered. We therefore propose in the following article to only deal with the early phase of the Federmesser groups and to present the data which enable us to relate it to the Magdalenian facies which immediately preceded it.

THE SITE OF LE CLOSEAU

The site of Le Closeau was discovered in the municipality of Rueil-Malmaison (Hauts-de-Seine) at less than 5 kilometres to the west of Paris (Figure 1). During an archaeological survey carried out along the route of the A86, two occupation levels attributable to the Federmesser groups (Azilian) were found in a surface area of about 400 square metres. There were four distinct concentrations of lithic artifacts, one of which also contained faunal remains, as well as hearths, and there were already suggestions of a diachronic difference between them (Bodu 1995; Kildea 1996).

This discovery was followed by two exhaustive seasons of excavation which were carried out from June 1995 to January 1996 and from March 1996 to October 1997 (Table 1). The total surface excavated, which covered 25,000 square metres, authorised an approach pertinent to the prehistoric settlement of this area (a total of 53 activity loci were found over this vast area). Following these two excavations, three other levels were found which completed the panorama of the occupation of the river channel (Figure 2). The first of these can be attributed to the Federmesser groups, while the other two are still problematic. Therefore

Figure 1: Location and topography of the site of Le Closeau

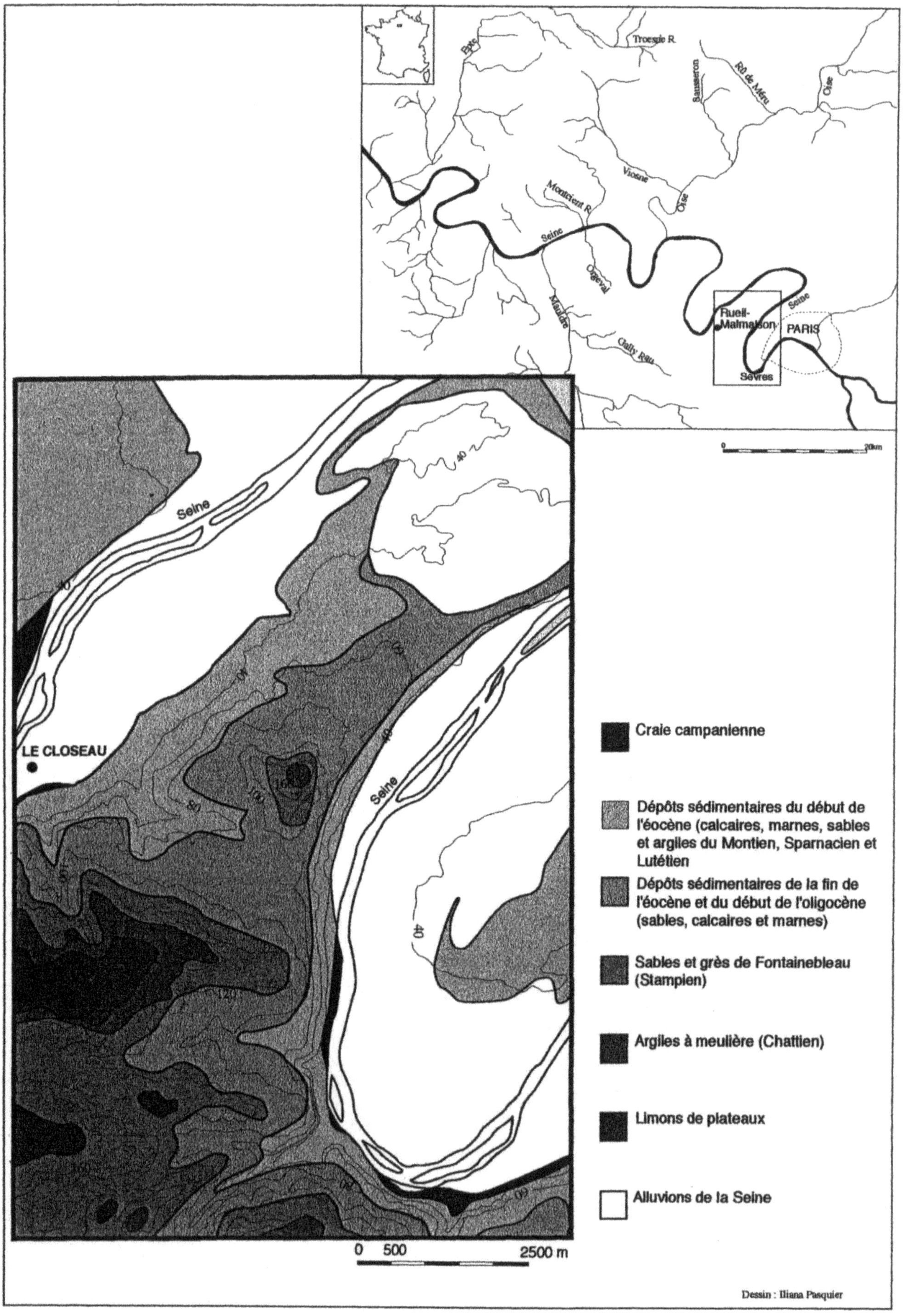

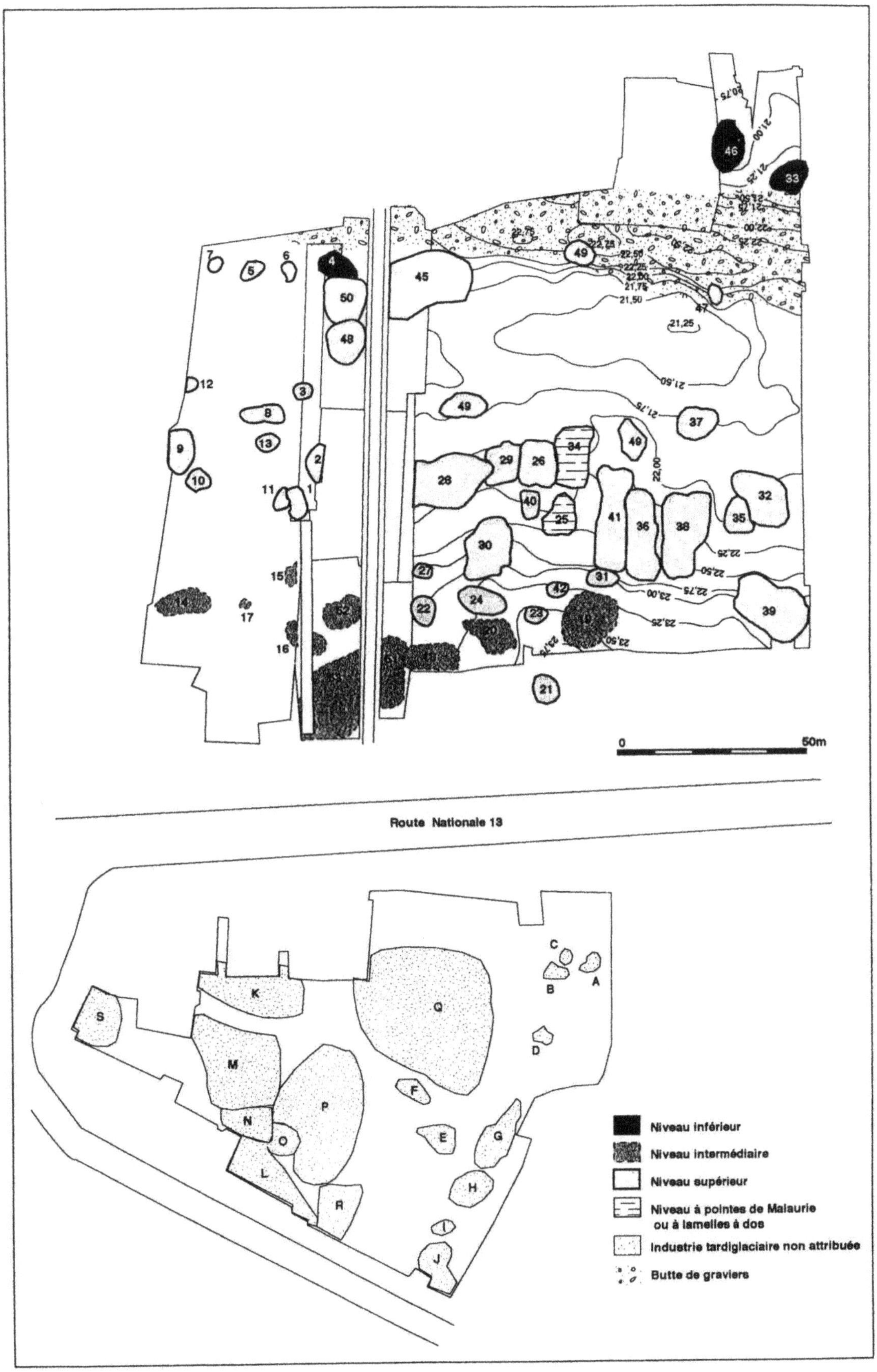

Figure 2: The site of Le Closeau. Plan of the loci according to their stratigraphic location

Table 1. Description of the different archaeological occupations at Le Closeau

Excavation season	Excavation surface area	Cultural attribution	Levels	Stratigraphic position of the industries
Le Closeau 1994	400 m²	Federmesser	lower	below the grey sand
			upper	summit of the grey sand
Le Closeau 1995	8000 m²	Federmesser	intermediate	?
			upper	summit of the grey sand
Le Closeau 1997	11,000 m²	Federmesser	lower	below the grey sand
			intermediate	?
			upper	summit of the grey sand
		with Malaurie points	?	summit of the grey sand
Sud RN13 1997	5000 m²	?	?	?

so far there are three Azilian levels at Le Closeau and two assemblages which are less clear.

The three levels were identified on the basis of their intrinsic characteristics as well as on the context of their discovery and the radiometric dates. The lower level lies under a pedogenised sand of a grey colour which is attributed to the Allerød, while the middle level lies in the centre of this sandy horizon. The last Azilian level lies on top of the sand. In this position, it seems to be relatively contemporary with a locus of very particular microliths (locus 25) which have been called Malaurie points and which have been found at certain sites in the Ile-de-France (Hinout 1985).

It was in this ancient channel of the river Seine, surrounded by hills with relatively steep slopes, the Jonchère hills, that the Palaeolithic remains have been fossilised (Figure 1). To the north a 10 metre wide mound composed of granulometrically different elements (ranging from small gravel to blocks weighing tens of kilograms) protects the site. A slow sedimentation constrained by this gravel bar explains the good state of preservation of the prehistoric settlements. Beyond this, a vast depression which could correspond to the ancient river bed of the Seine, had equally attracted the Palaeolithic populations. However, the settlement there seems to have been less important and more limited in space.

The oldest level was only found in the northern part of the site, while the intermediate level is concentrated on the southern slope of the channel. The upper level, which is the most extensive at the site, is essentially located at the centre of the channel although a few contemporary loci were found on the northern slope. At lower altitudes, towards the bottom of the channel, one observes a hiatus in the human occupation. The presence of a certain type of malacofauna characteristic of wet environments and notably of spring resurgences, explains the disinterest of the prehistoric occupants for this place.

One of the two assemblages with technological and typological characteristics which differ from those of the Azilian was found in the centre of the river channel like the loci of the upper level. In contrast, the second one is found at the foot of the Jonchère hills, on a sort of vast terrace, next to the flint outcrop zones.

THE LOWER LEVEL AT LE CLOSEAU: REMINISCENCES OF THE MAGDALENIAN?

Three loci appertain to the lower level: 4, 33 and 46. Two of these (4 and 46) correspond to real habitation units, while the third one, locus 33, can be better classified as a technical structure or annex unit. Its proximity and the origin of some of the lithic artifacts suggest that it was functionally related to locus 46.

Although it is not possible at the moment to establish the strict synchronicity of these structures, their common stratigraphic position, the similarities of the technological and typological characteristics of the industries and their spatial organisation are all homogeneous. So far it has only been possible to date one of these units; this is locus 4, which has two pre-Allerød dates (Table 2).

Table 2. Dates of the different levels at the Federmesser site of Le Closeau

Federmesser level	Material dated	Date number	Date BP	Calibrated date BC
lower	bone	OxA 5680 (Lyon 166)	12,090±90	12,511 to 11,846
	bone	OxA 6338 (Lyon 313)	12,050±100	12,474 to 11,785
intermediate	charcoal	AA 21677 (Lyon 358)	11,240±80	11,412 to 11,013
upper	charcoal	Lyon 7190	10,470±110	10,695 to 10,038
	charcoal	Lyon 7189	10,670±110	10,880 to 10,371
	charcoal	Lyon 206 OxA	10,650±75	10,812 to 10,422
	bone	OxA 6337 (Lyon 312)	10,840±110	11,047 to 10,568

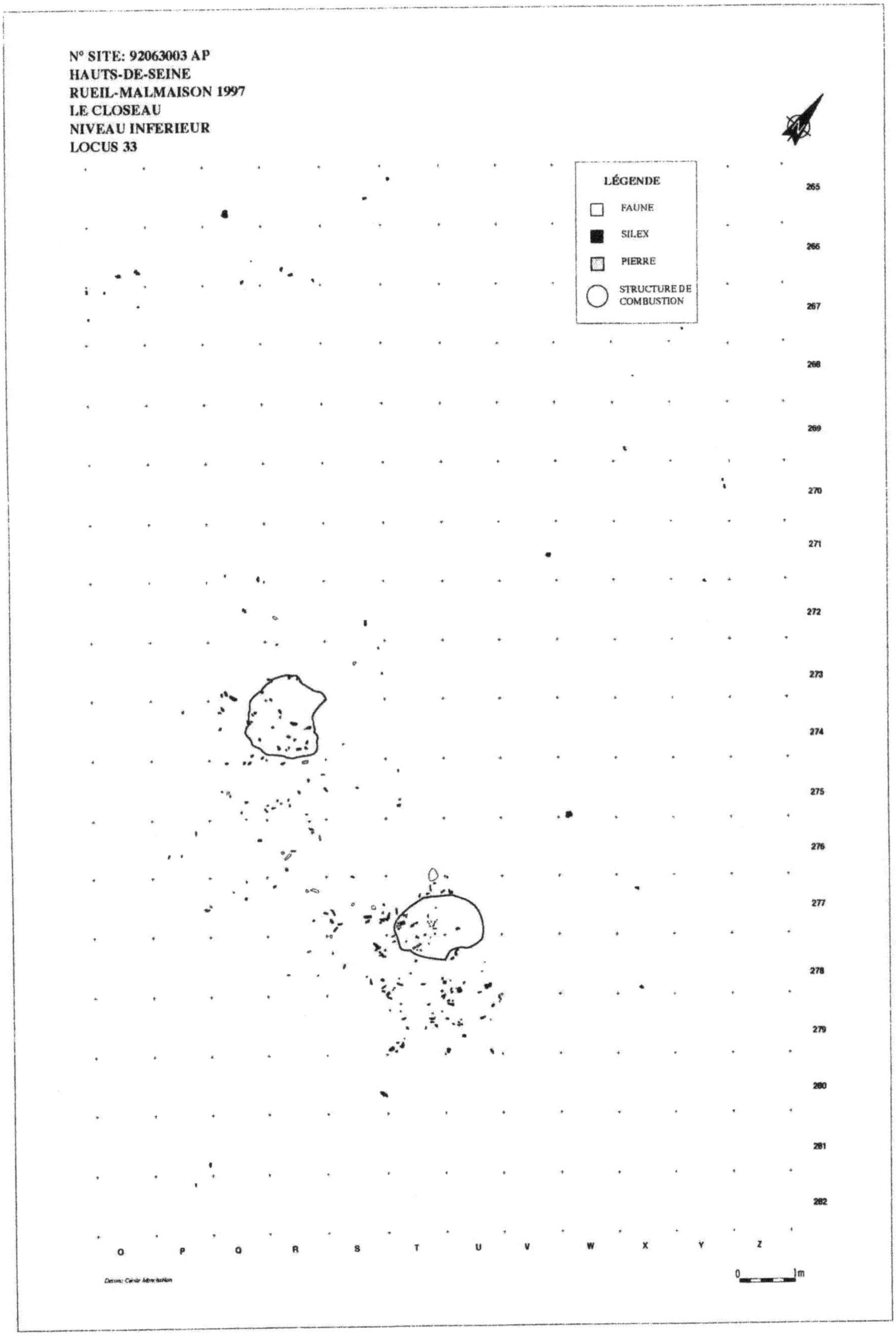

Figure 3: Lower level. Locus 33. Plan

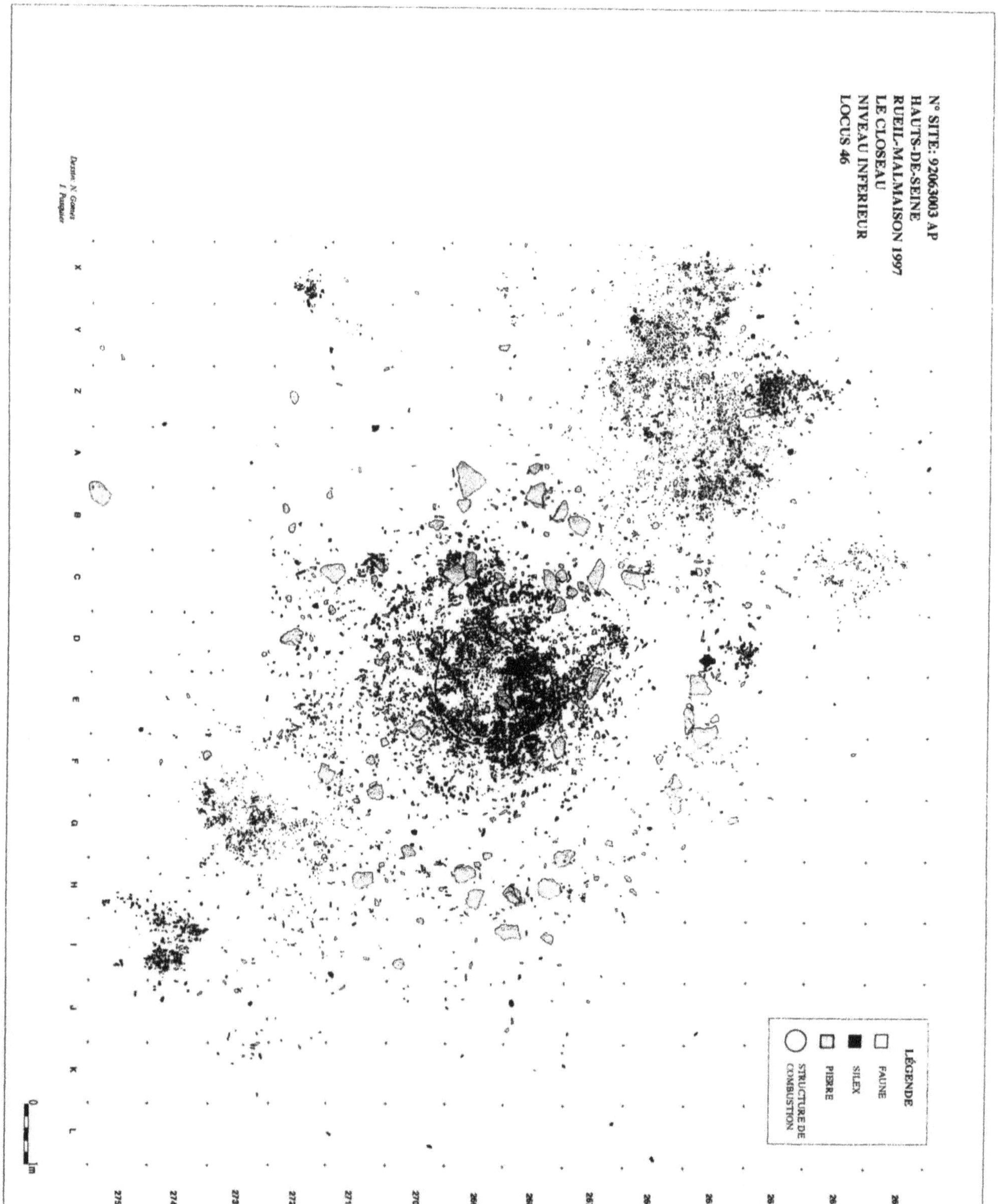

Figure 4: Lower level. Locus 46. Plan

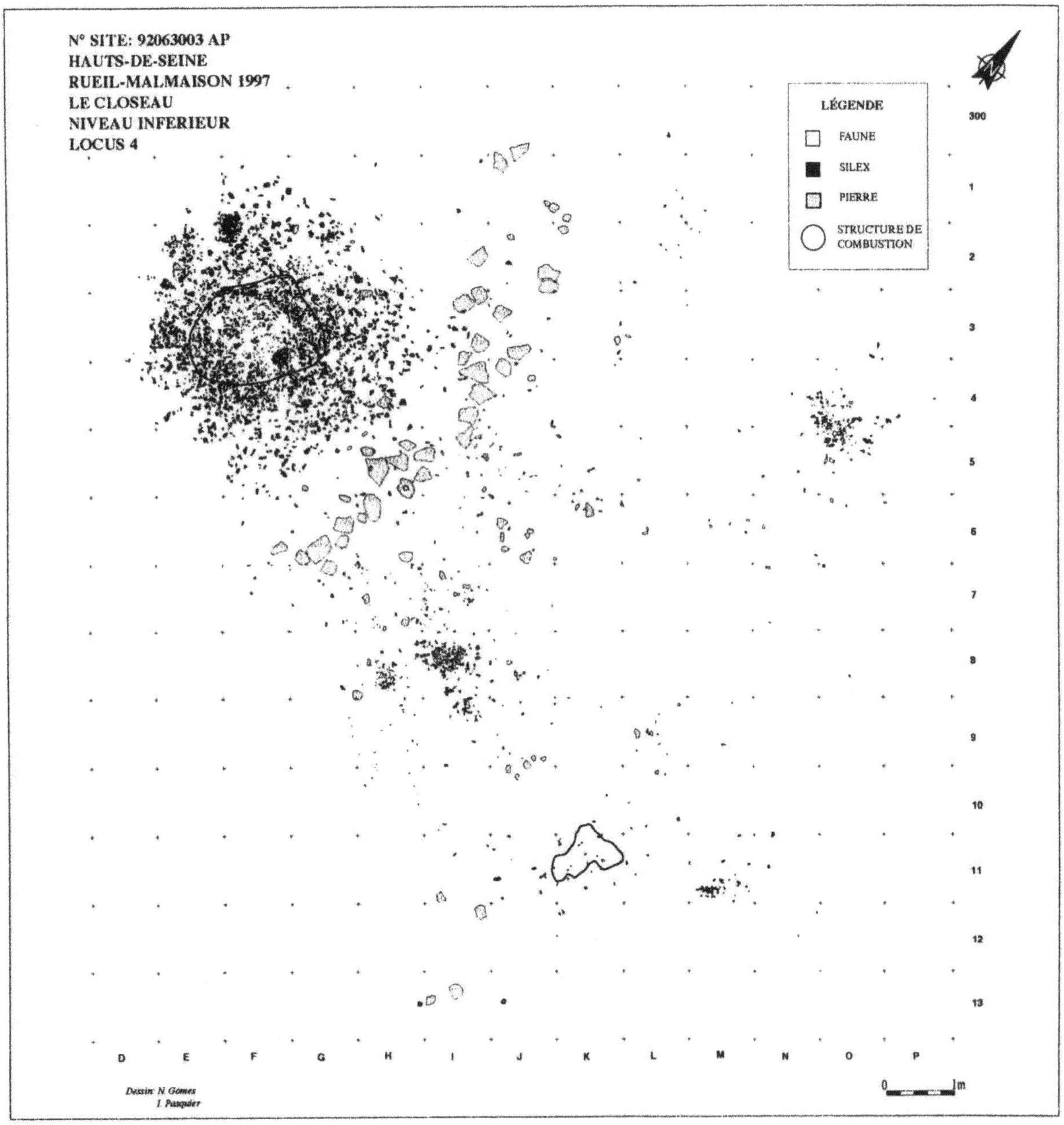

Figure 5: Lower level. Locus 4. Plan

SPATIAL CHARACTERISTICS

Unit 33 is characterised by a weak spatial organisation. 250 artifacts, of which the most part is made of flint (221 artifacts), were found concentrated around two very thin patches of ashes without any stone structure (Figure 3). The surface area of this concentration is only half a square metre. All this suggests a very ephemeral nature of this occupation which can be considered as an annex structure.

In contrast, the habitation units are remarkable for their dense spatial organisation. Both of them have an arrangement of large slabs of stone (sandstone, limestone, buhrstone) which enclose the central space organised around a hearth. This structure either takes the form of a real circle of stones (Figure 4) with a diameter of nearly six metres (locus 46), or else it is limited to an alignment of blocks as is the case of the edge of locus 4 (Figure 5). In this case one can without doubt explain this morphology by the partial destruction of the stone cordon by an ancient colluviation. Between this artificial boundary and the concentration of artifacts which surround the hearth, there is a relatively empty space close to the stone border. This may have been a circulation area or may equally have been a rest area or a store of perishable material (skins).

The structural blocks weigh many hundreds of kilos (300 for locus 4 and 600 for locus 46). The first is made up of about 30 blocks while the second is made up of about 40. Some of the individual blocks weigh about 50 kilos. This general mass would have prohibited long distance transport, which suggests that they were procured from the elongated bar of gravel which lies less than 10 metres away.

Surrounded by the blocks, which probably served as elements for securing a cover which has since disappeared, there was evidence of various different activities, such as flint knapping, use of the flint and bone tool kit, the remains of food, fire etc. The latter was most frequently lit directly on the ground. There are no real arranged hearths but rather the utilisation of a slight depression (although this may have formed during the cleaning-out after the fire, which is documented by the pieces of burnt bone found outside of the circle of stones). In the case of locus 46, some of the large blocks had been carried close to the hearth and may have been used as seats, anvils or other elements of site furniture (Figure 4).

A large quantity of lithic artifacts and bone remains was found outside of the central space, at a distance of between two to four metres from the stone border. The former appear in the form of dense concentrations and sparser patches which is suggestive of both a knapping scatter and discard areas. The bones, which are spatially associated with the flint, consist of numerous small fragments of burnt bone which may have resulted from the cleaning-out of the hearths. In locus 46 these areas are located symmetrically on both sides of the hearth (Figure 4).

Without risking to make too hasty assumptions, it is possible to highlight the existence of certain similarities between the habitation units in the lower level at Le Closeau and certain Magdalenian structures known in the Paris basin, especially at Etiolles where unit W11 had a similar vast circle of stones around a central area and exterior activity areas situated on both sides (Figure 5) (Julien *et al.* 1988). The difference in date between the lower level at Le Closeau and the Magdalenian at Etiolles is in the order of one thousand years. Comparisons with other Azilian sites or sites which have been attributed to the Federmesser groups are less satisfactory though. The spatial aspect has often not been sufficiently well described and when it has been one rarely finds any structures which are comparable to those of the lower level at Le Closeau (De Bie 1997). At the sites in the north of France, J.-P. Fagnart indicates a centripetal organisation within a perimeter of lesser dispersion (Fagnart 1997). According to this author, the finds were centred around a hearth of which the traces were not always detectable. After the grouping of certain kinds of finds, one equally finds the presence of specialised activity areas. But it is rarely, if ever, a marked delimitation of space by stone blocks. The most pertinent comparisons with this level at Le Closeau are therefore those with certain Magdalenian sites, comparisons which can be extended to different aspects of the lithic industry and to a lesser extent to the fauna.

THE FAUNA

The main characteristics of the fauna found in the lower level at Le Closeau will be briefly presented. Only very few fragments (about ten) were found in locus 33. Apart from the species which are also represented in the two large units (horse, wolf), there was also a fragment of a bovid bone, a species which has not been found elsewhere in the lower level at the site.

Loci 4 and 46 furnished a dense concentration of bones, either burnt or not, which were in a good state of preservation despite a particularly high rate of breakage (just under 4000 fragments bigger than one centimetre in each locus). Found around the hearths, in unit 46 they formed a real ring with a width of 30 centimetres. In the middle of locus 4 Anne Bridault identified the presence of horse (of which the bones are most common), as well as red deer, wild boar and a hare (Bridault in Bodu 1995). Céline Bemilli, who is in charge of the analysis of the fauna of locus 46, identified the same faunal spectrum, and she also discovered the presence of a small number of individuals for each species (a maximum of three horses), as well as the presence of a wolf and above all of bone remains of a lion (fragments of the forelimbs and hind limbs). This latter species is exceptional in Late Glacial faunas in the Paris basin. Its presence has been documented in some Final Magdalenian (La Vache - Pailhaugge 1996) and Azilian sites (La Tourasse - Harlé 1894) in the Pyrenees in the form of bones or art (the frieze of lions at La Vache - Buisson and Delporte

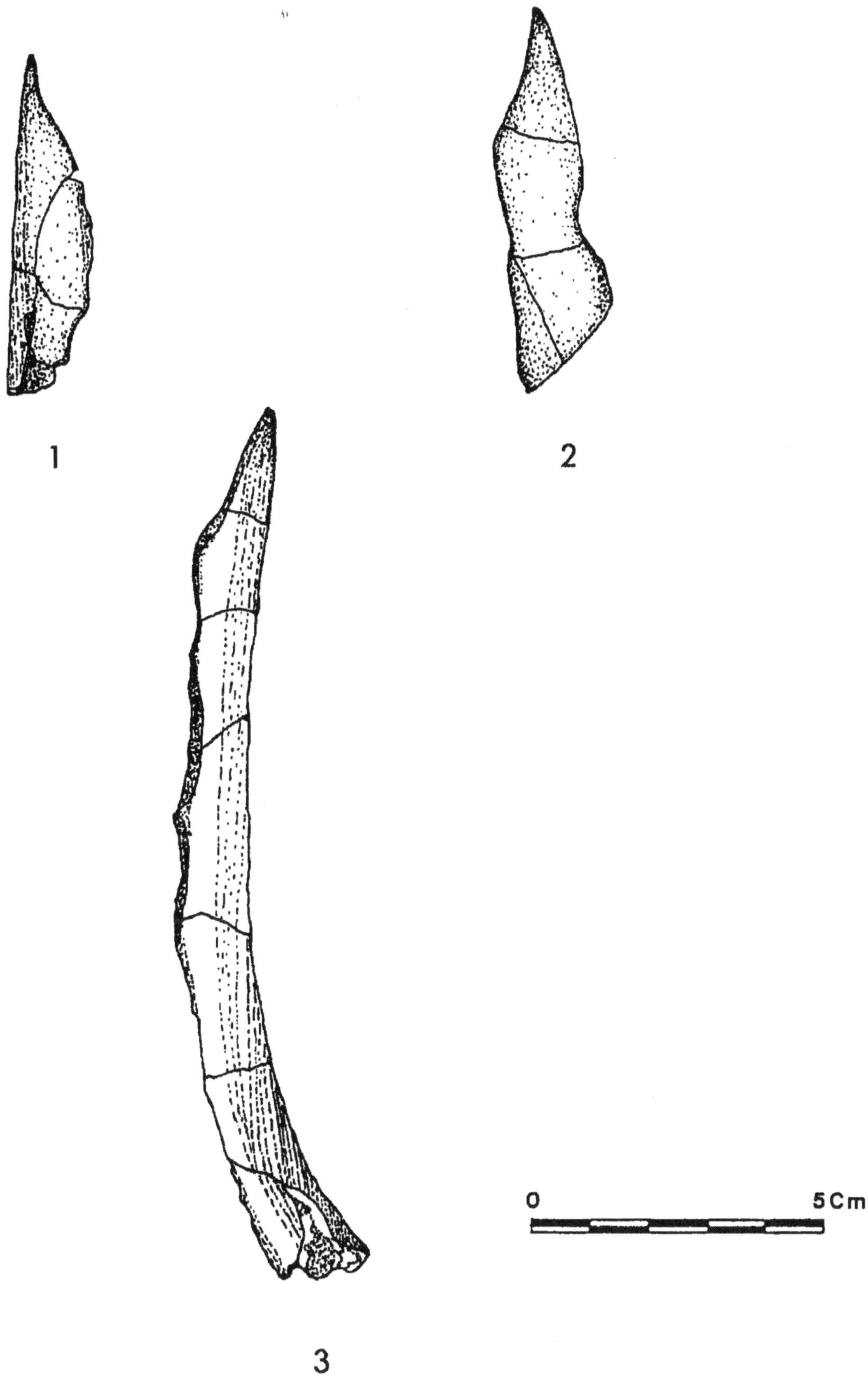

Figure 6: Lower level. Locus 46. Points made on ribs

1988). At the moment, the lion at Le Closeau corresponds to the latest evidence of the presence of this animal in a Palaeolithic archaeological context, although its existence has been documented until around 10,000 BP, but in palaeontological sites (Argant in Guérin and Pathou-Mathis). The mode of introduction and the acquisition of this animal remains to be seen (in the form of ornaments, as an animal skin, an animal which had been hunted by the group at Le Closeau or acquired by means of exchange, and so on).

On the basis of the small number of individuals found around the two large hearths and their species attribution, one can hypothesise that the site was a hunting station which was occupied for a fairly short period of time and which did not specialise in hunting any particular species. The diversity which characterises the hunted animals (at least five species) is more similar to the short-term Magdalenian hunting stations found on the edge of the Paris basin (Yonne valley), and in particular the site of Arcy-sur-Cure (Schmider *et al.* 1996), than to the large monospecific hunting camps (reindeer, horse) such as those illustrated at Pincevent (Leroi-Gourhan and Brézillon 1972) or the sites of the Grand-Canton (Marolles-sur-Seine) (Alix *et al.* 1993).

The quality of the Magdalenian bone industries is well known, and in contrast the bone industries at Epipalaeolithic sites are often quite impoverished. Apart from flat Azilian harpoons made of red deer antler and a few points, there are very few objects among the most recent assemblages (Fagnart 1997). The discoveries at Le Closeau complete this picture of a restricted toolkit and confirms the monotonous character of the Late Glacial bone industries. No worked bones were found in locus 4, while four points made on deer or horse ribs were found in the middle of locus 46. These have been only very roughly worked at one end of each rib by means of scraping and subsequently a slight smoothing (Figure 6). We cannot exclude the possibility that the latter task was carried out using a very coarse grained block which was found at the periphery of locus.

CHARACTERISTICS OF THE LITHIC INDUSTRY

In the middle of the two dense structures of the lower level the quantity of knapped flint amounts to about 2300 artifacts in each assemblage. Locus 33 is much poorer, with just over 200 artifacts. Apart from having a similarly structured spatial organisation the two large loci, 4 and 46, have the same density of lithic artifacts. If one only takes into account the retouched tools, these account for about 4% of each assemblage (3.8% in locus 46 and 4.3% in locus 4). When one also considers the unretouched blanks which show traces of use retouch, the majority of which are blades, the percentages rise to 15.5% in locus 4 and 18% in locus 46.

In locus 33, the annex unit of locus 46, the retouched toolkit represents about 5% of the lithic industry, which is a slightly higher percentage than in the two large loci. If one also considers the blades and flakes with cutting edges which could potentially have been used, the percentage rises to 42%. It is worth pointing out that at locus 33 there is only a small amount of waste material and that without doubt it was a place where blanks were used, whether they were retouched or not. This fact could explain the high utilitarian tendency of the lithic artifacts discovered there which makes the percentage so high.

THE TOOLKIT

If one considers only the retouched tools one finds similar tendencies in the two large loci of the lower level. There are about one hundred classic tools in both locus 4 (108) and locus 46 (81). Endscrapers are the dominant type, accounting for nearly 39% of the tools in locus 4 and about 60% in locus 46. In the latter there is an under representation of burins (2.46%) while in locus 4 these are better represented (19.44%). The abundance of this tool type in locus 4 may be related to the existence of fragments of red deer antler which were found in the same area.

Locus 33 is characterised by a low percentage of tools (a dozen) which are dominated by pointed backed bladelets (8), two truncations, one burin and one composite tool (burin/endscraper). Half of the microliths has clear flake scars on the cutting edges, and one of them has residues of resin preserved on the backing retouch. The other four points may correspond to production failures, because of the breakage types and because they are made of the same type of flint as the few blocks which were knapped here.

In the large loci pointed backed bladelets are the second most common tool type, after the endscrapers (36.11% in locus 4 and 32% in locus 46). These consist almost exclusively of double pointed backed bladelets which are well known in Azilian assemblages in south-west France (Célérier 1983). The refitting of a large number of fragments made it possible to reduce their actual quantities: instead of a total of 39 fragments there are 18 pointed backed bladelets in locus 4 while the 26 elements in locus 46 can be reduced to slightly over 20. The quantity seems to be similar in the two units and one could question whether this does not signify the existence of the same kind of hunting toolkit in the two loci.

Macroscopic impact traces are visible on some of the pieces (breaks with spin-offs, burin-like scars). These were confirmed by microwear analyses which document the presence of polish and chipping on the cutting edges. The location of these stigmatas and the presence of residues of "resin" suggest that the backed points were fixed axially in the haft. However, it is still necessary to determine whether they were arrows or spear points. Analysis with a scanning

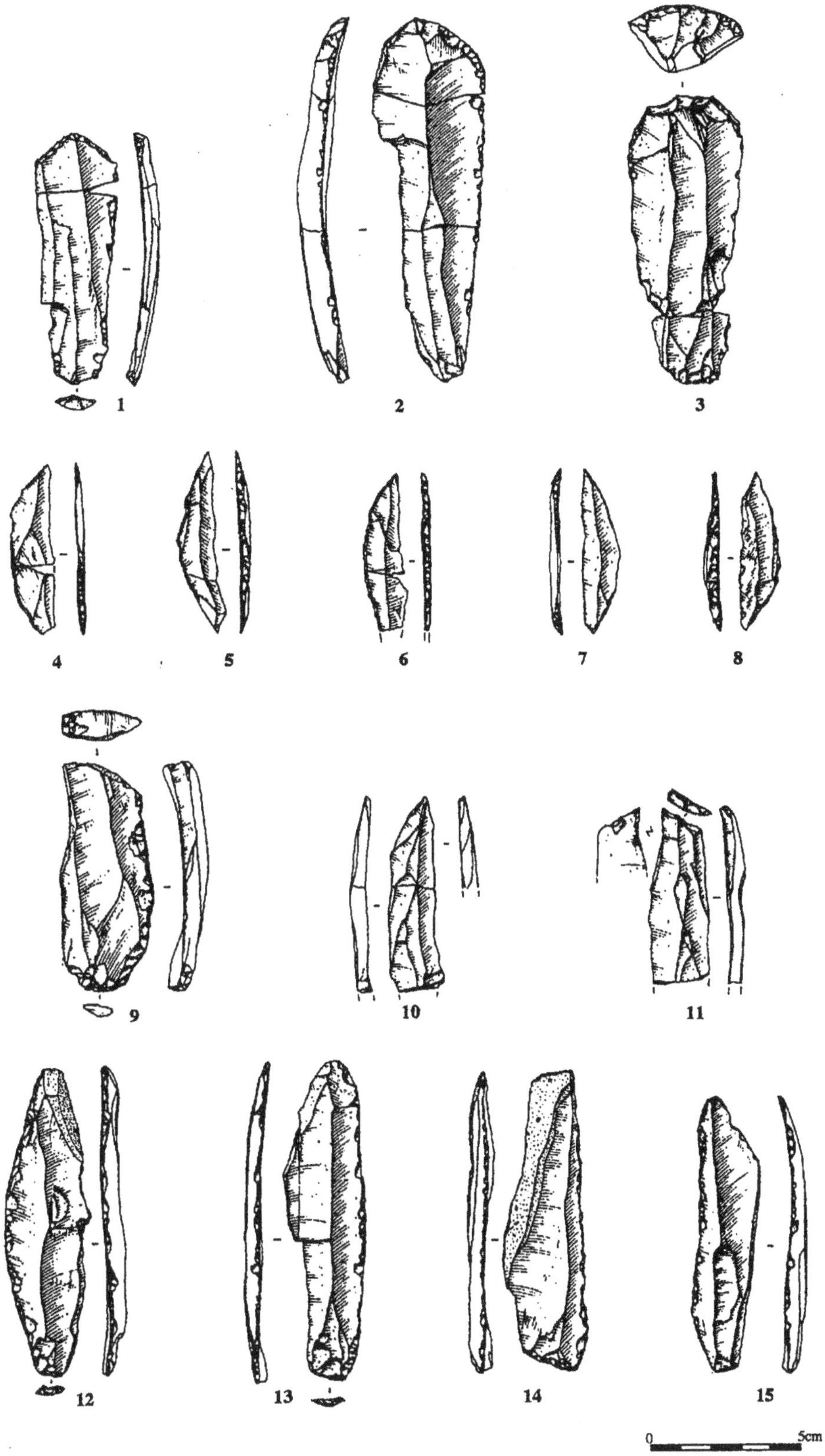

Figure 7: Lower level. Locus 4. Toolkit. 1- 3 endscrapers; 4-8 backed points; 9-11 burins; 12-15 utilised and retouched blades

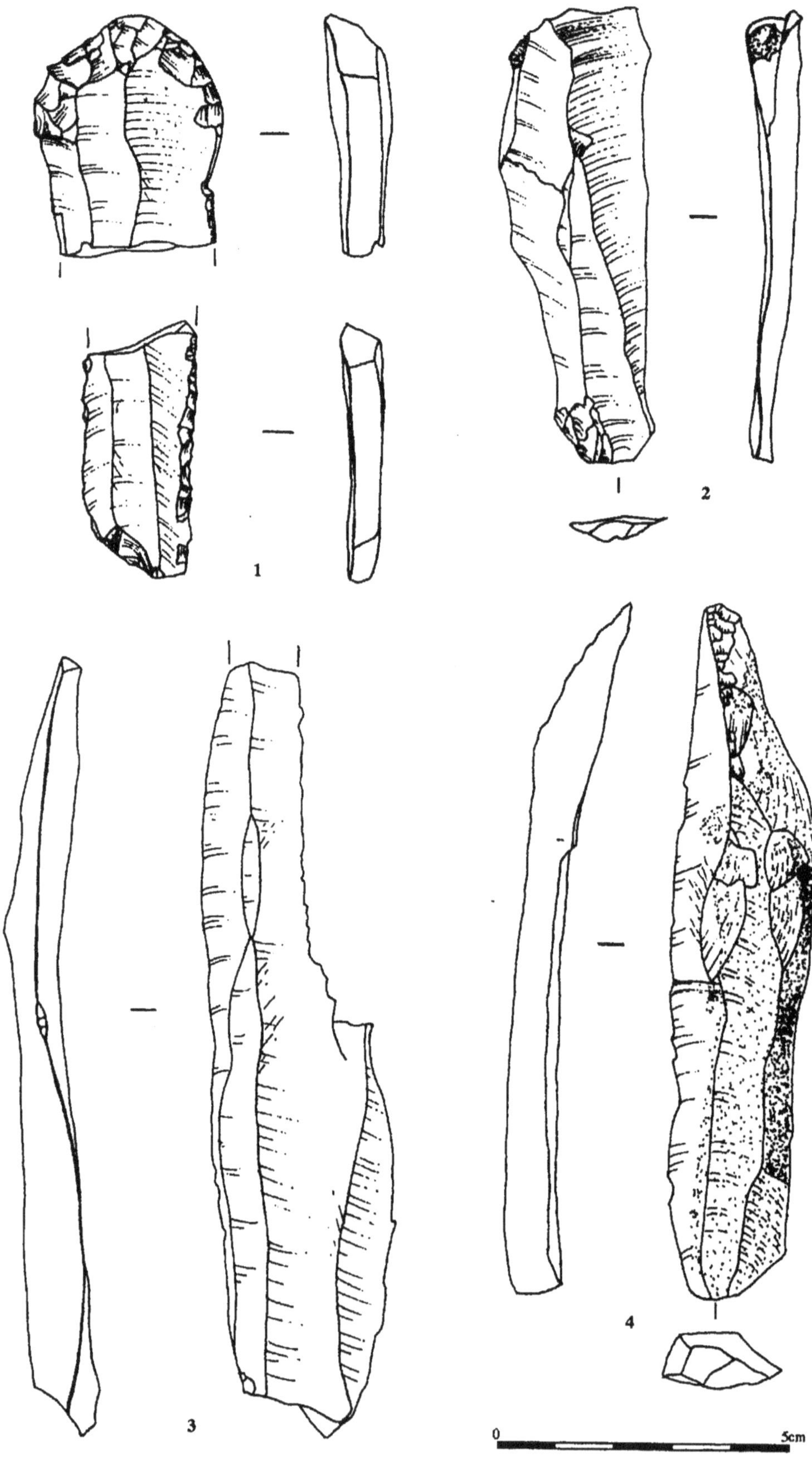

Figure 8: Lower level. Locus 46. Artifacts made of Tertiary flint. 1 endscraper on retouched blade; 2-4 unretouched blades

electron microscope made it possible to define the main components of the residual traces of resin. There is quite a high percentage of carbon as well as a high proportion of organic matter, while the black colour is due to the presence of manganese.

Apart from the three main tool types, there are other artifacts which have been retouched. There is a weak but recurrent presence of composite tools, the majority of which are burins/endscrapers in each of the three loci (6 in locus 4, 4 in locus 46 and 1 in locus 33) as well as a circumstantial presence of truncations (two in locus 33).

Table 3. Main tool types in the loci of the lower level

	Locus 4	Locus 33	Locus 46
Endscrapers	42	0	49
Burins	21	1	2
Backed points	39	8	26
Composite tools	6	1	4

The most numerous "tools" are made on blades, laminar flakes and flakes with "utilised" edges although they do not have real retouch (between 200 and 300 artifacts in each locus). This aspect was already clear after the study of the artifacts from locus 4 and it was confirmed by the analysis of those from locus 46. The blades, which are sometimes very regular, have almost systematic low angle removals on one side of the two cutting edges, the most characteristic of which can be compared with scalar retouch. It does not seem that the mode of obtaining this type of retouch was related to only one use. On the contrary, it seems that it may have been a rejuvenation of those parts of the cutting edge which had been previously damaged, using a soft hammer (wood, antler, bone).

A preliminary microwear analysis showed that some of these blades had, as is frequently observed, been used for cutting animal matter although traces of cutting non woody plants were also detected. In locus 4 these traces were in fact located in the same position as the consecutive traces which resulted from working hard animal matter. In contrast, it seems that the retouch which affected the majority of the laminar products was that formed by cutting plants, after their initial use.

Without doubt this evokes a certain longevity and relatively measured use of these blanks. Finally there is also evidence of working mineral matter on some of the cutting edges of the blades. The matter in question has been identified: it was an iron oxide and its black colour was due to the presence of magnetite. The reason why they used these blades to scrape magnetite is however difficult to discern.

THE TECHNICAL PROCESSES

The raw material

The observation of the cortical surfaces makes it possible to hypothesise that the flint was collected from sources close to the habitation areas. The surrounding area and in particular the Jonchère hills contain a large quantity of secondary flint of Campanian age. This is good quality raw material which made it possible to carry out quite sophisticated phases of core preparation for blade production. The prehistoric flint knappers probably procured the flint on the foothills in the slope scree which formed following colluviation phenomena. One cannot also exclude the possibility that some flint was procured from the alluvium on the valley bottom since some of the blocks have a smoothed and rolled cortex. Another argument which confirms that the lithic raw material was procured locally is the fact that tested blocks were abandoned at the two large loci. The testing was not carried out there however, as the impossibility of refitting the flakes from this first phase in the core preparation sequence shows. The disinterest in these transported nodules, which were of satisfactory quality, is undoubtedly related to the abundance of raw material and to its proximity.

Some blanks made of an allochthonous Tertiary flint were found in both of the large loci in the lower level. These consist of one blade in locus 4 and at least four blades in locus 46 (Figure 8). Although the Champigny limestone which outcrops towards the summit of the hills contains some flint of a colour which varies from brown to beige, it occurs in the form of slabs which do not seem to correspond to the nodules from which the transported blades were made. These few exogenous blanks therefore document the circulation of objects over longer distances which have yet to be determined. This behaviour is documented in the majority of the Magdalenian groups in Ile-de-France and the south of the Paris basin, who generally procured their raw material from local secondary flint deposits but who also imported finished products (blades, tools) made of Tertiary flint (Bodu 1993). In contrast, this behaviour is unknown in the more recent phases of the Federmesser groups. Once again, the best comparison for this lower level at Le Closeau seems to be that of the regional Magdalenian series.

The aims of the reduction sequence

The analysis of the retouched tools and utilised unretouched blanks clearly shows that the aim of the reduction sequence was the production of blades. There are no bladelets which can be said to have been intentionally manufactured. The blades were detached using a soft stone hammer and they have a very rectilinear profile. Their length ranges between 15 and 5 centimetres. The smaller ones were destined to be transformed into microliths, while some of the larger ones were used to make endscrapers, although the majority of them were used unretouched. Burins were made on both good quality blades as well as on more irregular laminar flakes.

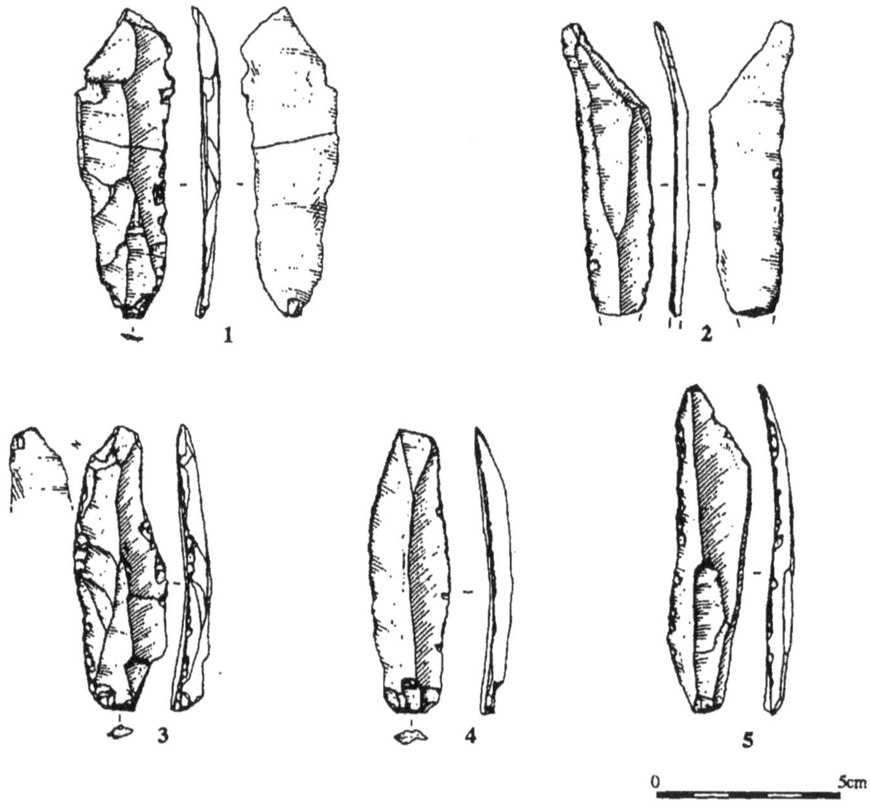

Figure 9: Lower level. Locus 4. Unretouched blades

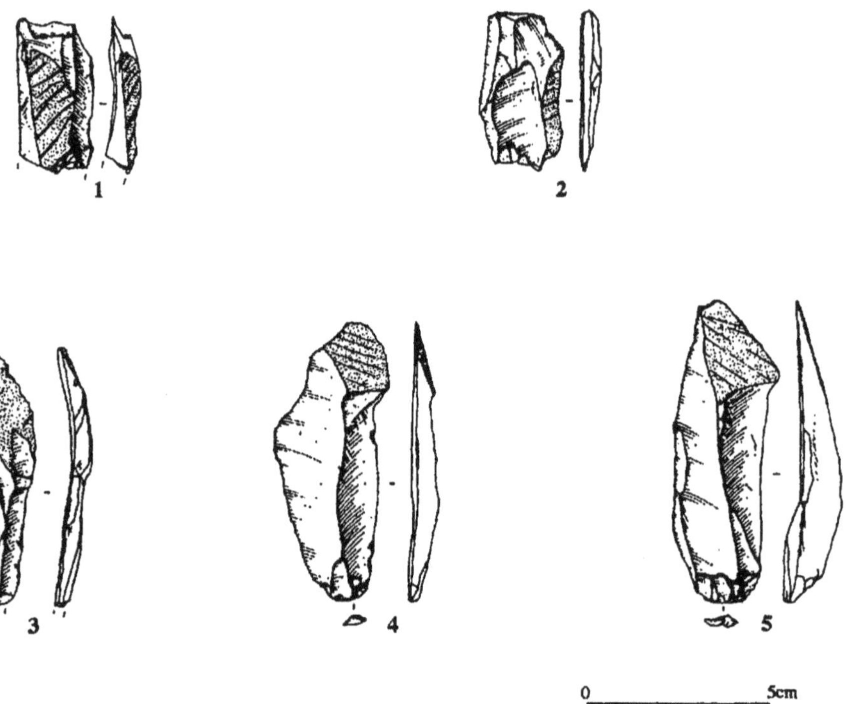

Figure 10: Lower level. Locus 4. Engraved cortex

The reduction sequence was not oriented towards the production of flakes but this did not exclude the provisional use of this type of blank for some activities. The end of the exploitation of the blocks also shows the removal of flakes using a less tangential gesture that that used for the removal of the blades, although it is difficult to understand the purpose of these elements which may have been intentional products but may equally have been production failures resulting from the loss of the convexities necessary for blade production.

The technique

A soft stone hammer was used throughout the reduction sequence, a characteristic which contrasts markedly with what is known about Magdalenian knapping. The latter was characterised by the use of two types of hammer, soft organic and soft stone, in precise moments during the reduction sequence. In the lower level at Le Closeau a soft stone hammer, without doubt made of sandstone like the one found in locus 46, was used for the removal of blades as well as for that of flakes, although a more tangential gesture was used for the former. An abrasion of the overhang of the striking platform sometimes associated with facetting of the latter made it possible to aim the blows close to the edge and thus to obtain relatively thin and standardised blades.

Core preparation

The phases of core preparation have been documented by the presence of cortical and semi-cortical flakes, crests and sub-crests. The negative flake scars on some of the cores testify to an adjustment of the convexities which concerned above all the blade flaking surface but also sometimes affected the lateral and dorsal sides of the blocks. As a result the care dedicated to the creation of crests is often less pronounced than that attested in the Magdalenian series. The arrises are more sinuous, and the negative bulbs deeper. Refittings showed that some of these operations were not carried out at the locus itself, but perhaps at the raw material procurement area. The majority of the blocks were not extensively prepared since their natural roundness was favourable for knapping, as is suggested by the existence of numerous cortical and semi-cortical first blades. It should be pointed out that the degree of investment made in this phase of the reduction sequence is much higher than that in the more recent Azilian levels at Le Closeau (Bodu 1995).

Core reduction

The majority took place using blocks with two opposed striking platforms (two thirds of the cores in locus 4). The blades were detached from both ends of the cores according to a rhythm which varied as a function of the hazards of knapping and the evolution of the tabular morphology. It was not a strictly alternating series of blade removals, but rather a compromise between this and removals which favoured one of the striking platforms. The use of a stone hammer in fact, for the purposes attempted, prohibited an exclusively unipolar core reduction in the same way as it prohibited a strictly alternate removal of blades (Figure 9). During the reduction process frequent controls were made of the convexities (self-maintenance), since more or less deep flake scars accumulated on the blade flaking surface. Side blades accompanied the flakes removed transversally to the laminar axis (neo-crests). The surfaces of the striking platforms were sometimes rejuvenated by means of the removal of a whole tablet or partial tablets, as the regular facetting of the platforms of some of the flakes or blades was damaging.

Discard

The sizes of the blocks which were discarded during core reduction are relatively variable (between 5 and 15 centimetres long) although there is a predominance of those with a length comprised between 8 and 10 centimetres. The last flake scars correspond to blades of about 6 centimetres in length which were used to make microliths. Sometimes a violent blow aimed at the interior of the striking platform resulted in the removal of a last thick flake of which the function has not been identified (blank?) and which disfigured the blade flaking surface.

Engraved cortex

One of the most particular aspects of the lithic industry is the presence of "engravings" on the cortex of various blanks which refitting has shown to appertain to the same block. These consist of four or five parallel lines spaced a few millimetres apart (Figure 10). The refitting of the group shows that the lines had been engraved before the block was knapped. The scraping of the surface of a block may enable one to remove thick cortex, although the examples from Le Closeau do not have thick cortex and the regular lines therefore cannot be attributed to this type of gesture. In addition, the motifs of the engravings are particularly widespread in Epipalaeolithic art, and in particular on pebbles (Baffier 1992). Therefore one cannot exclude the possibility that the engraved cortex at Le Closeau represents a form of artistic expression. This hypothesis could be confirmed by the identification of quite particular engravings. These are the figurative traces representing the simple outline of an unidentified animal's head and an eye surmounted by an eyebrow. These engravings revoke the Epipalaeolithic level at Pincevent where an engraving representing a horse's head was found on the cortex of a flake (Baffier in Gaucher 1996). The engraved elements at Le Closeau therefore enrich this "artistic" ensemble which was more symbolic than technical.

CONCLUSIONS

The industry, the dates obtained in locus 4 and the stratigraphic position place the assemblage in the Allerød and at the extreme end of the Bølling. Fruitful comparisons have been made with the site of Gouy situated in Seine-Maritime (Valentin 1995) where the radiocarbon dates suggest an ancient phase of the Federmesser groups (12,050±130 BP). These assemblages can be cautiously compared with the industries from the levels attributed to the ancient phase of the Federmesser groups discovered in the Somme valley (Fagnart 1997). Finally, because of its spatial, technological and economic characteristics we have compared the lower level at Le Closeau which was discovered in 1994 with regional Magdalenian assemblages. The two new concentrations (loci 33 and 46) confirm the idea of a close relationship between the last Magdalenian populations and the first groups of Azilians and the possibility of a direct continuity between the two groups is seen as a topic for future research.

Various different aspects plead in favour of a similarity between the first groups of Azilians and the last of the Magdalenians. As regards the percussion techniques however there are some differences. We recall that the Magdalenians, who used a soft organic hammer for blade manufacture, developed the use of the soft stone hammer during some of the phases of core reduction. This is the case at some of the final Magdalenian sites in the Paris Basin (Marsagny, Cepoy) where blade production using a soft organic hammer was followed by the removal of some blades using a soft stone hammer which were destined to be made into angular backed points (Valentin 1995). The use of a soft stone hammer by the Magdalenians documents a knowledge and skill of a procedure which was continued and developed by the Azilians.

These differences do not however detract from the similarities noted. The main characteristics of the reduction sequence are the same, both in the mode of raw material procurement as well as in the core preparation and core reduction phases. It is above all at the level of the economic treatment of the blanks that one notes the closest similarities. In both cases the different types of blades were destined to be made into the same type of tool: the endscrapers tended to be made on the more regular blades, the smallest blades (or bladelets in the case of the Magdalenians) were oriented towards the production of microliths, while the less standardised blanks were transformed into burins. As is the case in the Magdalenian, a certain number of blades was used unretouched for cutting different types of matter and the backed knives which are so common in the more recent Federmesser assemblages are in fact absent at Le Closeau. Finally, a similar behaviour was observed as regards the economy of the allochthonous raw material: both the Magdalenians and the early Azilians at sites in the centre of the Paris Basin frequently imported finished products made of Tertiary flint, a behaviour which is absent or much less frequent in the more recent Azilian populations.

The similarities seem to be even more pertinent if one considers the modes of spatial organisation. The marked structuring of the two units at Le Closeau evoke those of certain Magdalenian sites. We have suggested similarities with certain units at Etiolles which are dated to the Bølling and we have shown that such a spatial organisation, which was so onerous in terms of time and energy, does not exist in any contemporary or more recent sites of the Federmesser groups tradition.

Therefore, despite the slight differences in the knapping techniques, one can propose as a working hypothesis that there was a direct filiation between certain Magdalenian groups and the first Azilians, and in fact the radiocarbon dates demonstrate a relative contemporaneity of the two cultures.

Having shown this, it is necessary to work on the reasons behind this putative evolution (adaptation to environmental or climatic constraints, cultural choice etc.). It already seems that one cannot accept either one or the other of these reasons rather that it is necessary to evoke a combination of all the phenomena in order to explain this long development. Another question arises and it is tempting to answer it in terms of environmental determinism: the relationship between the early Azilian level and the more recent Azilian levels, between which one observes a technical rupture and at the same time a change in the economic and spatial aspects. If this rupture turns out to be significant, one could explain it as being the result of the slow climatic amelioration during the Allerød which evidently played an important role in the faunal and vegetal composition and thus on the availability of these resources and the techniques which the hunter-gatherers developed as a response. One should not forget, however, in order to underline the importance of cultural choices on the evolution of techniques, that a change in the fauna and similarly in the vegetation is perceptible in the early phase of the Azilian, and that this seems to have resulted in only very slight changes in the technical behaviour.

One of the numerous directions of research which the site offers is therefore the analysis of the relationship between the different Azilian levels at Le Closeau, of which only the early phase has been presented here. In future research it will be necessary to establish the characteristics of each of the groups and to propose an evolutionary scheme which takes into consideration all the various different cultural aspects.

REFERENCES

Alix, P., A. Averbouh, L. Binter, P. Bodu, A. Boguszewski, C. Cochin, V. Deloze, P. Gouge, V. Krier, C. Leroyer, D. Mordant, M. Philippe, J.-L. Rieu, P. Rodriguez & B. Valentin

1993 Nouvelles recherches sur le peuplement magdalénien de l'interfluve Seine-Yonne: Le Grand Canton et Le Tureau des Gardes à Marolles-sur-Seine (Seine-et-Marne). *Bulletin de la Société Préhistorique Française* 90 (3): 196-218.

Baales, M. & M. Street
1996 Hunter-gatherer behavior in a changing Late Glacial landscape: Allerød archaeology in the central Rhineland. *Journal of Anthropological Research* 52: 218-316.

Baffier, D.
1992 Le Mésolithique. In J. Garanger (ed.) *La Préhistoire dans le Monde*, 516-524. Paris: Nouvelle Clio P.U.F.

Bodu, P.
1993 *Analyse typo-technologique du matériel lithique de quelques unités du site madgalénien de Pincevent (Seine-et-Marne). Applications spatiales, économiques et sociales.* Unpublished doctoral thesis, Université de Paris I.
1995 Le Closeau, Rueil-Malmaison (Hauts-de-Seine). *Document final de Synthèse de diagnostic, A.F.A.N., S.R.A. d'Ile-de-France, D.D.E. des Hauts-de-Seine.*
1996 Le Closeau, un gisement à Federmesser sur les bords de la Seine. Rueil-Malmaison (Hauts-de-Seine). *Document final de Synthèse de sauvetage urgent, A.F.A.N., S.R.A. d'Ile-de-France, D.D.E. des Hauts-de-Seine.*

Buisson, D. & H. Delporte
1988 Intérêt d'un raccord pour l'authentification d'une oeuvre d'art. *Bulletin de la Société Préhistorique Française* 85 (1): 4-6.

Célérier, G. & E.H. Moss
1983 L'abris-sous-roche de Pont d'Ambon à Bourdeilles (Dordogne). Un gisement magdalénien-azilien. Micro-traces et analyse fonctionelle de l'industrie lithique. *Gallia Préhistoire* 26 (1): 81-108.

De Bie, M.
1997 L'industrie à Federmesser de Rekem (Belgique): datation, technologie lithique et répartition spatiale. In J.-P. Fagnart & A. Thevenin (eds.) *Le Tardiglaciaire en Europe du Nord-Ouest.* Actes du 119è Congrès National des Sociétés Historiques et Scientifiques (Amiens 1994), 381-396. Paris: C.T.H.S.

Fagnart, J.-P.
1997 *La Fin des Temps Glaciaires dans le Nord de la France. Approches Archéologique et Environnementale des Occupations Humaines du Tardiglaciaire.* Mémoires de la Société Préhistorique Française 24.

Fosse, G., B. Valentin & C. Billard
1997 Une occupation des groupes à Federmesser en Haute-Normandie: le gisement du Cornet. In J.-P. Fagnart & A. Thevenin (eds.) *Le Tardiglaciaire en Europe du Nord-Ouest.* Actes du 119è Congrès National des Sociétés Historiques et Scientifiques (Amiens 1994), 245-255. Paris: C.T.H.S.

Gaucher, G.
1996 *Fouilles de Pincevent. Le Site et ses Occupations Récentes. L'environnement Tardi et Post-Glaciaire et les Témoins Postérieurs au Magdalénien.* Mémoires de la Société Préhistorique Française 23.

Guérin, C. & M. Pathou-Mathis
1997 *Les Grands Mammifères Plio-Pleistocènes d'Europe.* Paris: Masson.

Hinout, J.
1985 Le gisement épipaléolithique de la Muette I, commune de Vieux-Moulin. *Bulletin de la Société Préhistorique Française* 82: 377-388.

Harlé, E.
1894 Restes d'élan et de lion dans une station préhistorique de transition entre le Quaternaire et les temps actuels à Saint-Martory. *Revue de Comminges* 9: 402-406.

Julien, M., F. Adouze, D. Baffier, P. Bodu, P. Coudret, F. David, G. Gaucher, C. Karlin, M. Larriere, P. Masson, M. Olive, M. Orliac, N. Pigeot, J.-L. Rieu, B. Schmider & Y. Taborin
1988 Organisation de l'espace et fonction des habitats magdaléniens du Bassin parisien. In *De la Loire à l'Oder. Les Civilisations du Paléolithique Final dans le Nord-Ouest Européen. Actes du Colloque de Liège 1985.* British Archaeological Reports International Series 444 (i): 81-123.

Kildea, F.
1996 *Etude du Matériel Lithique du Niveau Récent du Site à Federmesser du site du Closeau à Rueil-Malmaison (Hauts-de-Seine). Approche Technologique, Typologique et Spatiale de Six Unités d'Occupation.* Mémoire de Maîtrise, Université de Paris I.

Leroi-Gourhan, A. & M. Brézillon
1972 *Fouilles de Pincevent. Essai d'Analyse Ethnographique d'un Habitat Magdalénien (la section 36).* VIIe supplément à Gallia Préhistroire.

Pailhaugue, N.
1996 Faune et saisons de chasse de la salle Monique, grotte de "La Vache" (Alliat, Ariége). In H. Delporte & J. Clottes (eds.) *Pyrénées Préhistoriques, Arts et Sociétés.* Actes du 118è Congrès National des Sociétés Historiques et Scientifiques (Pau 1993), 173-191. Paris: C.T.H.S.

Schmider, B., B. Valentin, D. Baffier, E. David, M. Julien, Arl. Leroi-Gourhan, C. Mourer-Chauvire, T. Poulain, A. Roblin-Jouve & Y. Taborin
1996 L'abri du Lagopède (fouilles Leroi-Gourhan) et le Magdalénien des grottes de la Cure (Yonne). *Gallia Préhistoire* 37 (1): 55-114.

Valentin, B.
1995 *Les Groupes Humains et Leurs Traditions au Tardiglaciaire dans le Bassin Parisien. Apports de la Technologie Comparée.* Unpublished doctoral thesis, Université de Paris I.

PUTTING HUMPTY TOGETHER AGAIN: PRELIMINARY OBSERVATIONS ON REFITTING STUDIES IN THE EASTERN MEDITERRANEAN

Nigel Goring-Morris, Ofer Marder, Angela Davidzon & Fawzi Ibrahim

INTRODUCTION

A comprehensive program of refits being conducted on 24 chipped stone assemblages from short term encampments in a restricted area of the western Negev Desert, Israel, has resulted to date in about 8500 conjoined artefacts. The targeted assemblages represent the various typo-technologically defined local Epipalaeolithic industries dated between ca. 20,000-10,000 BP. The paper presents the background to the project, as well as some preliminary results and observations on the refits (detailed technological and intra-site spatial studies are in progress).

Though incomplete, refitting rates for assemblages commonly exceed 15% and reach up to 32%, with maximal individual refitted blocks in sites often exceeding 50 items. Temporal tendencies in conjoin rates have been recognised which may reflect changing mobility, curation and discard patterns. The observed trends also probably correlate to some degree with changes in raw material procurement practices.

Three general reduction sequence schemes have been identified. One, though numerically rare, is a constant feature in most assemblages, involving the procurement and off-site production of blanks for massive tools from nearby outcrops, using the block-on-block technique. Another concerns the occasional appearance in a few assemblages of reduction sequences specifically geared to the production of medium-sized tools, especially scrapers. The third *chaîne opératoire*, by far the most common and with many, often chronologically vectored, variations focused on the manufacture of blanks for microlithic tools and, as by-products, blanks for medium-sized tools.

Developments in this last general reduction sequence scheme are suggested to reflect changing conceptual approaches to tool manufacture, away from a predetermined configuration of the microlithic blank to an emphasis on achieving the desired form by subsequent modification of the blank by more invasive retouch and application of the microburin technique. Contrary to the superficial appearances of a decline in knapping proficiency, this is actually shown to be a sophisticated development, leading to increasing efficiency and economising practices.

Chipped stone refitting studies and concomitant reconstruction of reduction strategies have been widely adopted over the past two decades as analytical tools in the study of lithic assemblages in both the Old and New Worlds, including the southern Levant. The impact and contribution of such studies are manifold and varied (e.g. Bodu *et al.* 1991; Cziesla *et al.* 1990; Hofman & Enloe 1992; Karlin *et al.* 1992; Karlin & Julien 1995; Marks 1983; Pelegrin *et al.* 1988; Ploux 1991; Schlanger 1995; Tixier 1976; Villa 1982, 1991; Volkman 1983). For a variety of pragmatic and other reasons, such studies commonly focus on single assemblages or roughly contemporary assemblages from a limited geographical area. While site or horizon specific dynamics are often demonstrated, it is rarely obvious to what extent these reflect vectored patterns through time and space (Note 1).

Over the past 25 years dozens of occupation sites have been excavated in the Negev and Sinai Deserts dating to the Epipalaeolithic period, ca 20-10,000 BP uncalibrated (summarised in Goring-Morris 1987, 1994, 1997). In this arid region of the southern Levant the impact of the Pleniglacial and its immediate aftermath were marked, resulting in major changes in the landscape which affected the distributions, densities and patchiness of floral and faunal resources (Goring-Morris & Goldberg 1990; Baruch & Goring-Morris in press; Horowitz 1977; Tchernov *et al.* 1986). The material culture residues of the human populations in the area during this period display a major increase in diversity and tempo of change in comparison to the preceding Upper Palaeolithic. This variability is reflected especially, but by no means solely, in standard techno-typological analyses of the lithic assemblages. Coupled with stratigraphic considerations, radiometric dating and other patterns in the material record (e.g. hearth types, marine mollusc assemblages, architecture, groundstone tools, etc.), these have enabled seriation of the microlithic assemblages and assignment to the numerous cultural entities identified in the region during the Terminal Pleistocene (Table 1). Based also on all these variables, within and between the sites, models regarding changes in mobility, settlement patterns and subsistence have been proposed (Goring-Morris 1987, 1989, 1991). While many of the Early and Middle Epipalaeolithic entities are relatively mobile and largely restricted to the lowland dunes of northern Sinai and the western Negev, Late Epipalaeolithic entities indicate seasonal movements from residential basecamps in the central Negev Highlands to the lowlands.

Many of the sites are small-scale, *in situ* or partially deflated occupations in the western Negev *erg* (dunefields). The number, size and nature of these sites provided a unique opportunity for an ongoing program of lithic conjoins (Note 2). To date refitting studies have been systematically

Table 1: Scheme of Terminal Pleistocene socio-cultural developments in Negev and Sinai (listed assemblages have been refitted)

14C Chronology*	habitual use of mbt	no use of mbt
	LATE EPIPALAEOLITHIC	
10,750-10,100	**Harifian** Shunera XXIV; Shunera IX	
11,500-10,750	**Late Natufian** Shunera XVIII; Givat Hayil I	
12,500-11,500	***Terminal Ramonian*/Early Natufian** Nahal Sekher 23; Shunera VII; Azariq XV	
	MIDDLE EPIPALAEOLITHIC	
13,250-12,500	***Ramonian*** Nahal Lavan 1003; Nahal Lavan 1009; Shunera XXI	
14,000-13,250 14,500-14,000 *14,500-14,000* ~15,000-14,500	***Classic Mushabian*** Azariq XII; Azariq VIIB; Har Qeren VIII; ?Shunera II? ***Proto-Mushabian*** Shluhat Qeren II; Azariq X	**Geometric Kebaran** Azariq XVIII; Shunera XIIB; Azariq VIIA, Azariq II **Intermediate Kebaran/ Geometric Kebaran**
	EARLY EPIPALAEOLITHIC	
16,500-15,500 ~16,500-14,500	***Nizzanan*** Hamifgash IV	**Late Kebaran** Shunera XVII
~18,500-16,500		**Early Kebaran** Azariq IV
~20,000-19,000 ~19,500-16,000	***Nebekian?***	**Masraqan** Azariq XIII; Shunera XVI
	UPPER PALAEOLITHIC	
~25,000-20,000		**Arqov/Divshon**
~35,000-25,000		**Early Ahmarian (Lagaman)** Nahal Nizzana XIII

* BP uncalibrated

conducted on 24 different assemblages. With but four exceptions (Marder 1994), detailed technological studies of the *chaînes opératoires* and their spatial correlates are still incomplete, as indeed are some of the conjoining efforts. In consequence the present paper is a brief overview of the background to and the aims of the project, followed by presentation and discussion of some of the more obvious, but inevitably tentative, results achieved to date.

BACKGROUND

The Western Negev Landscape

The western Negev and northern Sinai is a low-lying area rising incrementally from sea level up to elevations of 350 metres, and is punctuated by a series of gently rounded anticlines. Some 60 kilometres inland, the central Negev Highlands, also comprising a series of dissected anticlines and synclines, begin to rise up, eventually reaching elevations up to about 1000 metres some 100 kilometres south-east of the coast (sea levels during the Terminal Pleistocene were lower, so that the coastal plain was then 20-30 kilometres wider).

The western Negev and north Sinai *erg* swiftly penetrated the area north-eastward from the Nile Valley during the cold and dry conditions of the Pleniglacial, corresponding to the Early Epipalaeolithic, ca. 20,000-14,500 BP. The dunefields then stabilised as conditions rapidly ameliorated during the Middle Epipalaeolithic, ca. 14,500-12,500 BP. Since then deflation and re-mobilisation of the dune crests have occurred at various times, e.g. during the Younger Dryas cold event ca. 11,000-10,000 BP, though much of the *erg* today consists of the consolidated Terminal Pleistocene sands. The initial incursion of dunes into the region blocked the nahal (= wadi = seasonal drainage) systems flowing from the Negev Highlands towards the Mediterranean, resulting in local ponding and the creation of at least seasonal playas. Once filled with fluvial and/or limnic silty deposits the nahals breached the dunes, leading to down-cutting (in some instances the process of in-filling is still incomplete and local ponding still occurs in winter). It is in such favoured ecotones around the playas that major clusters of Epipalaeolithic occupations are commonly situated, almost always specifically located on or in the sands rather than adjacent loess and silt surfaces.

Bedrock exposures of Eocene limestones are common in this part of the western Negev, around the 20 kilometres long Qeren-Rogem anticline, often providing ready access to bands of chert and flint, which occur in the form of cortical nodules of variable size, homogeneity and quality (larger nodules atop Har Qeren reach diameters up to 75 centimetres diameter). All the studied sites are located within less than 2-3 kilometres of such sources. Conglomeratic flint is also present in small quantities, the sources of which are still undetermined, but probably from Neogene or later outcrops around Nahal Besor. Somewhat more distant (3-25 kilometres) to the south-east, in the central Negev Highlands and foothills, bedrock exposures additionally include extensive outcrops of Campanian, Senonian, Turonian and Cenomanian limestones. Some contain smaller quantities of medium-sized flint nodules and fractured bands, the latter including distinctive blue/grey/brown translucent chalcedony, which rarely exceeds 7-8 centimetres in maximum dimension. The chalcedony can also be found in smaller quantities in secondary contexts in the major wadi beds or in pedestaled Early and Middle Pleistocene terraces.

Aims and provisos

The aims of the refitting project, which are ultimately planned for some 30 assemblages, have been manifold. In common with many similar studies, at the level of the individual assemblage, they include reconstruction of the various *chaînes opératoires*: the investigation of raw material procurement strategies; chipped stone reduction sequences; and the dynamics of intra-site and off-site activity patterning and relative mobility, as reflected by discard patterns for the various stages of the different reduction sequences. Perhaps the most significant contribution of the present study concerns the number of assemblages involved. This permits the investigation of diachronic and synchronic variability and vectored developments with regards all these variables over the course of approximately 10,000 ^{14}C years from a limited geographical region.

The study has been planned to provide at least two assemblages from each of the traditionally defined Epipalaeolithic socio-cultural entities defined in the Negev, a goal that generally has been achieved (Table 1). The targeted assemblages were recovered from the southern and south-eastern edge of the western Negev *erg* (dune-fields), along a SW to NE axis of approximately 40 kilometres. All the sites derive from broadly similar geomorphologic settings and all but three of the assemblages derive from two restricted areas, respectively encompassing less than 20 km^2 and 5 km^2, and located approximately 15 kilometres from each other, while the others are located in broadly analogous settings. Har Qeren VIII, located on the northern flank of the Har Qeren anticline, is the only site some distance from a major drainage system.

Assemblages for the current study derive from six areas where concentrations of Late Quaternary prehistoric sites have been investigated (Figure 1):

1. The Nahal Nizzana - Nahal Lavan confluence area (including the Nahal Nizzana, Azariq, Nahal Lavan, and Shluhat Qeren assemblages): 11 sites (Goring-Morris 1987; Marder 1994).
2. The Shunera Dunes area adjacent to Wadi-es-Sid: 9 sites (Goring-Morris 1987, 1993, in preparation; Marder 1994).

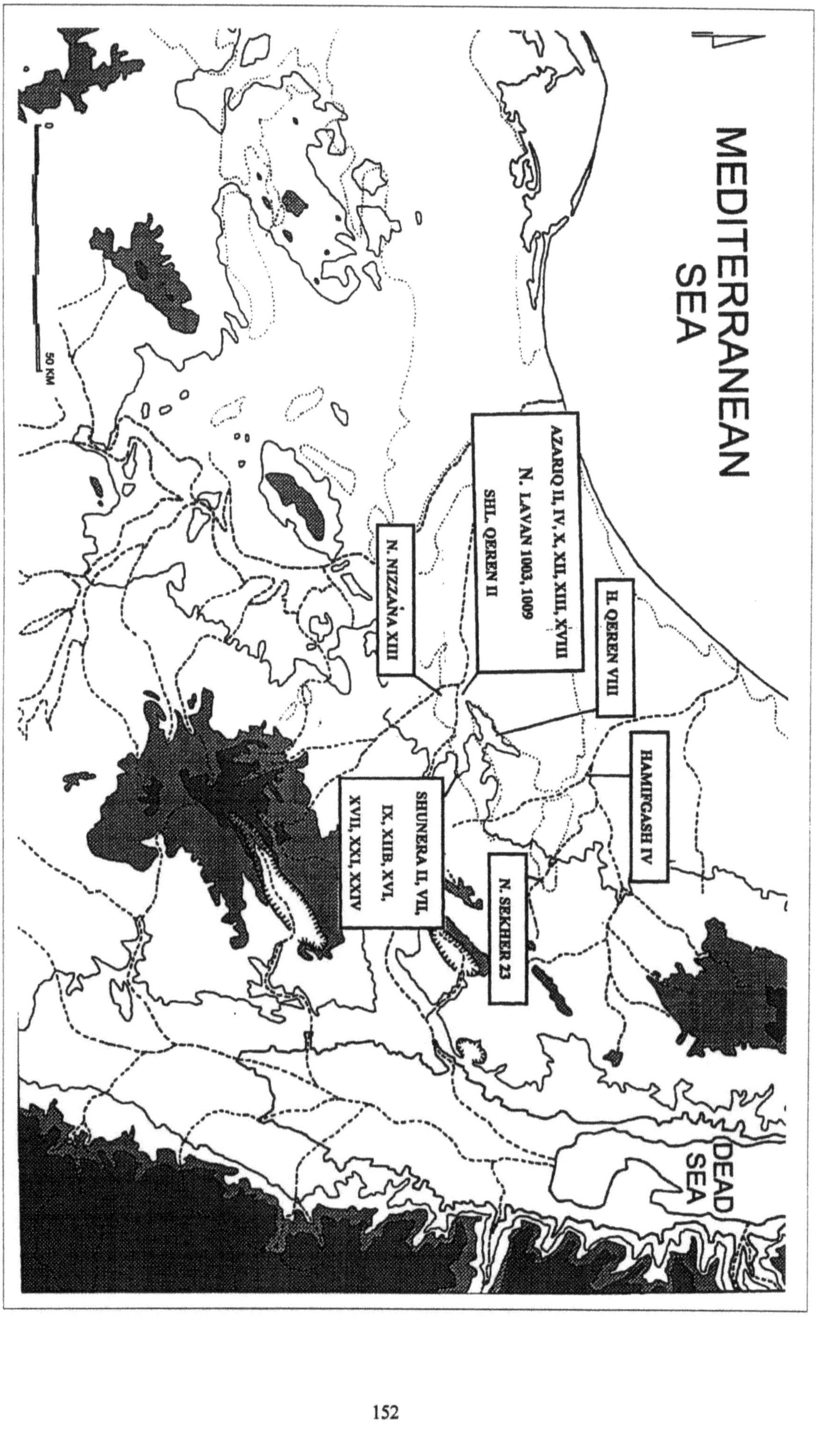

Figure 1: Map of the western Negev showing the locations of the studied assemblages. Dotted lines indicate dunefields and contours at 300 metre intervals

3. Givat Hayil, adjacent to Nahal Pakuah: 1 site (Goring-Morris 1997, in preparation).
4. The Nahal Sekher area: 1 site (Goring-Morris 1987, in preparation; Marder 1994).
5. The Nahal Beersheva - Nahal Besor confluence area (Hamifgash): 1 site (Goring-Morris 1987, in preparation).
6. The Har Qeren area: 1 site (Goring-Morris & Rosen 1987).

The selection was intended to neutralise as far as possible raw material availability for the chipped stone assemblages as a factor in the outlined investigations at the regional level. Such changes in raw material usage as are documented are thus more likely to reflect considered choices of the knappers during the course of their seasonal or annual rounds. Naturally, the impact of documented climatic changes should be taken into consideration regarding the biotic resources, especially plants, exploited from the sites studied, since these may well have varied during the course of the Epipalaeolithic (Baruch & Goring-Morris in press). The major portion of most lithic assemblages appear to be associated with hunting and prey processing gear.

Another matter to be taken into consideration is that some entities, such as the Masraqan, Geometric Kebaran, and Mushabian, display a relatively limited range of site and assemblage sizes and variability in terms of tool classes. By contrast other entities, such as the Ramonian and, most especially, the Natufian and Harifian, with large residential basecamps in the highlands and smaller logistical camps in the lowlands, are far more complex. Attempting refitting programs on assemblages from basecamp sites, however, is not practically feasible, since they commonly include lithic assemblages numbering hundreds of thousands of artefacts, or more.

REFITTING STUDIES

The refitted assemblages

To date significant refitting studies have been conducted on 24 different assemblages often, but not always, with considerable success. As can be seen from Table 2, all these sites were located on or in sand dune, sandy surface or silty sand floodplain settings. The vast majority were found deflating to various degrees from relatively flat surfaces, commonly with a large proportion of the sites still *in situ* immediately below the surface; this is attested by the presence of hearths, the pristine state of most artefacts, and the presence of slight carbonate encrustations on the underside of artefacts. In only a few instances deflation was more intense, when most of an assemblage was exposed on the surface. Sometimes deflation was not horizontal, but rather had scalloped out part of the consolidated fossil dunes, either in the form of miniature mesas or buttes (Nahal Lavan 1003, Nahal Lavan 1009, Shunera XXIV), or as a small 'step' in the slope of a dune (Shunera IX) so that part of the site remained *in situ* and part was completely deflated, scattering artefacts down-slope. In several instances sites were only found due to the path of tracked vehicles which had disturbed the biological crust on the surface and brought artefacts to the surface (Shunera XIIB, Shunera XVII). One site (Nahal Sekher 23) was exposed in a bulldozer trench in the side of an eroded dune, whilst another was exposed by local gullying adjacent to a major wadi channel (Hamifgash IV).

Standard retrieval techniques were based on a ¼m² grid over each site, with all sediment sieved through 2-3 mm mesh. Occasionally, where spatial integrity was seemingly less well preserved, the grid size was increased to 1 m² (Givat Hayil I). In a few instances, when extensive deflation onto compacted dune surfaces had occurred, all visible material was collected without a grid, with (Shunera XXIV) or without sieving (Shunera IX, Nahal Lavan 1003, Nahal Lavan 1009). These do not appear to have appreciably biased the results presented below (especially when taking into account the fact that the *in situ* portions have still not been excavated in the last three cited instances).

All but one of the sites can be assigned to the Terminal Pleistocene Epipalaeolithic period, ca. 20,000-10,000 BP. The single exception is an Upper Palaeolithic Lagaman (Early Ahmarian) assemblage (Nahal Nizzana XIII), probably predating the others by at least 10-15,000 years. It was found deflating from sandy silts unconformably underlying dune sands and has been included within the present study for comparative purposes, since a portion of the assemblage is microlithic.

All of the studied sites were small open-air occurrences, in which the only preserved constructional features were usually one or more hearths and associated ash dumps. The occupation horizons, generally thickest in the immediate vicinity of hearths, rarely exceeded 15 centimetres and most commonly varied from 5-8 centimetres in thickness. Material culture remains were limited to the chipped stone assemblages, as well as rare workslabs, limestone tools and hammerstones. Variable quantities of ornamental non-terrestrial shells, some modified, were present, as were small amounts of ochre. Though bone splinters of prey are sometimes present, bone preservation in these sandy sediments is poor. Ostrich eggshell fragments are sometimes present in variable quantities. Distribution maps and the presence of micro-chips indicate that the spatial integrity of artefacts is commonly excellent, with little evidence for subsequent post-depositional movement; after abandonment most sites appear to have been buried under conditions of low energy aggradation, usually aeolian sands.

Though estimates of occupation length are highly speculative, it is likely that all were of relatively short duration, perhaps varying from a day or two (Shunera XVII, Nahal Lavan 1003, Har Qeren VIII, Shunera XXIV) up to a

Table 2. Refitted assemblages, geomorphological settings and retrieval techniques

PERIOD/Entity/Site	Setting	Preservation	Collection method	Estimated Areal Extent	% collected*
UPPER PALAEOLITHIC					
1. N.Nizzana XIII	Silt/sand floodplain	extensively deflated	$1m^2$- sieved	$50m^2$	75
EARLY EPIPALAEOLITHIC					
Masraqan					
2. Azariq XIII	Sand-dune	largely deflated	¼m^2- sieved	$75m^2$	95
3. Shunera XVI	Sand-dune	slightly deflated	¼m^2- sieved	$50m^2$	95
Kebaran/Nizzanan					
4. Azariq IV	Sand-dune	partially deflated	¼m^2-sieved	$25m^2$	100
5. Hamifgash IV	Silt/sand floodplain	partially eroded	¼m^2- sieved	$30m^2$	75
6. Shunera XVII	Sandy flood plain	slightly deflated	¼m^2- sieved	$9m^2$	95
MIDDLE EPIPALAEOLITHIC					
Geometric Kebaran					
7. Azariq XVIII	Inter-dunal slope	largely deflated	¼m^2-sieved	$50m^2$	85
8. Azariq II	Sand/silt flood-plain	largely deflated	¼m^2- sieved	$75m^2$	80
9. Shunera XIIB	Sandy floodplain	slightly deflated	¼m^2- sieved	$25m^2$	95
Proto-Mushabian					
10. Shluhat Qeren II	Interdunal slope	partially deflated	¼m^2-sieved	$75m^2$	50
11. Azariq X	Sand-dune	slightly deflated	¼m^2-sieved	$50m^2$	85
Mushabian					
12. Azariq XII	Sand-dune slope	partially deflated	¼m^2-sieved	$150m^2$	75
13. Har Qeren VIII	Sand-dune	largely deflated	¼m^2-sieved	$20m^2$	95
14. ?Shunera II	Sandy floodplain	slightly deflated	¼m^2-sieved	$20m^2$	95
Ramonian					
15. Nahal Lavan 1003	Sand-dune	extensively deflated	complete coll.	$10m^2$	75
16. Nahal Lavan 1009	Sand-dune	extensively deflated	sieving & coll.	$25m^2$	75
17. Shunera XXI	Sand-dune	partially deflated	¼m^2-sieved	$75m^2$	50
LATE EPIPALAEOLITHIC					
Terminal Ramonian/ Early Natufian					
18. Nahal Sekher 23	Sand-dune	partially truncated	¼m^2-sieved	$25m^2$	95
19. Shunera VII	Sandy surface	partially deflated	¼m^2-sieved	$125m^2$	95
20. Azariq XV	Sandy surface	partially deflated	¼m^2-sieved	$35m^2$	85
Late Natufian					
21. Shunera XVIII	Sand-dune	mostly disturbed	sieved	$25m^2$	85
22. Givat Hayil I	Interdunal basin	extensively deflated	$1m^2$	$125m^2$	55
Harifian					
23. Shunera XXIV	Sand-dune	extensively deflated	complete coll.	$10m^2$	75
24. Shunera IX	Sand-dune	extensively deflated	complete coll.	$150m^2$	35

* gross estimate

maximum of a few weeks (Shluhat Qeren II, Azariq XII, Shunera VII). The likelihood of repeated visits by the same group to a site seems improbable in most instances, though Givat Hayil I and Shunera IX could be exceptions.

All the tool assemblages are absolutely dominated by microlithic forms (non-geometric, geometric and point classes). Other, standardised tool classes occurring in low, but variable frequencies include scrapers, notches and denticulates, burins, truncations, perforators, and heavy duty tools: some demonstrate vectored patterning through time in terms of presence/absence or their typological attributes (Goring-Morris 1987).

The microlithic blanks in many Early and Middle Epipalaeolithic entities (Masraqan, Kebaran, Geometric Kebaran) were shaped by retouch of one form or another (fine/Ouchtata, semi-abrupt, abrupt). However, beginning already in the later part of the Early Epipalaeolithic (Nizzanan), becoming predominant and even exclusive before the end of the Middle Epipalaeolithic (Proto-Mushabian, Mushabian, Ramonian), and continuing through the entire Late Epipalaeolithic (Terminal Ramonian, Natufian, Harifian) in the region, microliths were habitually manufactured using the microburin technique in addition to other more specific backing techniques (abrupt, Helwan, and bipolar).

Conjoins

Since refitting attempts have not been exhausted in many of the assemblages, the numbers presented below should be viewed as minimal rates (Table 3). Even when considered complete, we have often found that, after an interval, further conjoins can be accomplished during detailed technological studies. We have found it beneficial to rotate the refitting crew from one assemblage to another and back again.

Targeted assemblages in the study vary in size from 340 to 6811 chipped artefacts. These counts include cores, debitage, microburins and formal tools, but exclude debris, defined here either as chips or chunks, commonly comprising 40-80% of total assemblages. Some 8500 refits have been accomplished to date, with conjoins for individual assemblages ranging from 4 to 1125, corresponding to refitting rates from 0.1% to 32.2% (Table 3). The majority attain rates well in excess of 200 items and 10% (included in the refitting rates are broken, conjoinable pieces, as well as formal conjoins of individual artefacts). Overall, the refitting rate for the 24 assemblages combined is 12.5%.

The sizes of individual conjoined blocks range from 2 items to 131, with the greatest number of refits in assemblages commonly exceeding 30 items. All other matters being equal, the study seems to indicate that optimal assemblage sizes for conjoins range up to a maximum of about 5000-6000 items (Figure 2). Since theoretical permutations of potential conjoins are exponential, they become unmanageable with larger assemblages, even when taking into account variability in raw materials by colour, texture etc.

Of the 11 sites where studies are considered to be more or less complete, the success rate commonly varies from 15-23% of the cores, tools, debitage and microburins. The highest rate was achieved for the Early Upper Palaeolithic Lagaman (=Early Ahmarian) assemblage of Nahal Nizzana XIII, with 32%. The results thus obtained compare favourably with those for other studies in the Levant (and Europe), especially given the microlithic nature of the vast majority of assemblages in the project (Note 3).

The present conjoins include 652 tools and microburins. Most rates for individual assemblages vary from 4-20%, though the overall range is from zero to an exceptional high of 46% at Nahal Nizzana XIII. Conjoinable tools often include scrapers, burins (sometimes with spalls), truncations, and notches and denticulates. In several assemblages the conjoins include microliths (finely retouched bladelets, arch backed bladelets, *lamelles scalènes*, Ramon points, lunates, Harif points and others), La Mouillahs or *piquant trièdres* and their microburins fitting onto cores or refitted blocks, though these are relatively rare. In part this derives from a combination of the microliths' sizes and the fact that subsequent retouch or backing considerably modified the original shape of the blank. In Ramonian and Terminal Ramonian sites, for which data are available, we found only 1-2% of microliths could be refitted. Another possible factor may involve the relatively greater mobility of microliths out of sites in relation to other larger and heavier items.

In assemblages where the microburin technique was habitually applied in microlith manufacture, conjoin rates for these items were most successful for Ramonian assemblages. This reflects the highly standardised use of single, distal microburins for microliths in the Ramonian, in contrast to the Mushabian where less patterning is displayed with regards which extremity of the blank to which the microburin technique was applied. In Natufian and Harifian assemblages two microburins (proximal and distal) were commonly produced during the manufacture of the lunate-triangle form (Goring-Morris 1987). In general distal microburins were more easily fitted than proximal examples, due to their more distinctive morphological shapes.

Chronological trends

In addition to the above noted correlations with regards site size and the success of conjoins, certain diachronic trends in refitting rates also appear to be indicated. This is reflected both when individual assemblages are ordered chronologically and, more obviously, when the results are pooled by entity (Tables 3-4 and Figures 3- 4).

The most successful conjoin result of the entire programme was achieved on the Upper Palaeolithic Lagaman (Early Ahmarian) assemblage of Nahal Nizzana XIII (32%).

Table 3. Current state of refitting studies, with details of assemblage sizes and refitting rates of various artefact categories (December 1997).

	RAW COUNTS				REFITS			
PERIOD/Entity/Site	Cores	Waste	Tools & Mbt	**Total ‡**	Tools & Mbt	**% of Tools & Mbt**	Total items refitted	**Total Refitting Rate**
UPPER PALAEOLITHIC								
N.Nizzana XIII	20	1127	39	**1163**	18	**46.2**	375	**32.2**
EARLY EPIPALAEOLITHIC								
Masraqan								
Azariq XIII	40	3639	1188	**4867**	41	**3.5**	>1125	**23.1**
Shunera XVI	35	4181	243	**4459**	49	**20.2**	>975	**21.9**
Kebaran/Nizzanan								
Azariq IV	13	1622	199	**1834**	45	**22.6**	440	**24.0**
Hamifgash IV*	30	1543	565	**2138**	22	**3.9**	214	**10.0**
Shunera XVII	1(3†)	529	407	**925**	12	**2.9**	170	**18.4**
MIDDLE EPIPALAEOLITHIC								
Geometric Kebaran								
Azariq XVIII	13	2543	519	**3075**	34	**6.6**	278	**9.0**
Azariq II	18	1514	389	**1921**	16	**4.1**	137	**7.1**
Shunera XIIB	5	1224	331	**1560**	5	**1.5**	58	**4.1**
Proto-Mushabian								
*Shluhat Qeren II**	15	4910	1379	**6304**	27	**2.0**	344	**5.5**
*Azariq X**	37	2415	535	**2987**	7	**1.3**	118	**4.0**
Mushabian								
Azariq XII*	53	4421	1688	**6162**	52	**3.1**	835	**13.6**
*Har Qeren VIII**	*3*	*653*	*22*	***678***	-	***0.0***	*43*	***6.3***
Shunera II*	24	3773	580	**4377**	53	**9.1**	860	**19.6**
Ramonian								
Nahal Lavan 1003*	7	229	104	**340**	8	**7.7**	72	**21.2**
Nahal Lavan 1009*	12 (9†)	742	645	**1399**	28	**4.3**	230	**16.4**
Shunera XXI*	25	4026	1719	**5770**	88	**5.1**	585	**10.2**
LATE EPIPALAEOLITHIC								
Terminal Ramonian/ Early Natufian								
N. Sekher 23*	26	1095	514	**1635**	32	**6.2**	360	**22.0**
Shunera VII*	41	2035	755	**2831**	36	**4.8**	446	**15.8**
Azariq XV*	12(13†)	2408	320	**2740**	46	**14.4**	438	**16.0**
Late Natufian								
*Shunera XVIII**	*11*	*681*	*212*	***904***	*1*	***0.5***	*69*	***7.6***
*Givat Hayil I**	*44*	*4863*	*1904*	***6811***	-	***0.0***	*4*	***0.1***
Harifian								
Shunera XXIV*	8	563	154	**725**	17	**11.0**	100	**13.8**
*Shunera IX**	*46*	*1705*	*450*	***2201***	*15*	***3.3***	*206*	***9.4***
TOTAL REFITS:			14848	**67807**	**652**	**4.4**	**>8478**	**12.5**

Notes:
Italics - sites where refitting is considered largely incomplete;
‡excluding debris (chips and chunks);
† based on refits;
* assemblages in which the microburin technique is habitually used in microlith manufacture.

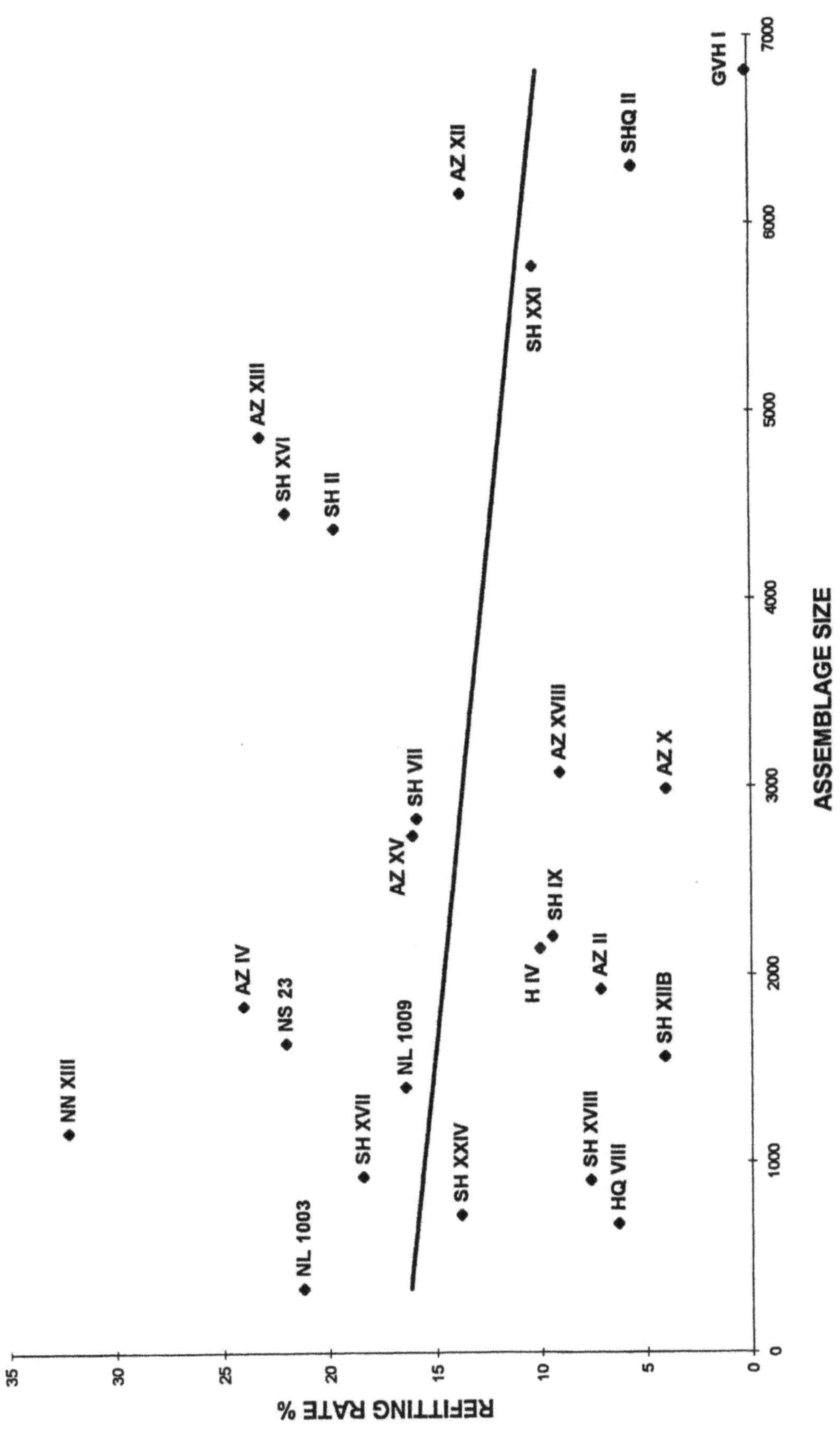

Figure 2: Conjoin rates in refitted assemblages, arranged by assemblage size from smallest (left) to largest (right). Note the trend for rates to increase as assemblages decline in size.

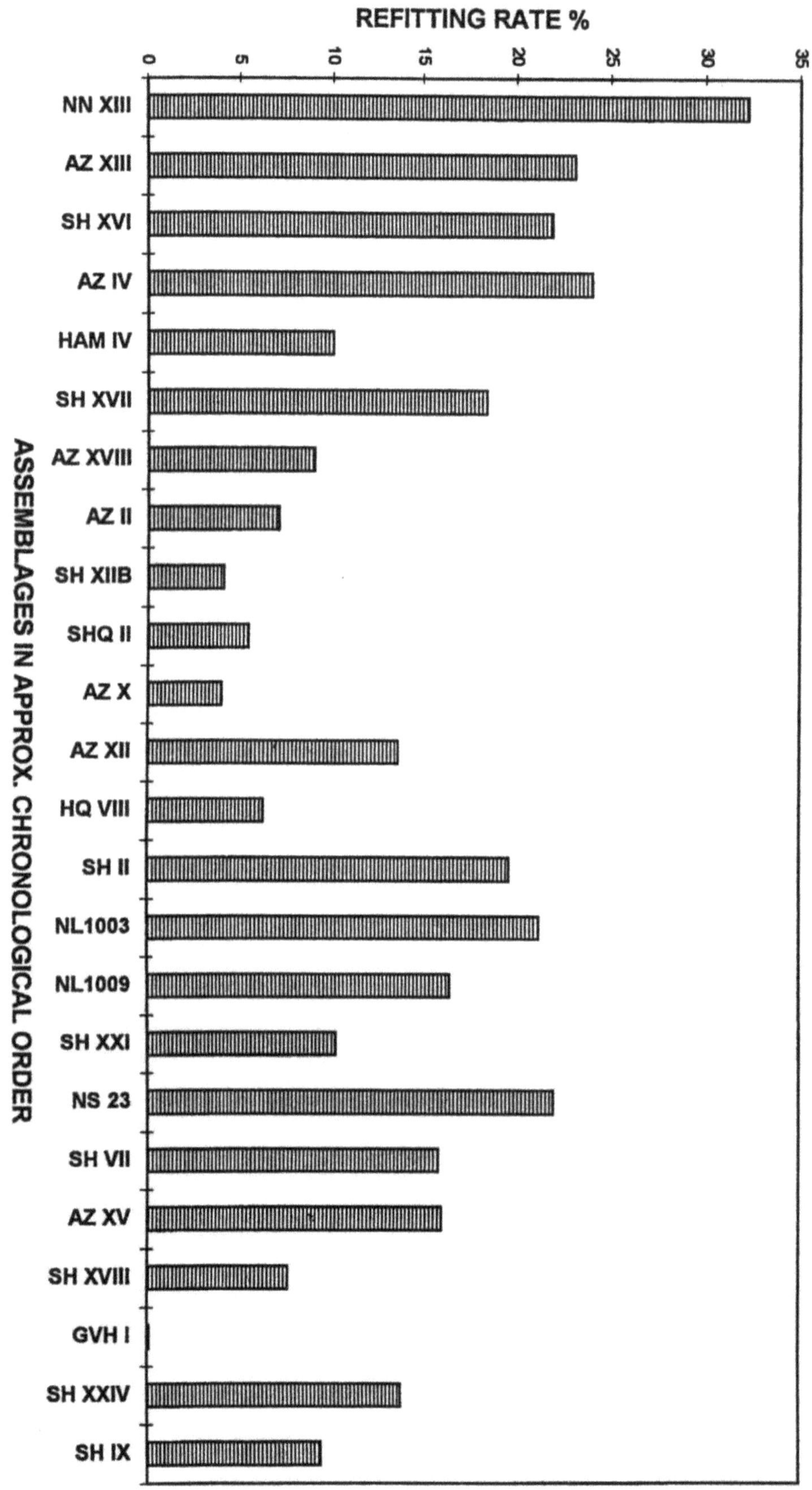

Figure 3: Conjoin rates in refitted assemblages, arranged in approximate chronological order from early (left) to late (right)

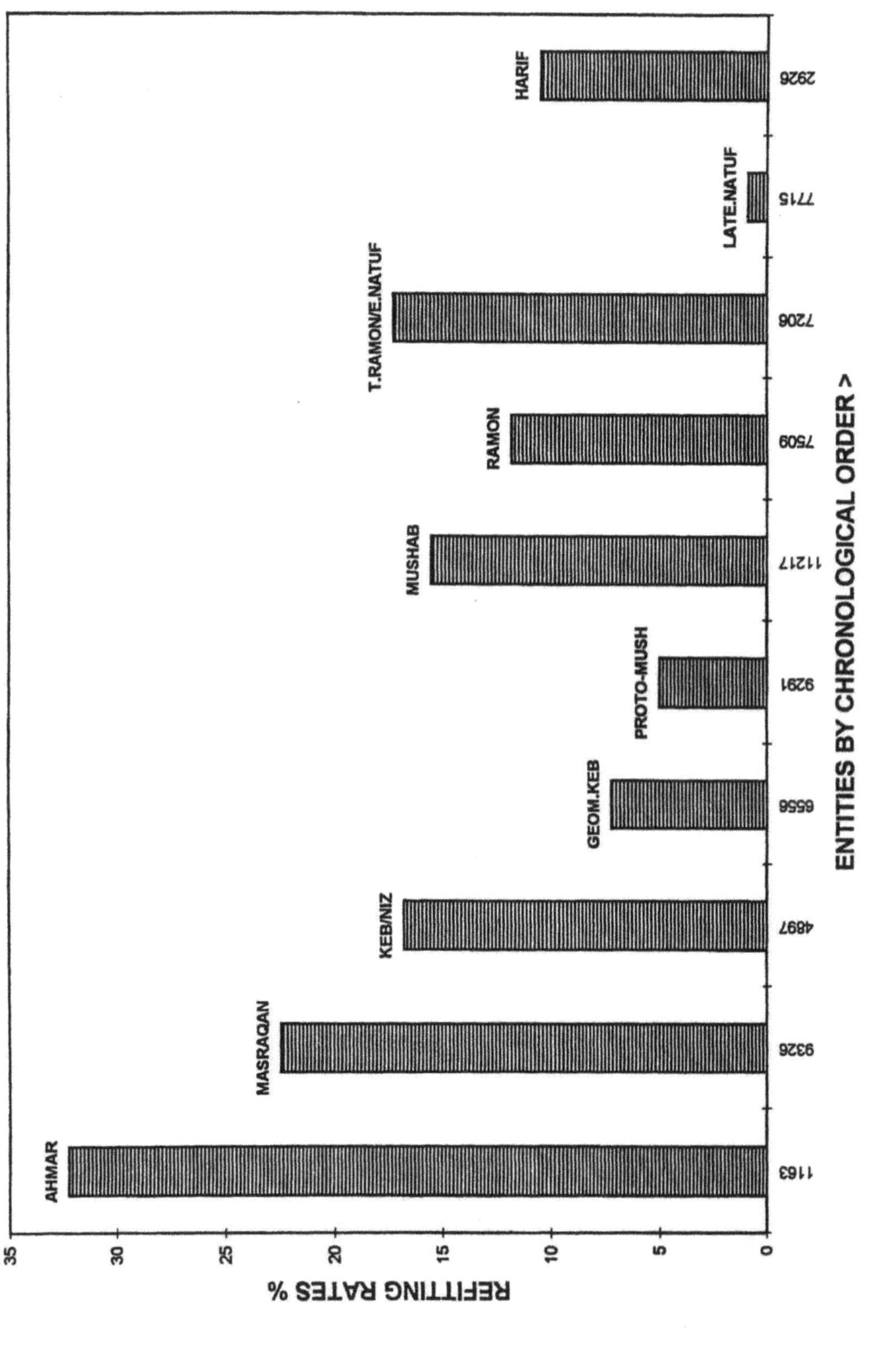

Figure 4: Conjoin rates grouped in chronological order by entity from early (left) to late (right).
Rates were computed by counting all artefacts and refits in the various assemblages assigned to each phase

Table 4. Current state of refitting rates according to traditionally defined socio-cultural entities (December 1997).

PERIOD/Entity	RAW COUNTS Cores	Waste	Tools & Mbt	Total ‡	CONJOINS Tools & Mbt	% of Tools & Mbt	Total items refitted	Total Refitting Rate
UPPER PALAEOLITHIC								
Lagaman	11	1127	39	**1163**	18	**46.2**	375	**32.2**
EARLY EPIPALAEOLITHIC								
Masraqan	75	7820	1431	**9326**	90	**6.2**	>2100	**22.5**
Kebaran/Nizzanan	44	3694	1171	**4897**	79	**6.7**	824	**16.8**
MIDDLE EPIPALAEOLITHIC								
Geometric Kebaran	36	4057	1239	**6556**	55	**4.4**	473	**7.2**
Proto-Mushabian	52	7325	1914	**9291**	34	**1.8**	462	**5.0**
Mushabian	80	8847	2290	**11217**	105	**4.6**	1738	**15.5**
Ramonian	44	4997	2468	**7509**	124	**5.0**	887	**11.8**
LATE EPIPALAEOLITHIC								
T.Ramonian/E. Natufian	79	5538	1589	**7206**	114	**7.2**	1244	**17.3**
Late Natufian	55	5544	2116	**7715**	1	**0.04**	73	**0.9**
Harifian	54	2268	604	**2926**	32	**5.3**	306	**10.5**
TOTAL REFITS:					**652**	**4.4**	**>8478**	**12.5**

Note: refitting rates for entities are computed using total artefact and conjoin counts for individual assemblages in each entity.

Preliminary publication of refitting studies on other Upper Palaeolithic Lagaman/Early Ahmarian assemblages in Sinai also indicate relatively high success rates, e.g. the Abu Noshra and Qadesh Barnea sites (Phillips 1991; Gilead & Bar-Yosef 1993 and personal communications), as well as the early Upper Palaeolithic assemblages of Boqer Tachtit (Volkman 1977). This would seem to indicate that the Nahal Nizzana XIII assemblage, though limited in size, is nevertheless representative and reflects broader behavioural patterns in addition to size factors.

A gradual decline in conjoin rates is apparent for the Early Epipalaeolithic assemblages, from the Masraqan (23%) and the probably slightly later and rather heterogeneous Kebaran and Nizzanan assemblages (17%). The dip continues and is emphasised for the Middle Epipalaeolithic Geometric Kebaran (7%) and "Proto-Mushabian" (5%) assemblages. However, rates subsequently pick up markedly again with the Mushabian (16%), Ramonian (12%) and at the beginning of the Late Epipalaeolithic with the Terminal Ramonian and Early Natufian (17%). An especially low trough occurs with the subsequent Late Natufian (1%), though for the Harifian rates again pick up (11%).

Explanations for such trends undoubtedly include a combination of several variables, including the preservation and integrity of the studied assemblages, as well as raw material preferences and their relative amenity to refitting (see below). Yet, tentatively, the evidence would appear to indicate that this patterning has broader connotations and displays significant correlations with the general adaptations and mobility levels of the various socio-cultural entities based on other criteria (see Goring-Morris 1987: Chapter XI). Thus Terminal Ramonian/Early Natufian, Late Natufian and Harifian assemblages represent relatively small, task specific occupations of much larger, logistically organised adaptations which encompassed large and variegated ranges - the studied assemblages derive from only a small fraction of their total annual (or seasonal) rounds, and do not include the major residential basecamps. This is notwithstanding contrasting patterns of seasonality between the Late Natufian and Harifian (Goring-Morris 1987). Various lines of evidence, by contrast, indicate that Geometric Kebaran (and probably Proto-Mushabian) adaptations in the Negev and Sinai were highly mobile and commonly restricted to lowland settings. The Lagaman, Masraqan, Mushabian and Ramonian display intermediate patterns of small-scale aggregation and dispersal, whether primarily in the lowlands or including the highlands as well (for the Ramonian). The refitting rates may thus reflect the degree to which artefacts (including blanks and cores, as well as tools) were moved from the place of production within sites to their discard localities, whether lost or abandoned 'off-site' or at subsequent occupation sites.

In this context it is worth recalling that refitting rates for microliths in all assemblages were low, so that the above-described tendencies are more likely to reflect the mobility

Table 5. Distributions of refitted blocks with more than 5 conjoins

Site	n	Number of refits on block 5-9	10-19	20-29	30-39	40-49	50+	Maximum refits on a single block
Early Ahmarian								
N.Nizzana XIII	14	5	5	1	1	1	1	80 (91)
Early Epipalaeolithic								
Azariq XIII	42	8	13	5	6	6	4	65
Shunera XVI	38	7	14	7	5	3	2	131
Azariq IV	23	15	6	2	-	-	-	24
Hamifgash IV	10	7	2	-	-	1	-	41
Shunera XVII	9	6	1	1	1	-	-	36
Proto-Mushabian								
Shl. Qeren II	23	16	7	-	-	-	-	15 (29)
Azariq X	7	6	1	-	-	-	-	11
Geometric Kebaran								
Azariq XVIII	12	12	-	-	(1)	-	-	9 (30)
Azariq II	5	5	-	-	-	-	-	9
Shunera XIIB	1	-	-	1	-	-	-	29
Mushabian								
Azariq XII	32	16	13	1	-	1	1	51
Shunera II	40	17	14	3	2	1	3	60
Har Qeren VIII	1	-	1	-	-	-	-	9
Ramonian								
Shunera XXI	34	21	10	2	1	-	-	33
N. Lavan 1003	3	3	-	-	-	-	-	9
N. Lavan 1009	12	8	3	1	-	-	-	20
Terminal Ramonian/Early Natufian								
N. Sekher 23	19	11	6	1	-	-	1	70
Shunera VII	18	14	3	-	1	-	-	30
Azariq XV	10	nd	3	2	2	-	3	64
Late Natufian								
Shunera XVIII	0	-	-	-	-	-	-	4
Givat Hayil I	0	-	-	-	-	-	-	2
Harifian								
Shunera XXIV	3	2	1	-	-	-	-	10
Shunera IX	7	5	1	-	-	-	1	60

Note: numbers in parentheses relate to conjoined blocks that almost refit but for which direct contact was not achieved.

of the non-microlithic classes. The mobility and discard patterns of the microlithic components of assemblages are more likely to be reflected by the restricted microburin indices for those entities in which it was a characteristic technique for microlith manufacture (see Goring-Morris 1987, 1997; Goring-Morris & Avner 1986). In some assemblages many microliths display microburin scars attesting to its systematic application in their manufacture, yet the microburins (the waste by-products) are either absent or only occur in below expected frequencies - and thus the microliths discarded in the site were made elsewhere, as at Haj I. Conversely, in other assemblages the microburins appear in greater profusion relative to the recovered microliths - the microliths made on the site were discarded elsewhere, as is apparent at Givat Hayil I.

Table 5 presents details concerning distributions of the number of items fitting onto individual blocks or cores within the assemblages. The maximum number of items refitting onto a single block commonly exceeds 30 and the maximal number is 131. While many refits comprise only 2-4 items (not listed in Table 5), it can be seen that sample sizes of the more complete blocks (commonly 5-30 items in a block) are sufficient to conduct systematic analyses of the *chaînes opératoires* and reduction sequences for each assemblage. Not included here in the counts are many instances when several refitted blocks obviously derive from the same nodule but actual conjoins of the blocks have yet to be achieved.

CHAÎNES OPÉRATOIRES

Raw material procurement patterns

Preliminary analyses indicate patterned changes in raw material procurement, and on-site and off-site production of blanks for different tool classes, e.g. heavy duty, as opposed to medium-sized and microlithic artefacts. These include initial decortication and nodule splitting, platform preparation, blank production and the choice of blanks for modification into retouched tools, as well as subsequent re-sharpening. These variables provide evidence concerning the standardisation of knapping technologies within cultural units, as well as vectored developments in specific technological aspects through time. Furthermore the studies enable investigation of patterning in the dynamics of site formation by the occupants concerning the movement of lithic artefacts and raw material into, through and out of the sites, and at what stages they were discarded, and hence introduced into the archaeological record. These also provide information concerning the relative mobility of the various groups in the region.

The choice of raw materials in most entities was frequently from local flint and chert sources readily available in the immediate vicinity of the sites (50 metres - 2 kilometres distant). Sometimes the shape and texture of the raw material was carefully selected, as with the use of fine-grained, discoidal nodules at Azariq XIII (Figures 5-7). Nodules on occasion display evidence of having been tested at source to assure its quality, whether in the form of a notch on the end of the nodule from Azariq XIII (Figure 6) or by partial decortication.

In other assemblages the choice was made with surprisingly little regard for the quality of the flint or shape of the nodule, whether as a general pattern, at Shunera XVI, or with respect to part of an assemblage, as at Azariq XII, where a couple of microlithic cores were knapped on chalky chert, which was unlikely to produce any viable blanks. In some cases, notwithstanding obvious internal flaws in the form of hairline cracks, thick and irregular cortex and/or differential textures, nodules continued to be knapped and reduced even though they were unlikely to produce usable blanks, e.g. Shunera XVI, Azariq XII, and Shunera II. A possible explanation could involve differing levels of expertise or even the attempts of apprentices (Ploux 1991).

Cherty, rough textured nodules for the massive tools that comprise a minor but constant component of almost all these Epipalaeolithic assemblages were commonly knapped at source, and blanks and retouched tools were then transported back to the sites (Figures 8-10, 13). Outcrops of massive nodules are to be found throughout the Har Qeren - Har Rogem and Givat Ivha areas. The only exceptions to this pattern were noted at Shunera XVI, Shunera XXI, and Har Qeren VIII, all located within 50-100 metres of such outcrops. At Shunera XVI one such core, measuring 19 x 18 x 10 centimetres, upon reconstruction derived from a huge nodule some 45 x 30 x 12 centimetres in size (Figure 11). However, a minimally retouched scraper, a truncation and a couple of other less formally retouched massive items were the only tools deriving from this specific reduction sequence. Another massive nodule had been roughed out on-site but the core seemingly had been removed.

Another raw material occurring in sometimes significant quantities is translucent chalcedony, which was imported from more distant sources (>5 kilometres). Although occurring sporadically in earlier contexts, starting with some Ramonian assemblages, e.g. Shunera XXI, chalcedony is more systematically used, especially in the Terminal Ramonian/Early Natufian and Late Natufian, as at Nahal Sekher 23, Shunera VII, Azariq XV, Givat Hayil I, and Shunera XVIII. In the Harifian its overall use declines, though it continues to be used for certain tool categories at Shunera IX and Shunera XXIV. The presence of small caches of unmodified (or minimally tested or fractured) chalcedony was noted in various sites, as at Shunera VII and Nahal Sekher 23. Individual chunks of such raw material usually vary in weight from 100-300 grams, i.e. fist sized and smaller. We found that, in contrast to the other types of flint used, chalcedony was particularly difficult to conjoin, which likely may affect refitting rates.

The degree to which translucent and/or glossy chalcedony was used as a raw material is likely to be significant. It probably reflects the mobility levels and geographical movements of these groups (i.e. Ramonian through Harifian), as indicated by various other lines of evidence, especially if raw material procurement was embedded within general subsistence strategies. However, it should be noted that in these same assemblages, even when chalcedony was not used, there is a distinct emphasis on high quality lustrous flint (Figures 15, 16, 18 and see below). In addition to aesthetic, mobility and traditional factors, this may also reflect specific knapping properties of such raw materials for the application of characteristic Helwan retouch (a form of pressure retouch) in the Terminal Ramonian and Early Natufian.

Observations on reduction sequences

Observations during refitting indicate certain general chronological and cultural trends, while other redundant patterns are cross-cultural and remain relatively static. Three quite separate approaches to knapping can be defined for the refitted assemblages:

1. Already alluded to above, this reduction sequence concerns the production of massive, thick blade blanks for heavy duty tools. Due to size and weight constraints (the 'schlepp' factor) this knapping usually seems to have

taken place in the immediate vicinity of outcrops of huge nodules (many workshop sites, often attributable to specific periods but dating from at least the Middle Palaeolithic through Chalcolithic, were noted around the summit of Har Qeren, on Shluhat Qeren, Givot Ivha and Mitzpeh Shunera). The 'block-on-block' technique appears to have been extensively employed, with minimal platform preparation or decortication. Blanks or complete tools were then transported back to occupation sites for use (Figures 8, 9, 10, 13,). A degree of curation is indicated by re-sharpening and/or other modifications, sometimes from one massive tool type to another. This reduction sequence, however, is usually represented by merely a few items within the refitted assemblages and is temporally transgressive. A few exceptions to the above pattern can be cited, usually close to outcrops of massive nodules. Shunera XVI (Masraqan) includes relatively numerous massive tools and sometimes even the full reduction sequence (Figure 11), while Har Qeren VIII (Mushabian) is an ephemeral camp in which the production of massive items seems to have been the primary activity. Massive tools are also a major component at Azariq XV (Early Natufian).

2. An even less common reduction sequence seems to involve cores specifically tailored to the production of medium-sized tools, especially scrapers. Examples appear at Shunera II (Mushabian) and Shunera XXIV (Harifian), where series of scraper blanks were produced in sequence. This approach may also have been employed in some Geometric Kebaran assemblages. In general, however, no obvious temporal or spatial patterning has yet been detected.

3. By far the most common approach, however, involved various reduction sequences for the production of blanks for microlithic tools. In the process of initial striking platform preparation, rejuvenation and decortication larger by-products were the most common blanks for medium-sized tool manufacture. This approach is already manifest in the Lagaman and continues throughout the entire spectrum of refitted sites. In general, most cores are single platform varieties, with an extra platform used when necessary to surmount specific knapping problems. The Late Epipalaeolithic Late Natufian and Harifian are exceptions in that, increasingly, any appropriate surfaces are opportunistically (*ad hoc*) knapped.

Yet within this general uniformity, temporally and culturally specific trends can be discerned. These may be succinctly summarised as involving changing conceptual approaches to the fabrication of microliths:

The Upper Palaeolithic Lagaman and Masraqan display evidence for systematic predetermination in the dimensions and shapes of the blade/let blanks, which then required minimal modification to produce the desired attributes for the complete tools. Such a concept involved well-planned and relatively fastidious application in decortication, striking platform preparation and acute angle maintenance. Conceptually, this may even be viewed as the culmination of the Levallois approach.

However, starting with the later Early Epipalaeolithic (Kebaran/Nizzanan, as well as the Geometric Kebaran), the desired final shapes and sizes of the microlithic tools were increasingly achieved by the application of abrupt and more invasive retouch to the blade/let blanks. In consequence the necessity for specific configurations and standardisation of potential blanks declined. Since a wider range of blanks could be fashioned into microliths, blank production no longer needed to be so tightly controlled.

From the Nizzanan and Proto-Mushabian onwards an innovative technique in microlith production was introduced that became predominant during the rest of the Epipalaeolithic in the Negev and Sinai. This was the habitual use of the microburin technique in the shaping of blanks for pointed microlithic tools, which increasingly negated the necessity for size and shape standardisation of blanks. Cores were often still quite standardised in the Mushabian, Ramonian, and Terminal Ramonian/Early Natufian, though this novelty widened the pool of potential blanks.

Ultimately, by the Late Natufian and Harifian, as mean microlith dimensions declined, blank specifications became so irrelevant that almost any small, elongated flake (in addition to blade/lets) could easily be modified into a microlith.

The above trends are also accompanied by a shift from initial acute platform angles on narrow carinated cores in the Lagaman and Masraqan to wider angles later in the early Epipalaeolithic. A shift towards rectangular, sub-pyramidal and pyramidal cores ensues, which is most notable in the Ramonian and Terminal Ramonian/Early Natufian (Figures 14-18). However, during the Late Natufian and Harifian cores were increasingly amorphous and less standardised as knapping became more opportunistic.

Although the reduction sequence patterns outlined above could easily be interpreted as a temporally transgressive 'decline' in knapping abilities and concerns, superficially reflected in less standardised and aesthetically pleasing blank production, reality indicates that, to the contrary, this masks a sophisticated and increasingly economical approach to raw material use, as techniques became more efficient with regards microlithic tool production (see also Goring-Morris 1987).

Platform preparation commonly reduced the original volume of nodules by about one third to one half in the Lagaman (Nahal Nizzana XIII), as well as in many other entities (e.g.

Ramonian and Terminal Ramonian/Early Natufian). Often initial platform preparation precedes systematic decortication. A means of increasing efficiency was sometimes obtained by splitting the nodule into two halves and making two separate cores, with the striking platform along the 'split', as eloquently expressed at Azariq XIII (Figures 5-6). The production of two or even three cores from small nodules is found also in Mushabian, Ramonian, and especially Terminal Ramonian/Early Natufian assemblages (Figures 15-18).

Decortication was achieved by several techniques. It could either precede the production of blade/bladelet blanks, taking the form of lateral 'onion' peeling, especially if the nodule was discoidal, by the alternate removal of cortical flakes from each side. This approach appears to correlate with narrow carinated cores, where relatively elongate blade/let blanks were produced. It is notable in the Lagaman, the Masraqan and Kebaran (Figures 5-7). Another approach to decortication involves rough removal of the cortex, followed by subsequent decortication during the course of blade/let production. It was noted especially in some Ramonian, Late Natufian and Harifian assemblages. A similar approach has been described for some portions of Ahmarian assemblages at Abu Noshra in southern Sinai (Phillips 1991).

At Azariq XIII there are also numbers of cores with two striking platforms, either opposed, same side, or opposed, opposite side (Figure 6). The striking platforms are achieved by the use of core tablets and the intensive abrasion of the striking platform edge along the removal face, usually between each blank removal. These features are most marked in the Early Epipalaeolithic and decline, particularly during the Late Epipalaeolithic (e.g. Shunera XXIV, Shunera IX), when blank production appears to become more opportunistic (see below).

During the Early Epipalaeolithic (e.g. Azariq XIII, Azariq IV, Shunera XVI) bladelet production was achieved by the use of narrow removal surfaces, whether from (commonly sectioned) discoidal nodules, or by the use of large flakes functioning as cores - some of the latter in their abandoned state are traditionally classifiable as burins or lateral carinated scrapers. Here (at Azariq XIII, Azariq IV, and Shunera XVI) the removal of subsequent core tablets in the course of the reduction sequence is seemingly accomplished to readjust the angle of the striking platform, since the proximal ends of the bladelet blanks tend to be thicker than the distal tips. It is of interest to note that many such core tablets at Shunera XVI tend to be shattered. Indeed, while Azariq XIII and Shunera XVI, both assigned to the Masraqan, display numerous conceptual similarities in terms of initial decortication and platform preparation, the actual production of bladelet blanks at Shunera XVI is clearly inferior, less controlled and *ad hoc*, resulting in bladelets with thicker striking platforms. In an earlier technological study of blade/let production of the assemblages Wiseman (1993) identified the use of both hard hammer and punch technique at both sites. It will be of some interest to see whether this is confirmed by detailed studies of the *chaînes opèratoires*.

During the Middle and earlier part of the Late Epipalaeolithic (Mushabian, Ramonian, Terminal Ramonian/Early Natufian and some of the Late Natufian) cores are commonly prepared as pyramidal forms, sometimes around one side to produce a semi-circular striking platform, on others being circumferential (Figures 14-18). Abrasion of the removal surface between each removal is often intensive. In the Late Natufian and the Harifian the provision of striking platforms is less standardised and directed, with removals tending to be opportunistic from any available surface, and little abrasion of the removal surfaces.

Some cores were knapped to extremely small dimensions, sometimes producing numbers of bladelets towards the end of the reduction sequence which were clearly too small to be modified into retouched microliths (Shunera II, Nahal Sekher 23). In other instances refitted nodules seemingly produced no retouched tools whatsoever (micro-wear studies may help in evaluating whether such items were actually used). An interesting feature at Shunera XVI is the seemingly intentional breakage of large blades, often modified into truncations or burins, by a single blow on one of the dorsal *arrêtes* (Figures 9-10).

A common practice in some of the Middle Epipalaeolithic assemblages is the blunting of sharp ridges on lateral areas of the cores (see also Marder 1994). This appears to be carried out in order to facilitate handling of the core while knapping bladelet blanks following the initial 'setting up' of the core. Further into the reduction sequence it was necessary to remove such items, but notwithstanding superficial similarities, they do not appear to relate to the function normally ascribed to ridge blades.

Medium sized tools, such as scrapers, notches and denticulates, and retouched pieces commonly seem to be made on opportunistic flake or blade blanks deriving from initial decortication and/or platform preparation, prior to systematic bladelet blank production. Only rarely do they seem to result from separate, specific reduction sequences, e.g. Shunera II and Shunera XXIV. A considerable portion of items defined as single notches or oblique truncations are actually unfinished microliths - the microburin scar has not been applied for whatever reason.

As noted above, some limited success has been achieved with refitting microliths and, in those entities where they appear, together with conjoinable microburins, especially distal forms which are more distinctive than proximal variants. A few instances of microliths refitting onto cores were also accomplished (Figures 14, 15b, 17). These enable the reconstruction of the original dimensions of the blanks from which the microliths were fabricated. While in the Early Epipalaeolithic these were obviously on narrow, elongated bladelet blanks, there is a marked decline in standardization through the Middle and Late Epipalaeolithic when blanks are shorter and squatter so that some approximate elongated flakes rather than bladelets (distal shape thus becomes virtually irrelevant). It is surmised that this tendency reflects diachronic changes in the approach to microlith manufacture.

In the Early Epipalaeolithic there is considerable investment in 'setting up' the core and its platform, in order to produce bladelet blanks which closely approximate the final size and shape of the microlith, and subsequent retouch is minimal. By contrast from the Middle Epipalaeolithic the final shape is primarily achieved by invasive backing and in some entities the systematic use of the microburin technique. The dimensions and form of the final product is thus achieved by investment in subsequent modification of the blank. This expands the range of potential blanks, so that almost any elongated item will suffice, and less initial investment in 'setting up' the core is required. The result appears more opportunistic but could also be viewed as ultimately more efficient in achieving the desired product.

Heat treatment

In most of the studied assemblages a small proportion of the artefacts displays evidence of exposure to high temperatures in the form of colour changes, hairline cracks, potlids, and shattering (Figure 6). In the vast majority of instances such alterations were clearly accidental, resulting from the artefacts chance proximity to hearths during the course of the occupation. However, during the Terminal Ramonian/Early Natufian (Nahal Sekher 23 and Azariq XV) there appears to be some evidence to indicate experimentation with intentional heat treatment of raw material to alter the physical and mechanical knapping properties of the flint. A waxy texture, with lustrous pink to purple colour was often achieved, often approximating chalcedony. In some instances this occurred on blanks after removal from the (unheated) core, while in others partially flaked cores were further knapped after heat treatment. However, the presence of characteristic uncontrolled heat by-products in the form of potlids, hairline fractures and shattering were also noted, indicating that such attempts were often not always particularly successful.

Intentional heat treatment of flint has also been reported from the Early Natufian site of Wadi Hammeh 27 (Edwards & Edwards 1990). This pyrotechnical experimentation should be viewed within the context of early lime-plaster production in the Negev and (Sinai from it least the Geometric Kebaran at Lagama North VIII (Kingery *et al.* 1988) for use as mastic in the hafting of microliths. Indeed, at Shunera VII a small quantity of lime-plaster was recovered adjacent to one of the hearths (Goring-Morris in preparation).

DISCUSSION

Traditional classification versus functional classification of artefacts

As noted above, reconstructions clearly demonstrate that for most sites the reduction sequence, particularly in the production of bladelet blanks, was primarily achieved by removals from a single striking platform. However, this is not always reflected in the abandoned core, resulting in the traditional classification system clearly at odds with the actual state of the core during most of the knapping sequence, e.g. seemingly "opposed platform cores" are in reality single platform through the vast majority of the reduction sequence, except for thinning of the base of the core, removal of arching on the removal surface, or final attempts to remove a few extra blanks at right angles to the main striking platform.

Another clearly similar phenomenon concerns items which functioned as core tablets based on the refits and reduction sequence, yet are often not easily identifiable as such by traditional classification criteria (and would normally be defined as regular flakes). So too, with the ridge blades and the above-mentioned use of lateral blunting, where the apparent aim was to dull sharp edges around the periphery of the core, rather than being concerned with the initial process of preparing the removal surface prior to the systematic removal of blade/let blanks. For the Ramonian assemblages it was found that, after refitting, some 2.9-4.3% of conjoined items changed definition from flakes to blades or vice versa (these are all broken items).

The sizes of the tool assemblages also change following refitting of broken items; thus two broken endscrapers, upon refitting, could transpire to represent a single double endscraper, or a seeming dihedral burin which clearly functioned primarily as a core, producing numerous microlithic blanks.

This is not to claim, however, that traditional typological categories of either debitage or tools are meaningless. Rather, more stringent criteria and sub-type definitions are needed for various debitage categories; indeed, for local Upper Palaeolithic assemblages, Ferring (1980) has demonstrated that approximate reconstructions of different reduction sequences can be achieved without actual refitting. With regards tools, the above cited instances are hardly likely to appreciably alter the general typological configuration of an assemblage.

Differential patination of broken items and refits

An interesting phenomenon noted in several of the sites concerns the different patination of both refitted broken artefacts as well as regular conjoins. In some instances this is easily explained as the result of accidental burning of items, or possible intentional heat treatment. In other cases, however, this is clearly not the case, but rather appears to reflect post-occupation taphonomic processes including differential exposure to the elements, or staining by the sediments.

Mobility of artefacts

Many of the sites included in the present study have been almost totally excavated. Some conjoined blocks have been completely reconstructed back to their original nodules, with only a few items missing. In others larger or smaller portions are missing, whether the core itself (e.g. at Shunera XVII, Azariq XV), or the more cortical outer regions of the original nodule (Azariq XIII, Shunera XVII). In other instances single tools or items of debitage cannot be matched to any others from the same assemblage, as for example a transverse burin with conjoinable spalls at Shunera XVI (Figure 12). These observations probably reflect the movements of the knappers, their raw materials, and finished products, whether blanks or retouched tools. Sometimes cores were roughed out or prepared elsewhere and introduced into the site where blanks were produced and the core eventually discarded. In other instances roughouts or initial preparation of nodules was conducted on-site and the core later removed and abandoned elsewhere. Intermediate situations also occur, with the initial stages of decortication missing, as is the core (Figure 18). Such results illustrate the dynamics of site formation processes and movements of raw materials, cores, tools (and debitage) through and between sites.

Largely irrespective of cultural and chronological affinity blanks for massive items (commonly in excess of 10 cm) such as scrapers, denticulates, battered pieces, truncations and burins, typically were produced off-site, presumably at nearby outcrops of suitable large nodules. These blanks were then introduced to the site, where they underwent modification into tools, sometimes being re-modified (resharpening or modification into other tools following breakage) on the basis of conjoinable retouching flakes and other elements (e.g. Shunera XVII, Shunera VII). This seems to reflect economising behaviour or what is known as the 'schlepp' effect. Only at Shunera XVI and perhaps Azariq XII, where large nodules were available in the immediate vicinity, is the whole reduction sequence for massive items present on-site.

The refitting of microliths has proved to be difficult in most sites, largely the result of their small size, and the commonly extensive modification by retouch to the initial shape of the blank. However, in numerous sites there would appear to be an excess of microliths relative to the refitted nodules and cores (Nahal Sekher 23). Also more microburins than microliths, even allowing for two mbt per tool (e.g. lunates at Shunera VII and Azariq XV; or as resharpening Krukowski variants at Nahal Sekher 23). In some instances the colour, texture and patination of some of the microliths do not appear to match those of refitted blocks, or indeed the un-refitted debitage. It seems likely that some of these microliths represent items produced previously elsewhere but discarded on-site. This again illustrates that the static artefact remains recovered from within individual sites were not themselves part of closed systems, but rather were part of dynamic and open systems, where items were mobile as the occupants moved from one site to another. This is valid even for very short term quite specialised camp sites, as represented by most of the occupations dealt with in this study - the majority seem to be primarily associated with hunting activities.

SUMMARY

It is clear that many, if not most, Epipalaeolithic assemblages in the Levant are not amenable to time-consuming refitting studies as such as those outlined above. Yet the judicial selection of assemblages for refitting can contribute to our understanding of the human behaviours behind these mute remains. The particular significance of the present project lies in its unprecedented scope - a large sample of assemblages representing all the numerous industries documented from a restricted geographical region over a relatively short timespan. As such it moves beyond the site-specific approach of many similar studies, and provides an opportunity to investigate both synchronic and diachronic variability at various levels. The study is already providing insights as to dynamic developments in knapping technologies, raw material exploitation, mobility patterns, the intrasite spatial organisation of activities, taphonomic processes, and many other related topics.

Acknowledgements

This research was supported by grants from the Israel Science Foundation founded by The Israel Academy of Sciences. The excavations and initial analyses of most of the assemblages described herein were conducted under the direction of one of the authors (ANG-M) in the framework of the Emergency Archaeological Survey of the Negev mostly during the years 1979-1984. The encouragement and support for the Prehistory Unit by the then head of the Survey, Dr. Rudolph Cohen, and Ofer Bar-Yosef are gratefully acknowledged. Artefacts were photographed by Gabi Laron, except for numbers 14 and 15 which were photographed by Klara Amit, and the artefact illustrations were drawn by Leonid Zaiger.

Notes

1. The first systematic refitting study in the Levant, at the Early Upper Palaeolithic site of Boqer Tachtit in the central Negev, was also aimed at investigating diachronic changes in reduction strategies and activity areas (Marks 1983).

2. Sporadic refittings have been conducted previously on Epipalaeolithic assemblages in the Levant, e.g. in Gebel Maghara, at Mushabi IV (Bar-Yosef 1977: Figure 84) and Lagama North VIII (ANG-M, personal observation).

3. Rates of 11%, 22%, 28% and 35% respectively were reported for the four levels of the macrolithic, transitional/Early Upper Palaeolithic site of Boqer Tachtit near Avdat, in the central Negev, seemingly 7% at the Late Mousterian site of Farah B in Nahal Besor, and 19% at the Late Upper Palaeolithic site of Meer in Belgium (Volkman 1983; Gilead 1988, 1995; Gilead & Fabian 1990; Cahen *et al.* 1979). By contrast, at the Mousterian site of Quneitra on the Golan Heights, the refitting rate was a mere 0.4% (and 0.1%, excluding broken items), primarily reflecting site function and taphonomic processes (Saragusti & Goren-Inbar 1990). It should also be noted that there are different definitions for debris in Middle Palaeolithic studies. The various different methods of counting refits should also be taken into account.

REFERENCES

Baruch, U.

1994 Late Quaternary pollen record of the Near East. In R. Kra & O. Bar-Yosef (eds.) *Late Quaternary Chronology and Paleoclimates of the Eastern Mediterranean.* Radiocarbon, University of Arizona: 103-120.

Baruch, U. & A.N. Goring-Morris

in press The aboreal vegetation of the central Negev Highlands in Epipalaeolithic times: evidence from archaeological charred wood remains. *Archaeobotany.*

Bar-Yosef, O.

1977 Mushabi IV. In O. Bar-Yosef & J.L. Phillips (eds.) *Prehistoric Investigations in Gebel Maghara, Northern Sinai.* Qedem 7: 184-193.

1987 Direct and indirect evidence for hafting in the Epipalaeolithic and Neolithic of the southern Levant. In D. Stordeur (ed.) *La Main et l'Outil. Manches et Emmanchements Préhistoriques,* Travaux de la Maison de l'Orient 15: 155-164. Lyon: Maison de l'Orient.

1991 Raw material exploitation in the Levantine Epi-Paleolithic. In A. Montet-White & S. Holen (eds.) *Raw Material Economies among Prehistoric Hunter-Gatherers.* University of Kansas Publications in Anthropology, Vol. 19: 235-250, Lawrence: Kansas.

Bar-Yosef, O. & A.N. Goring-Morris

1977 Geometric Kebaran occurrences. In O. Bar-Yosef & J.L. Phillips (eds.) *Prehistoric Investigations in Gebel Maghara, Northern Sinai.* Qedem 7: 115-148.

Bar-Yosef, O. & J.L. Phillips

1977 *Prehistoric Investigations in Gebel Maghara, Northern Sinai.* Qedem 7.

Bar-Yosef, O. & J. Vogel

1987 Relative and absolute chronology of the Epi-palaeolithic in the southern Levant. In O. Aurenche, J. Evin & F. Hours (eds.) *Chronologies in the Near East.* Oxford: British Archaeological Reports International Series 379: 220-245.

Bodu, P., C. Karlin & J. Pelegrin

1991 Processus techniques et chaînes opératoires. Comment les préhistoriens s'approprient un concept élaborépar les ethnologues. In H. Balfet (ed.) *Des Chaînes Opératoires Pourquoi Faire?* Paris: CNRS.

Cahen, D., L.H. Keeley & F. Van Noten

1979 Stone tools, tool kits, and human behaviour in prehistory. *Current Anthropology* 20: 661-683.

Cziesla, E., S. Eickhoff, N. Arts & D. Winter (eds.)

1990 *The Big Puzzle.* Studies in Modern Archaeology, Vol. 1. Bonn: Holos.

Edwards, P.C. & W.I. Edwards

1990 Heat treatment of chert in the Natufian Period. *Meditarch* 3: 1-5.

Ferring, C.R.

1980 *Technological Variability and Change in the Late Paleolithic of the Negev.* Unpublished Ph.D. Thesis, SMU, Dallas. Ann Arbor, University Microfilms International.

Gilead, I.

1988 Le site Mousterien de Fara II (Néguev séptentrional, Israel) et le remontage de son industrie. *L'Anthropologie* 92/3: 797-807.

1995 Problems and prospects in the study of the Levallois technology in the Levant. The case of Fara II, Israel. In O. Bar-Yosef & H.L. Dibble (eds.) *The Definition and Interpretation of Levallois Variability.* Monographs in World Archaeology, Vol. 23. Madison, Wisconsin: Prehistory Press: 79-91.

Gilead, I. & O. Bar-Yosef

1993 Early Upper Paleolithic sites in the Kadesh Barnea area, Northeastern Sinai. *Journal of Field Archaeology* 20: 265-280.

Gilead, I. & P. Fabian

1990 Conjoinable artifacts from the Middle Palaeolithic open air site Farah II, northern Negev, Israel. In E. Cziesla, S. Eickhoff, N. Arts, & D. Winter (eds.), *The Big Puzzle.* Studies in Modern Archaeology, Vol. 1. Bonn: Holos: 101-113.

Goring-Morris, A.N.

1987 *At the Edge: Terminal Pleistocene Hunter-Gatherers in the Negev and Sinai.* Oxford: British Archaeological Reports International Series 361.

1988 Trends in the spatial organization of Terminal Pleistocene hunter-gatherer occupations as viewed from the Negev and Sinai. In M.-C. Cauvin, O. Aurenche & P. Sanlaville (eds.) *Colloque Préhistoire Levant II: Les Processus de Changements Culturels. Paléorient* 14/2: 231-244.

1989 Developments in Terminal Pleistocene hunter-gatherer socio-cultural systems: A perspective from the Negev and Sinai Deserts. In I. Hershkovitz (ed.) *People and Culture in Change.* Oxford: British

Archaeological Reports International Series 508 (i): 7-28.

1990 Socio-cultural aspects of marine mollusc use in the Terminal Pleistocene of the Negev and Sinai regions of the southern Levant. In C. Hayes III (ed.) *Proceedings of the 1986 Shell Bead Conference. Selected Papers*. New York: Rochester Museum and Science Center, Research Records No. 20: 175-188.

1991 The Harifian of the southern Levant. In O. Bar-Yosef & F.R. Valla (eds.) *The Natufian Culture in the Levant*. Ann Arbor: International Monographs in Prehistory: 173-216.

1993 Negev: The Prehistoric Periods. In E. Stern (ed.) *The New Encyclopedia of Archaeological Excavations in the Holy Land*. New York: Israel Exploration Society and Simon & Schuster: 1119-1123.

1993 Shunera Dunes. In E. Stern (ed.) *The New Encyclopedia of Archaeological Excavations in the Holy Land*. Volume 4. New York: Israel Exploration Society and Simon & Schuster: 1380-1382.

1995 Complex hunter-gatherers at the end of the Paleolithic (20,000-10,000 BP). In T.E. Levy (ed.) *The Archaeology of the Holy Land*: 141-168. London: Leicester University Press.

1996 Square pegs into round holes: a critique of Neeley and Barton. *Antiquity* 70/267: 130-135.

1997 A Late Natufian campsite at Givat Hayil I, western Negev Dunes, Israel. *Journal of the Israel Prehistoric Society* 27: 43-61.

in prep. *Prehistoric Investigations in the Western Negev. Part 1: The Shunera Dunes, Givat Hayil, Nahal Sekher and Hamifgash*. 'Atiqot.

Goring-Morris, A.N. & U. Avner

1986 An Epipalaeolithic occurrence in eastern Sinai and some observations on the use of the microburin technique. *Mitekufat Haeven - Journal of the Israel Prehistoric Society* 19: 58-65.

Goring-Morris, A.N. & A. Belfer-Cohen

in press The articulation of cultural processes and Late Quaternary environmental changes in Cisjordan. *Paléorient* 23/2.

Goring-Morris, A.N. & P. Goldberg

1990 Late Quaternary dune incursions in the southern Levant: Archaeology, chronology and palaeoenvironments. *Quaternary International* 5: 115-137.

Goring-Morris, A.N. & S.A. Rosen

1987 Prehistoric Archaeology. In *Preliminary Safety Analysis Report. Nuclear Power Plant - Shivta Site. Appendix 2.5E, Volume 9/1. Late Cenozoic Geology in the Site Area. Attachment A*. Haifa: Israel Electric Corporation: 1-81.

Hofman, J.L. & J.G. Enloe (eds.)

1992 *Piecing Together the Past: Applications of Refitting Studies in Archaeology*. Oxford: British Archaeological Reports International Series 578.

Horowitz, A.

1977 Pollen spectra from two early Holocene prehistoric sites in the Har Harif (west central Negev). In A.E. Marks (ed.) *Prehistory and Paleoenvironments in the Central Negev, Israel. Volume II. The Avdat/Aqev area, Part 2, and the Har Harif*. Dallas: SMU Press: 323-326.

Karlin, C., P. Bodu, N. Pigeot & S. Ploux

1992 Some socio-economic aspects of the process of lithic reduction among groups of hunter-gatherers of the Paris Basin area. In A. Berthelet & J. Chavaillon (eds.) *The Use of Tools by Humans and Non-human Primates*. Oxford: Clarendon Press: 318-340.

Karlin, C. & M. Julien

1995 Prehistoric technology: A cognitive science? In C. Renfrew & E.B.W. Zubrow (eds.) *The Ancient Mind. Elements of Cognitive Archaeology*. Cambridge: Cambridge University Press: 152-164.

Kingery, D.W., P.B. Vandiver & M. Pickett

1988 The beginnings of pyrotechnology, Part II: production and use of lime and gypsum plaster in the Pre-Pottery Neolithic Near East. *Journal of Field Archaeology* 15: 219-244.

Leroi-Gourhan, A. & M. Brezillon

1966 L'habitation Magdalenienne de Pincevent près Montereau (Seine-et- Marne). *Gallia Préhistoire* 9: 263-385.

Marder, O.

1994 *Technological Aspects and Flint Industry of the Late Epipalaeolithic in the Negev and Sinai*. Unpublished M.A. Thesis, Hebrew University of Jerusalem.

Marks, A.E. (ed.)

1983 *Prehistory and Paleoenvironments in the Central Negev, Israel. Vol III, The Avdat/Aqev Area, Part 3*. Dallas: SMU Press.

Pelegrin, J., C. Karlin & P. Bodu

1988 Chaînes opératoires: Un outil pour le préhistorien. In J. Tixier (ed.) *Technologie Préhistorique*. Notes et Monographies Techniques No. 25. Paris: CNRS: 55-62.

Phillips, J.L.

1991 Refitting, edge wear and chaînes opératoires: A case study from Sinai. In *25 Ans d'Etudes Technologiques en Préhistoire*, XIe Rencontre Internationales d'Archéologie et Histoire d'Antibes. Juan-les-Pins: Editions APDCA: 305-317.

Ploux, S.

1991 Technologie, technicité, techniciens: méthodes de détermination d'auteurs et comportement technique individuel. In *25 Ans d'Etudes Technologiques en Préhistoire*, XIe Rencontre Internationales d'Archéologie et Histoire d'Antibes. Juan-les-Pins: Editions APDCA: 201-214.

Saragusti, I. & N. Goren-Inbar
1990 Conjoinable pieces from the lithic assemblage of Quneitra, Area B. In N. Goren-Inbar (ed.) *Quneitra: A Mousterian Site on the Golan Heights.* Qedem 31: 173-188.

Schlanger, N.
1995 Mindful technology: unleashing the chaîne opératoire for an archaeology of the mind. In C. Renfrew & E.B.W. Zubrow (eds.) *The Ancient Mind. Elements of Cognitive Archaeology.* Cambridge University Press: 143-151.

Singer, C.A.
1992 The 63 kilometer fit. In J.E. Ericson & B.A. Purdy (eds.) *Prehistoric Quarries and Lithic Production.* Cambridge: Cambridge University Press: 35-48.

Tixier, J.
1976 *Le Campement Préhistorique de Bordj Mellala, Ouargla,* Algérie. Paris: Editions de Cercle de Recherche et d'Etudes Préhistoriques.

Tchernov, E., T. Dayan & Y. Yom-Tov
1986 The Paleo-geography of *Gazella gazella* and *Gazella dorcas* during the Holocene of the southern Levant. *Israel Journal of Zoology* 34: 51-59.

Villa, P.
1982 Conjoinable pieces and site formation processes. *American Antiquity* 47: 276-290.
1991 From debitage chips to social models of production: The refitting method in Old World archaeology. *The Review of Archaeology* 12/2: 24-30.

Volkman, P.
1983 Boker Tachtit. Core reconstruction. In A.E. Marks (ed.) *Prehistory and Paleoenvironments in the Central Negev, Israel. Vol. III: The Avdat/Aqev Area, Part 2.* SMU Press, Dallas: 127- 188.

Wiseman, M.F.
1993 Lithic blade elements from the southern Levant: A diachronic view of changing technology and design process. *Journal of the Israel Prehistoric Society - Mitekufat Haeven* 25: 13-102.

Figure 5a-c: Azariq XIII (Masraqan). Thin, discoidal nodule sectioned down the long axis to produce two cores which were then laterally decorticated to produce bladelets from narrow striking platforms. Note the acute angle of the striking platforms. One core was firecracked after exhaustion. Scale in centimetres.

Figure 6: Azariq XIII (Masraqan). Thin, discoidal nodule sectioned down the long axis to produce two cores which were then laterally decorticated to produce bladelets from narrow striking platforms (one core has two striking platforms). The acute angle of the striking platforms can be seen, as well as a fire-cracked broken scraper on an initial core tablet. Scale in centimetres.

Figure 7: Azariq XIII (Masraqan). Thin, discoidal nodule decorticated to produce bladelets from a narrow striking platform. Note the use of core tablets to correct the acute angle of the striking platform. Scale in centimetres.

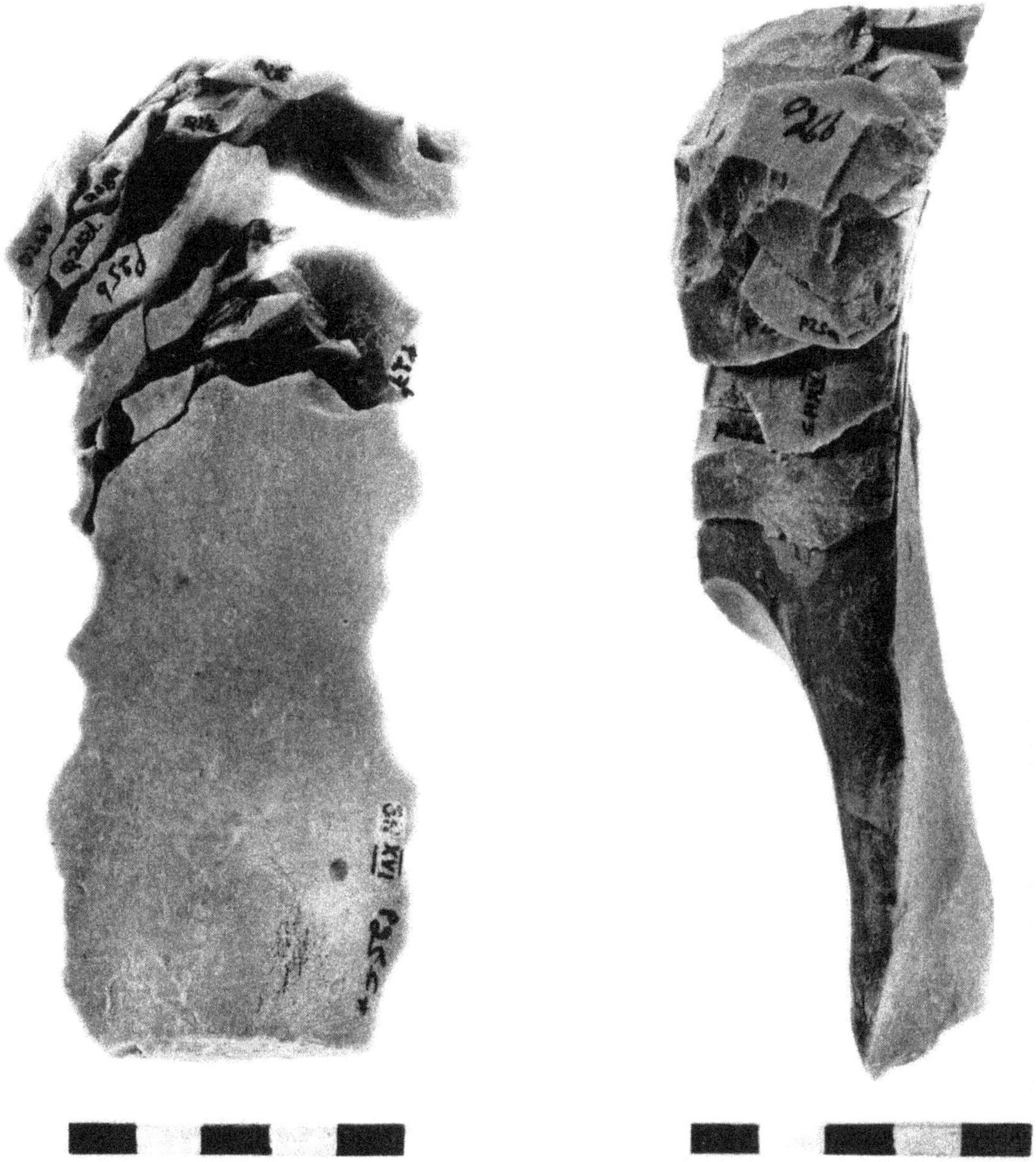

Figure 8: Shunera XVI (Masraqan). a, Ventral and b, profile views of denticulate on massive blade, showing intensive (?)re-sharpening of distal end. The core for the blank was not found on-site. Scale in centimetres.

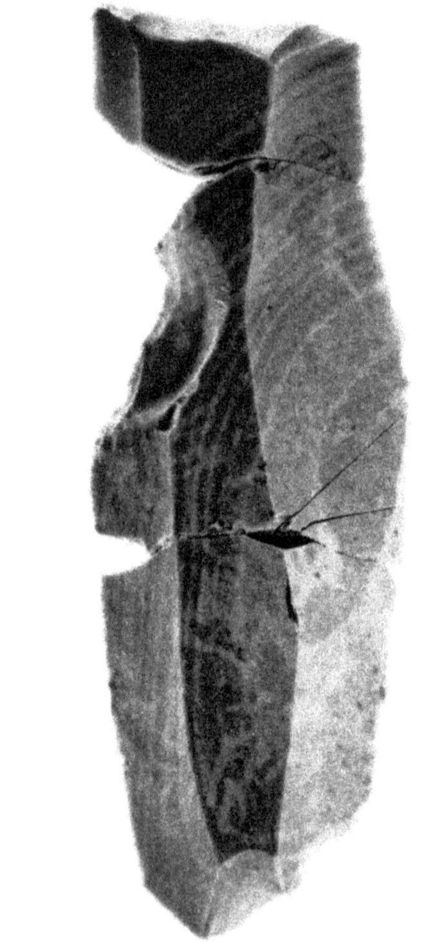

Figure 9: Shunera XVI (Masraqan). Dorsal view of multiple burin on a massive blade blank, showing intentional snap by a blow to the dorsal ridge, and burin spalls. The core for the blank was not found on-site. Scale in centimetres.

Figure 10: Shunera XVI (Masraqan). Double truncation on a massive blade. The blade was subsequently intentionally sectioned twice by blows to the dorsal ridges. The core for the blank was not found on-site. Scale in centimetres.

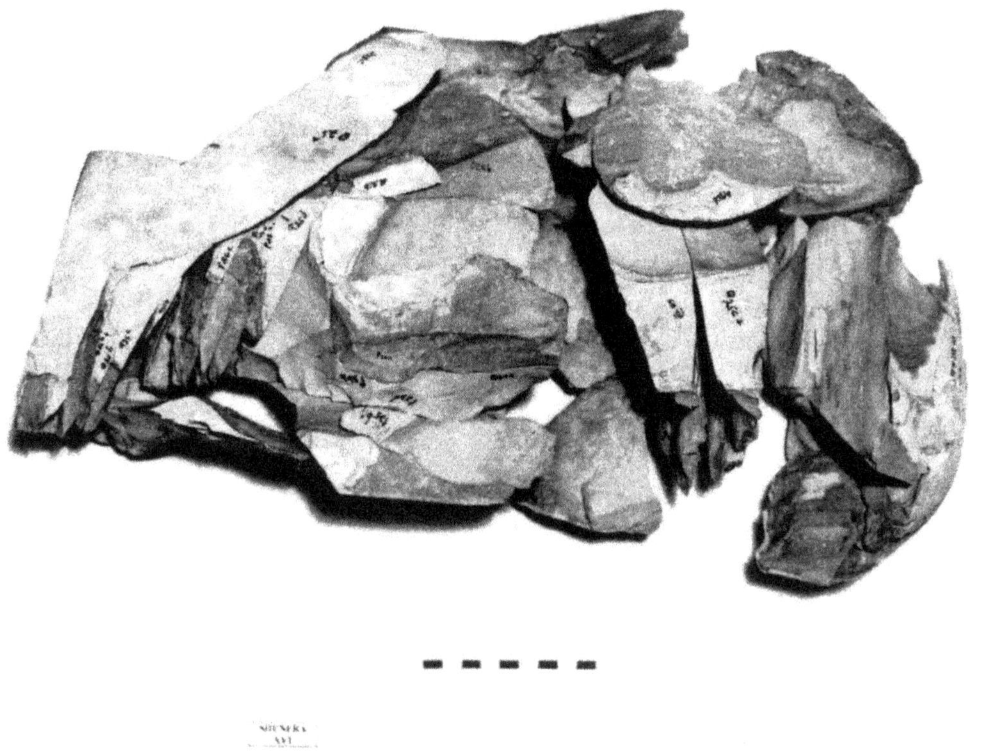

Figure 11: Shunera XVI (Masraqan). Huge discoidal nodule used to produce a series of large flakes and blades. Tools include a couple of massive scrapers, truncations and utilised pieces. Scale in centimetres.

Figure 12: Shunera XVI (Masraqan). Ventral view of transverse burin on a natural surface with multiple re-sharpening spalls. The core for the blank was not found on-site. Scale in centimetres.

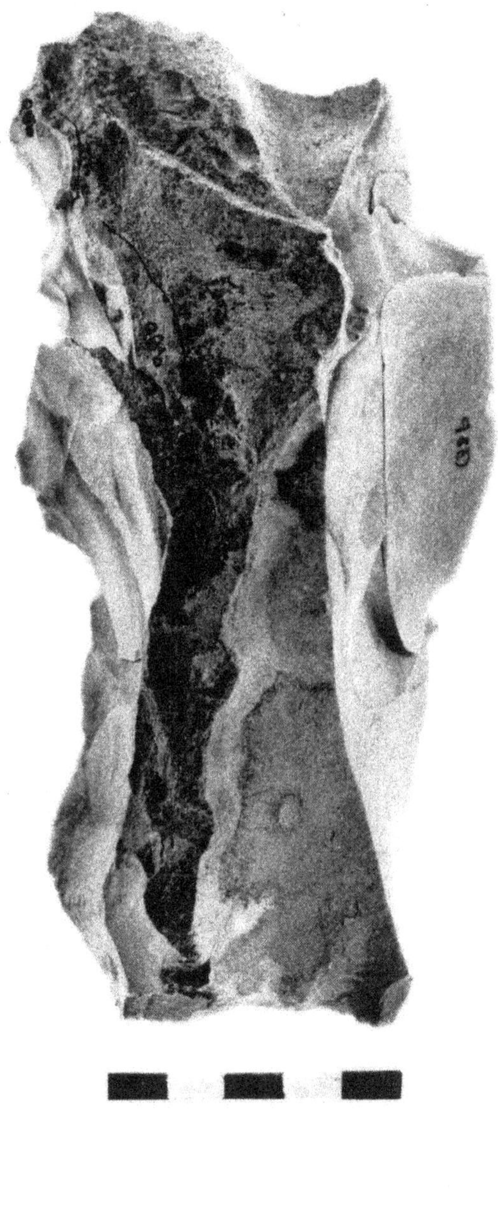

Figure 13: Shunera XVII (Late Kebaran). Massive denticulate on a cortical blade. Though re-sharpening was done on-site, the blank itself was produced off-site. Scale in centimetres.

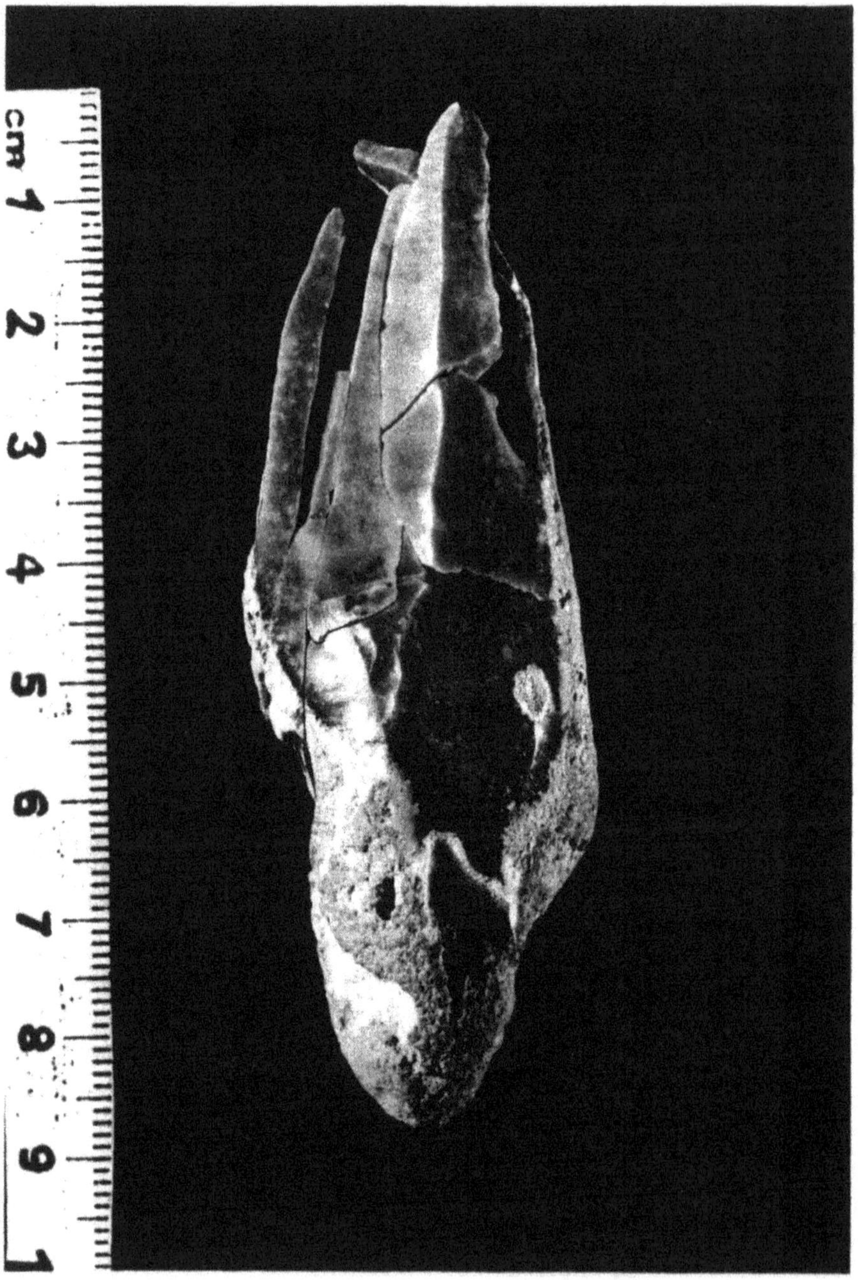

Figure 14: Nahal Lavan 1009 (Ramonian). Narrow single platform sub-pyramidal core. Note core tablet and Ramon point with its distal microburin. Initial decortication flakes missing. Scale in centimetres.

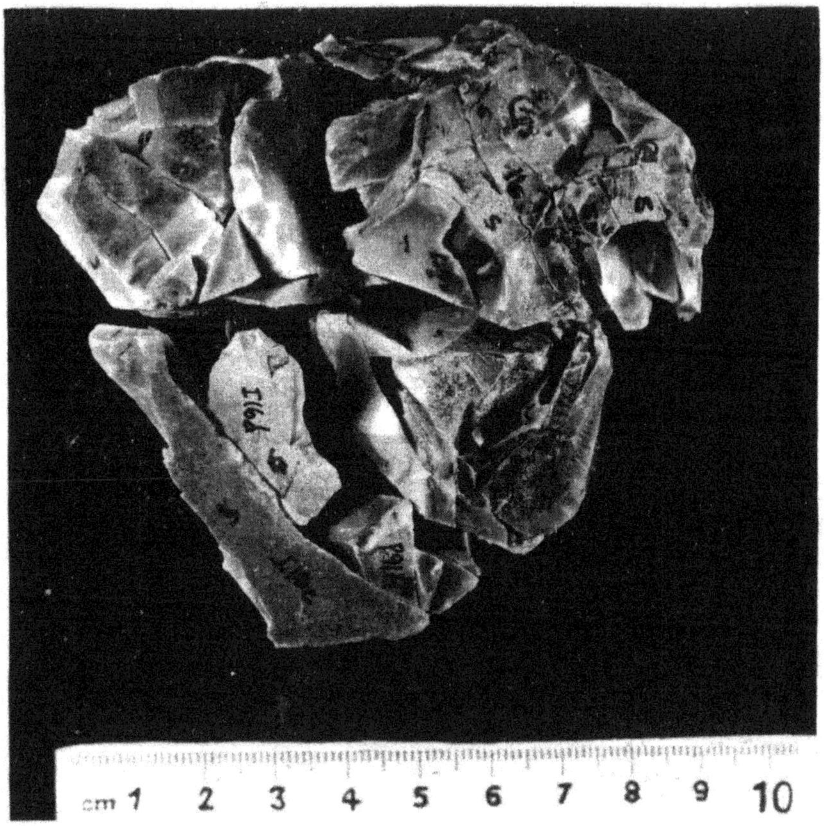

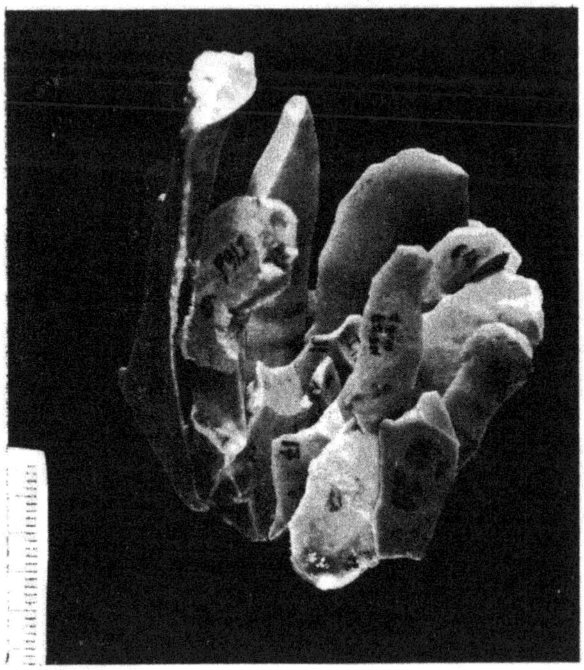

Figure 15a. Nahal Sekher 23 (Terminal Ramonian). Chalcedony nodule which produced three separate pyramidal cores; b, View of refits from one of the cores. Note interior cluster of six distal microburins, including one with refitted *piquant triède*. Scales in centimetres.

Figure 16: Azariq XV (Early Natufian). Large, amorphous nodule of variegated flint, used to produce two sub-pyramidal bladelet cores. Scale in centimetres.

Figure 17a, b: Azariq XV (Early Natufian). Single platform pyramidal core. The lateral view shows some of the bladelet removals including one of two helwan lunates refitted onto the core (right, centre). Scale in centimetres.

Figure 18: Azariq XV (Early Natufian). a, Exterior and b, interior and exterior views of refits from sub-pyramidal single platform core on high quality flint. The refits include a broken inversely retouched bladelet and two microburins, one of which is inverse (b, lower left). Neither the core nor initial decortication flakes were recovered.

www.ingramcontent.com/pod-product-compliance
Lightning Source LLC
La Vergne TN
LVHW071211210926
848482LV00015B/775